Waveform Design for Active Sensing Systems

A Computational Approach

With a focus on developing computational algorithms for examining waveform design in diverse active sensing applications, this guide is ideal for researchers and practitioners in the field. The three parts conveniently correspond to the three categories of desirable waveform properties: good aperiodic correlations, good periodic correlations and beampattern matching. The book features various application examples of using the newly designed waveforms, including radar imaging, channel estimation for communications, an ultrasound system for breast cancer treatment and covert underwater communications. In addition to numerical results, the authors present theoretical analyses describing lower bounds or limitations of performance. Focusing on formulating practical problems mathematically and solving the mathematical problems using efficient and effective optimization techniques, the text pays particular attention to developing easy-to-use computational approaches. Most algorithms are accompanied by a table clearly detailing iteration steps, and corresponding MATLAB codes are available on the companion website.

Hao He received his PhD from the Department of Electrical and Computer Engineering at the University of Florida, USA, in 2011. His student papers won awards at the IEEE 13th DSP Workshop & 5th SPE Workshop in 2009, and at the 2nd International Workshop on Cognitive Information Processing in 2010.

Jian Li is a professor at the Department of Electrical and Computer Engineering, University of Florida, USA, and a Fellow of IEEE and IET. She has published three books, four book chapters, and some 400 papers in archival journals and conference records. She is a co-author of the paper that received the M. Barry Carlton Award for the best paper published in IEEE Transactions on Aerospace and Electronic Systems in 2005.

Petre Stoica is a professor at the Department of Information Technology at Uppsala University, Sweden, a member of the Royal Swedish Academy of Engineering Sciences and the European Academy of Sciences, an honorary member of the Romanian Academy and a fellow of the Royal Statistical Society, IEEE and EURASIP. He has published 10 books, 15 book chapters, and some 700 papers in archival journals and conference records, and has won several awards of IEEE, IEE and EURASIP.

Waveform Design for Active Sensing Systems

A Computational Approach

HAO HE
University of Florida

JIAN LI
University of Florida

PETRE STOICA
Uppsala Universitet, Sweden

CAMBRIDGE
UNIVERSITY PRESS

CAMBRIDGE
UNIVERSITY PRESS

Shaftesbury Road, Cambridge CB2 8EA, United Kingdom

One Liberty Plaza, 20th Floor, New York, NY 10006, USA

477 Williamstown Road, Port Melbourne, VIC 3207, Australia

314–321, 3rd Floor, Plot 3, Splendor Forum, Jasola District Centre, New Delhi – 110025, India

103 Penang Road, #05–06/07, Visioncrest Commercial, Singapore 238467

Cambridge University Press is part of Cambridge University Press & Assessment, a department of the University of Cambridge.

We share the University's mission to contribute to society through the pursuit of education, learning and research at the highest international levels of excellence.

www.cambridge.org
Information on this title: www.cambridge.org/9781107019690

First published 2012

A catalogue record for this publication is available from the British Library

ISBN 978-1-107-01969-0 Hardback

Contents

Preface

The focus of this book is on developing computational algorithms for transmit waveform design in active sensing applications, such as radar, sonar, communications and medical imaging. Waveforms are designed to achieve certain desired properties, which are divided into three categories corresponding to the three main parts in the book, namely good aperiodic correlations, good periodic correlations and beampattern matching. The principal approach is based on formulating practical problems mathematically and then solving the problems using optimization techniques. Particular attention is paid to making the developed algorithms computationally efficient. Theoretical analysis that describes performance lower bounds or limitations is provided. Various application examples using the newly designed waveforms are presented, including radar imaging, channel estimation, an ultrasound system for medical treatment and covert underwater spread spectrum communications.

This book is a research monograph. Its backbone is a series of innovative waveform-design algorithms that we have developed in recent years. These algorithms address different specific problems of waveform design, yet the topics discussed are all centered around active sensing applications and the optimization techniques share similar ideas (e.g., iterative and cyclic procedures, incorporation of fast Fourier transforms, etc.). Notably, all these algorithms are computational approaches that reply on the implementation of computer programs, as opposed to classic waveform design approaches that are mostly analytical. By stitching these algorithms together in a book, we are able to tell a detailed story on various aspects of waveform design, within a consistent framework highlighting computational approaches.

The subject matter of this book falls into the field of electrical engineering, more specifically, into the area of signal processing. The main targeted readers are researchers as well as practitioners in industry who are interested in signal design for radar, sonar and communication systems. We have made a significant effort to present the algorithms in a rigorous and self-contained way, so that interested readers can learn the new waveform-design framework thoroughly. Besides newly developed algorithms, a considerable portion of the book presents tutorial-like materials; for example, we review existing waveforms, analyze properties of ambiguity functions and describe application scenarios. Little background knowledge is required beyond a basic understanding of signal processing and linear algebra. Therefore the book can also serve as an introduction to waveform design for active sensing applications. Readers can access demonstration

MATLAB programs that cover many examples illustrated in the book via the dedicated website http://www.sal.ufl.edu/book/.

Following Chapter 1, which contains an introduction, the other chapters are structured into four parts. In Part I (Chapters 2 to 8) we discuss *aperiodic* sequence design. The heart of the discussion is to achieve desired correlation properties for a single sequence or a sequence set. Revolving around the correlation concept, other topics discussed include frequency stopbands, ambiguity functions and receiver design. In Part II (Chapters 9 to 12) we discuss *periodic* sequence design, and this largely parallels the first four chapters of Part I. In Part III (Chapters 13 to 15) we describe array beampattern synthesis. Finally, in Part IV (Chapters 16 to 19) we study several application examples where the newly developed waveforms lead to an improved performance.

Research on waveform design has thrived since the 1950s, following a number of pioneering works such as Shannon's communication theory and Woodward's ambiguity function. A vast amount of literature has been devoted to this area. As such, we do not attempt to cover every aspect of it. In fact, most discussion in this book is limited to phase-coded waveforms and the correlation-related properties thereof, and this by no means represents all waveform models or application scenarios for active sensing. However, by focusing on modern computational methods we hope to bring in a new methodology and contribute from a perspective that is different from classic methods. It is conceivable that advances in computing power and optimization techniques will continue to herald new and improved approaches to waveform design.

We are grateful to Jun Ling, William Roberts, Xumin Zhu, Bin Guo and Yao Xie for providing relevant materials and helping with the simulations. We appreciate the professional assistance of Phil Meyler and Mia Balashova at Cambridge University Press. We would also like to thank the sponsors of our research: the Office of Naval Research, the US Army Research Laboratory and US Army Research Office, the National Science Foundation, the Swedish Research Council and the European Research Council. Finally, we are indebted to Yajie, to Jerry, Vivian and Lillian, and to Anca, for their understanding and moral support.

Notation

j	the imaginary unit; $j^2 = -1$
a^*	complex conjugate of scalar a
$\mathrm{Re}(a)$	real part of scalar a
$\mathrm{Im}(a)$	imaginary part of scalar a
$\|\|\mathbf{a}\|\|$	Euclidean norm of vector $\mathbf{a}$
$\mathbf{A}^*$	complex conjugate of matrix $\mathbf{A}$
$\mathbf{A}^T$	transpose of matrix $\mathbf{A}$
$\mathbf{A}^H$	conjugate transpose of matrix $\mathbf{A}$
$\mathrm{tr}(\mathbf{A})$	trace of matrix $\mathbf{A}$
$\|\|\mathbf{A}\|\|$	Frobenius norm of matrix $\mathbf{A}$
$\mathbf{A} \odot \mathbf{B}$	Hadamard (element-wise) product of two matrices
$\mathbf{A} \geq 0$	matrix $\mathbf{A}$ is positive semidefinite
$\mathbf{A} \leq \mathbf{B}$	matrix $\mathbf{B} - \mathbf{A}$ is positive semidefinite
δ_n	Kronecker delta; $\delta_n = 1$ if $n = 0$ and $\delta_n = 0$ otherwise
δ_{mn}	an extension of δ_n; $\delta_{mn} = 1$ if $m = n$ and $\delta_{mn} = 0$ otherwise
$\mathbf{I}_M$	identity matrix of dimension $M \times M$
$f(x) * g(x)$	convolution of two functions $f(x)$ and $g(x)$
$\lfloor x \rfloor$	largest integer less than or equal to (real-valued) x
$\arg(x)$	phase angle (in radians) of x

Abbreviations

AF	ambiguity function
BER	bit error rate
CA	cyclic algorithm
CAF	cross AF
CAN	cyclic algorithm-new
CIR	channel impulse response
CREW	cognitive receiver and waveform (design)
DFT	discrete Fourier transform
FFT	fast Fourier transform
ISL	integrated sidelobe level
IV	instrumental variable
MF	merit factor
MIMO	multi-input multi-output
MSE	mean-squared error
Multi-CAN	multi-sequence CAN
Multi-CAO	multi-sequence CA original
Multi-PeCAN	multi-sequence PeCAN
Multi-PeCAO	multi-sequence periodic CA original
Multi-WeCAN	multi-sequence WeCAN
PAF	periodic AF
PAR	peak-to-average power ratio
PeCAN	periodic CAN
PSL	peak sidelobe level
SAR	synthetic aperture radar
SCAN	stopband CAN
SNR	signal-to-noise ratio
UWA	underwater acoustic
WB-CA	wideband CA
WeCAN	weighted CAN
WeSCAN	weighted SCAN
WISL	weighted ISL
ZCZ	zero-correlation zone

1 Introduction

The goal of an active sensing system, such as radar or sonar, is to determine useful properties of the targets or of the propagation medium by transmitting certain waveforms toward an area of interest and analyzing the received signals. For example, a land-based surveillance radar sends electromagnetic waves in the direction of the sky, where objects such as airplanes can reflect a (usually very tiny) fraction of the transmitted signal back to the radar. By measuring the round-trip time delay, the distance (called the range) between the radar and the target can be estimated since the speed of propagation for radio waves is known (3×10^8 m/s). Additional target properties can be obtained by performing further processing at the receiver side; e.g., the speed of a target can be estimated by measuring the Doppler frequency shift of the received signal.

In 1904, a German engineer named Christian Hülsmeyer carried out the first radar experiment using his "telemobiloscope" to detect ships in dense fog by means of radio waves. As to sonar, Reginald Fessenden, a Canadian engineer, demonstrated this in 1914 using a sound echo device, though not successfully, for iceberg detection off the east coast of Canada. It was amongst several other experiments and patents said to be prompted by the 1912 Titanic disaster.

Radar and sonar underwent considerable development during the two world wars and later on spread into diverse fields including weather monitoring, flight control and underwater sensing. There are two factors that are critical to the system performance, namely the *receive filter* and the *transmit waveform*. The receive filter is used to extract from the received signals the information of interest, e.g., target locations in radar or sonar applications [Skolnik 2008] or channel conditions in communications [Proakis 2001]. The transmit waveform, not surprisingly, interplays with the receive filter. A good design of the transmit waveform lends itself to accurate parameter estimation and a reduced computational burden at the receiver.

Arguably the most commonly used receive filter is the matched filter, which maximizes the signal-to-noise ratio (SNR) in the presence of stochastic additive white noise [Turin 1960]. Examples of other well-known receive filters include the mismatched filter, the use of which is also called the instrumental-variable (IV) method [Ackroyd & Ghani 1973][Zoraster 1980][Stoica, Li & Xue 2008], the Capon estimator [Capon 1969], the amplitude and phase estimation (APES) algorithm [Li & Stoica 1996][Stoica et al. 1998][Stoica et al. 1999] and more advanced data-adaptive techniques such as the iterative adaptive approach (IAA) [Yardibi et al. 2010].

We concentrate our attention on transmit waveform design in this book. In particular, we are interested in synthesizing waveforms that have good correlation properties. In radar range compression, low auto-correlation sidelobes improve the detection performance of weak targets [Stimson 1998][Levanon & Mozeson 2004]; in code-division multiple access (CDMA) systems, low auto-correlation sidelobes are desired for synchronization purposes and low cross-correlations reduce interferences from other users [Suehiro 1994][Tse & Viswanath 2005]; and the situation is similar in many other active sensing applications such as ultrasonic imaging [Diaz et al. 1999]. An emitted probing waveform with low auto-correlation sidelobes maximizes the signal-to-noise ratio, when complemented by a matched filter at the receiver side, while significantly weakening the signals from adjacent range bins.

In addition to correlation properties, the problem of transmit beampattern synthesis is also considered in this book. A classical phased array steers narrow beams toward different angles by adjusting only the waveform phases across the antenna elements. In a modern MIMO (multi-input multi-output) system, however, waveforms can be chosen freely and this waveform diversity allows for more flexibility in beampattern synthesis. One example is hyperthermia treatment for breast cancer [Guo & Li 2008], where waveform diversity enables an ultrasonic focal point to be matched to the entire tumor region without impacting surrounding healthy tissues.

1.1 Signal model

Let $s(t)$ denote the transmitted signal, with t indicating time. Suppose that $s(t)$ consists of N symbols

$$s(t) = \sum_{n=1}^{N} x(n) p_n(t) \tag{1.1}$$

where $p_n(t)$ is the shaping pulse and $\{x(n)\}_{n=1}^{N}$ are the N symbols. The shaping pulse $p_n(t)$ (with duration t_p) can be an ideal rectangular shaping pulse

$$p_n(t) = \frac{1}{\sqrt{t_p}} \operatorname{rect}\left(\frac{t-(n-1)t_p}{t_p}\right), \quad n = 1, \ldots, N, \tag{1.2}$$

where

$$\operatorname{rect}(t) = \begin{cases} 1, & 0 \leq t \leq 1, \\ 0, & \text{elsewhere}, \end{cases} \tag{1.3}$$

or is equal to other pulse such as the raised-cosine pulse [Proakis 2001].

Note that the actual transmitted waveform is composed of the in-phase and quadrature components of $s(t)e^{j2\pi f_c t}$ where f_c is the carrier frequency. It is assumed that the signal demodulation has already been performed at the receiver side and thus the carrier term $e^{j2\pi f_c t}$ can be safely ignored in the analysis.

In practice, hardware components such as analog-to-digital converters and power amplifiers have a maximum signal amplitude clip. In order to maximize the transmitted power available in the system, it is desirable that the transmit sequences are unimodular or have low peak-to-average power ratios (PARs). In our design, we impose the following unit-modulus constraint whenever feasible:

$$x(n) = e^{j\phi(n)}, \quad n = 1, \ldots, N \tag{1.4}$$

where $\{\phi(n)\}$ are phases. Note that (1.1) combined with (1.4) provides a phase-coded signal representation. There are many other types of signal that are widely used or have been discussed in the literature; these include the well-known chirp waveform (see Section 1.3), discrete frequency-coded waveforms [Costas 1984][Deng 2004] and waveforms constructed from a particular set of functions such as the prolate spheroidal functions [Moore & Cada 2004] or the Hermite wave functions [Gladkova & Chebanov 2004]. In this book we have chosen to focus specifically on the phase-coded signal model, which serves as a practical and effective framework for designing waveforms with various desirable properties.

The waveform $s(t)$ is transmitted in the direction of a scene of interest and is reflected by various targets at different range locations. The reflected signals, which are time-shifted and weighted versions of $s(t)$, arrive linearly combined at the receiver side:

$$y(t) = \sum_k \alpha_k s(t - \tau_k) + e(t), \tag{1.5}$$

where τ_k is the round-trip time delay for the kth target, α_k is the coefficient related to the target reflection, for example the radar cross section (RCS), and $e(t)$ is the noise.

Suppose that we aim to estimate the coefficient $\alpha_{k'}$ by applying the filter $w(t)$ at the receiver. The estimated coefficient is given by

$$\hat{\alpha}_{k'} = \int_{-\infty}^{\infty} w^*(t) y(t) dt. \tag{1.6}$$

More precisely, according to a conventional convolution definition (1.6) is the receiver output at time instant 0 when $y(t)$ is the input and $w(-t)$ is the filter. However, we can simply refer to $w(t)$ as the filter in the receiver processing indicated by (1.6) without introducing any ambiguity in later discussions.

To determine an appropriate $w(t)$, we decompose $y(t)$ into three parts:

$$y(t) = \underbrace{\alpha_{k'} s(t - \tau_{k'})}_{\text{signal}} + \underbrace{\sum_{k \neq k'} \alpha_k s(t - \tau_k)}_{\text{clutter}} + \underbrace{e(t).}_{\text{noise}} \tag{1.7}$$

If there is no clutter and $e(t)$ is the zero-mean white noise then the matched filter $w(t) = s(t - \tau_{k'})$ will give the largest signal-to-noise ratio (SNR). A direct proof goes as follows:

$$\text{SNR} \triangleq \frac{\left|\int_{-\infty}^{\infty} w^*(t)\alpha_{k'} s(t-\tau_{k'})dt\right|^2}{\text{E}\left\{\left|\int_{-\infty}^{\infty} w^*(t)e(t)dt\right|^2\right\}} \tag{1.8}$$

$$= \frac{|\alpha_{k'}|^2 \left|\int_{-\infty}^{\infty} w^*(t)s(t-\tau_{k'})dt\right|^2}{\sigma_e^2 \int_{-\infty}^{\infty} |w(t)|^2 dt} \tag{1.9}$$

$$\leq \frac{|\alpha_{k'}|^2 \int_{-\infty}^{\infty} |s(t-\tau_{k'})|^2 dt}{\sigma_e^2} = \frac{|\alpha_{k'}|^2 \sigma_s^2}{\sigma_e^2} \tag{1.10}$$

where E denotes the expectation operator, and σ_e^2 and σ_s^2 are the noise power and signal power, respectively. Note that (1.9) is due to the white noise assumption $\text{E}\{e(t_1)e^*(t_2)\} = \sigma_e^2 \delta_{t_1-t_2}$ and that (1.10) results from the Cauchy–Schwartz inequality $\left|\int w^*(t)s(t-\tau_{k'})dt\right|^2 \leq \int |w(t)|^2 dt \int |s(t-\tau_{k'})|^2 dt$. The maximum value of the SNR in (1.10) is achieved if and only if the filter $w(t)$ is a scaled version of $s(t-\tau_{k'})$, which concludes the proof.

For the purpose of normalization, the matched filter $w(t)$ is chosen as $s(t-\tau_{k'})/\int |s(t)|^2$ and the corresponding estimate of $\alpha_{k'}$ in (1.6) is given by

$$\hat{\alpha}_{k'} = \frac{\int_{-\infty}^{\infty} s^*(t-\tau_{k'})y(t)dt}{\int_{-\infty}^{\infty} |s(t)|^2 dt}. \tag{1.11}$$

Besides boosting the signal component and suppressing the noise, the matched filter can also eliminate the clutter component (as easily seen from (1.7) and (1.11)) if

$$r(\tau) = \int_{-\infty}^{\infty} s(t)s^*(t-\tau)dt, \quad -\infty < \tau < \infty, \tag{1.12}$$

is zero for all $\tau \neq 0$. The function $r(\tau)$ as defined in (1.12) is called the *auto-correlation* of $s(t)$.

1.2 Design metrics

The previous section has outlined the benefit of small auto-correlation sidelobes $r(\tau)$ (for $\tau \neq 0$). For most practical cases, we need to focus only on a delay τ that is an integer multiple of the symbol length t_p. One reason is that in modern systems digital filtering is usually performed at the receiver side, that is, the integral in (1.11) is implemented as a summation of sampled signals. In addition, if a rectangular shaping pulse (see (1.2)) is used, the values of $r(\tau)$ can be obtained exactly by the linear interpolation of two neighboring auto-correlation samples [Levanon & Mozeson 2004, Chapter 6]:

$$r(\tau) = \frac{\tau - t_1}{t_p} r(t_2) + \frac{t_2 - \tau}{t_p} r(t_1), \tag{1.13}$$

where $t_1 = \lfloor \tau/t_p \rfloor$ and $t_2 = t_1 + t_p$. Such auto-correlations at integer multiple delays $\{kt_p\}_{k=-N+1}^{N-1}$ can be calculated for $k \geq 0$ as

$$
\begin{aligned}
r(kt_p) &= \int_{-\infty}^{\infty} s(t)s^*(t - kt_p)dt \\
&= \int_{kt_p}^{Nt_p} \sum_{n=k+1}^{N} x(n)p_n(t)x^*(n-k)p_n^*(t)dt \\
&= (N-k) \sum_{n=k+1}^{N} x(n)x^*(n-k) \int_0^{t_p} |p_n(t)|^2 dt \\
&= (N-k) \sum_{n=k+1}^{N} x(n)x^*(n-k).
\end{aligned} \tag{1.14}
$$

Correlations at negative delays can be obtained from $r(kt_p) = r^*(-kt_p)$. When shaping pulses other than rectangular are used, it can still be expected that $r(\tau)$ will be well controlled as long as $r(k)$ is sufficiently small.

It follows from the above discussions that the correlations of interest are given by

$$
r(k) = \sum_{n=k+1}^{N} x(n)x^*(n-k) = r^*(-k), \quad k = 0, \ldots, N-1. \tag{1.15}
$$

The above set $\{r(k)\}$ is called the auto-correlation of the discrete sequence $\{x(n)\}$. Note that the notation r is slightly abused in (1.12) and (1.15) in order to denote both continuous-time and discrete-time auto-correlations, yet a distinction can be made easily by examining the two different time variables.

For the set $\{r(k)\}_{k=-N+1}^{N-1}$ defined above, $r(0)$ is called the in-phase correlation and is always equal to the signal energy. All the other auto-correlations, i.e., $\{r(k),\ k = -N+1, \ldots, -1, 1, \ldots, N-1\}$, are collectively called the auto-correlation sidelobes. One of our main interests in Part I is the design of phase-coded sequences $\{x(n)\}$ whose auto-correlation sidelobes are as low as possible. Part I also discusses ambiguity function synthesis, which can be considered as a two-dimensional extension to correlation design.

More precisely speaking, the $\{r(k)\}$ defined in (1.15) constitute the *aperiodic* auto-correlation. The *periodic* auto-correlation of the sequence $\{x(n)\}$ is defined as

$$
\tilde{r}(k) = \sum_{n=1}^{N} x(n)x^*((n-k) \bmod N) = \tilde{r}^*(-k) = \tilde{r}^*(N-k), \qquad k = 0, \ldots, N-1, \tag{1.16}
$$

where "mod" is the modulo operator:

$$
p \bmod N = \begin{cases} p - \lfloor p/N \rfloor N, & p \text{ is not an integer multiple of } N, \\ N, & \text{otherwise.} \end{cases} \tag{1.17}
$$

The relationship between the aperiodic correlation (1.15) and the periodic correlation (1.16) can be easily obtained as follows:

$$\begin{aligned}\tilde{r}(k) &= \sum_{n=1}^{k} x(n)x^*(n-k+N) + \sum_{n=k+1}^{N} x(n)x^*(n-k) \\ &= \sum_{m=(N-k)+1}^{N} x(m-(N-k))x^*(m) + \sum_{n=k+1}^{N} x(n)x^*(n-k) \\ &= r^*(N-k) + r(k). \end{aligned} \tag{1.18}$$

Periodic correlations appear in many applications such as the synchronization of a CDMA (code division multiple access) system. Part II deals with the minimization of periodic correlation sidelobes.

As will be shown in Part III, waveform correlations can serve as a bridge connecting the underlying waveform to the desired beampattern of, e.g., an antenna array. In particular, the waveform diversity in a MIMO system leads to a flexible control of waveform correlations, which further leads to an agile transmit beampattern synthesis.

In the next section, we will review several well-known waveforms that have good auto-correlation properties, especially those that are phase-coded (1.4). For the sake of brevity, the word "periodic" is used explicitly whenever a periodic correlation (1.16) is referred to; otherwise, an aperiodic correlation, as defined in (1.15), is meant.

1.3 Review of existing waveforms

We start with the well-known chirp waveform. A chirp waveform is a linear frequency-modulated (LFM) pulse, whose frequency is swept linearly over a bandwidth B in a time duration T. Chirp signals have been widely used in radar applications since World War II, as they possess relatively low correlation sidelobes and are mostly tolerant to Doppler frequency shifts [Levanon & Mozeson 2004]. In addition, the power of a chirp signal is dispersed evenly throughout the frequency spectrum, which allows for high spectral efficiency.

A chirp signal $s(t)$ can be written as

$$s(t) = \frac{1}{\sqrt{T}} \exp\left(j\pi \frac{B}{T} t^2\right), \quad 0 \le t \le T, \tag{1.19}$$

where B/T is the chirp rate. Figure 1.1(a) shows the real part of $s(t)$ with parameters $T = 100$ s and $B = 1$ Hz. Figure 1.1(b) shows its auto-correlation function $r(\tau)$ (normalized by $r(0)$ and using a $20\log_{10}$ scale), where the peak sidelobe is -13.4 dB.

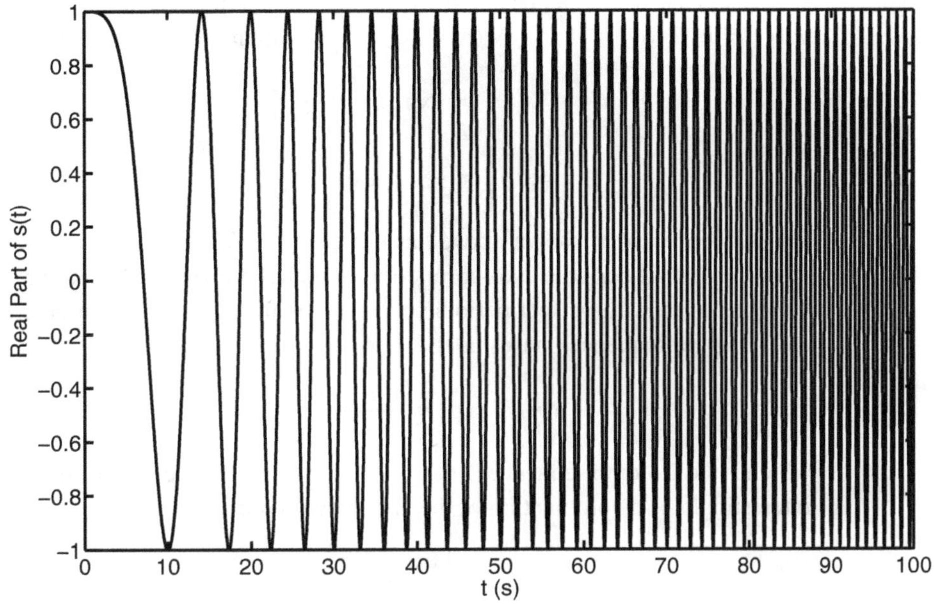

(a)

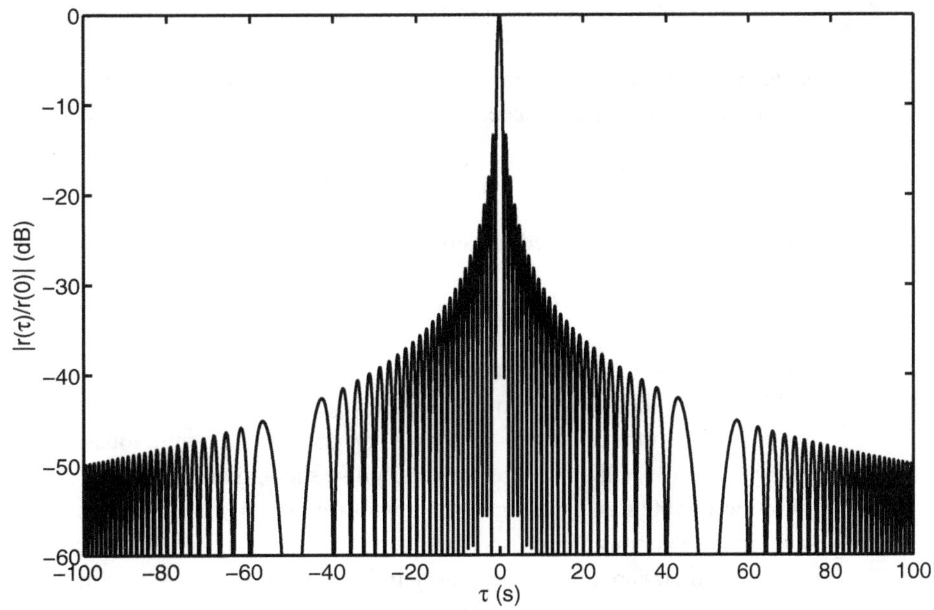

(b)

Figure 1.1 (a) The real part of $s(t)$ in (1.19) with $T = 100$ s and $B = 1$ Hz. (b) The auto-correlation function of $s(t)$.

Many phase codes can be derived from the chirp signal. We sample $s(t)$ at time intervals $t_s = n/B$ for $n = 1, \ldots, N$ $(N = BT)$ and obtain the following sequence:

$$\begin{aligned} x(n) &\triangleq s(nt_s) = \exp\left[j\pi \frac{B}{T}\left(\frac{n}{B}\right)^2\right] \\ &= \exp\left[j\pi \frac{n^2}{BT}\right] = \exp\left[j\pi \frac{n^2}{N}\right], \quad n = 1, \ldots, N. \end{aligned} \tag{1.20}$$

The sequence $\{x(n)\}$ shown above has perfect *periodic* auto-correlations if N is even, meaning that all periodic auto-correlation sidelobes are zero: $\tilde{r}(k) = 0$ for $k \neq 0$.

A sequence with perfect periodic correlations for any odd N can be constructed by changing the sequence phases in (1.20) to

$$x(n) = \exp\left[j\pi \frac{n(n-1)}{N}\right], \quad n = 1, \ldots, N, \tag{1.21}$$

which is the Golomb sequence [Zhang & Golomb 1993]. The Chu sequence [Chu 1972], interestingly, is a combination of the above two sequences:

$$x(n) = \begin{cases} \exp\left[jQ\pi \frac{n^2}{N}\right], & N \text{ even}, \\ \exp\left[jQ\pi \frac{n(n-1)}{N}\right], & N \text{ odd}, \end{cases} \tag{1.22}$$

where Q is any integer that is prime to N. As expected, the Chu sequence has perfect periodic correlations for any (positive) integer N.

Besides the Golomb and the Chu sequences, there are many other phase-coded sequences whose phases are quadratic functions of n, such as the well-known Frank sequence and the P4 sequence. The Frank sequence is defined for $N = L^2$ as

$$x((m-1)L + p) = \exp\left[j2\pi \frac{(m-1)(p-1)}{L}\right], \quad m, p = 1, \ldots, L. \tag{1.23}$$

The P4 sequence is defined for any length N as

$$x(n) = \exp\left[j\frac{2\pi}{N}(n-1)\left(\frac{n-1-N}{2}\right)\right], \quad n = 1, \ldots, N. \tag{1.24}$$

Both the Frank and P4 sequences have perfect periodic correlations.

Figure 1.2 shows the auto-correlation $r(k)$ of the P4 sequence of length $N = 100$. Note that from the sequence $\{x(n)\}$ we can construct a continuous-time waveform $s(t)$ using (1.1). We will choose rectangular pulse shaping and set $t_p = 1$ s so that the signal duration is 100 s and the signal bandwidth is roughly 1 Hz (i.e., $1/t_p$); these are the same parameters as those used in Figure 1.1. The real and imaginary parts of the so-constructed $s(t)$ are shown separately in Figure 1.3(a). The auto-correlation $r(\tau)$ of this $s(t)$ is shown in Figure 1.3(b). The peak sidelobe is −26.3 dB, which is much lower than that of the chirp waveform in Figure 1.1(b).

The auto-correlation properties of the Golomb, Chu or Frank sequences are similar to those of the P4 sequence and are omitted here.

Another widely used sequence is the maximum length sequence (m-sequence) [Proakis 2001], which is one of the most commonly known PN (pseudo-noise)

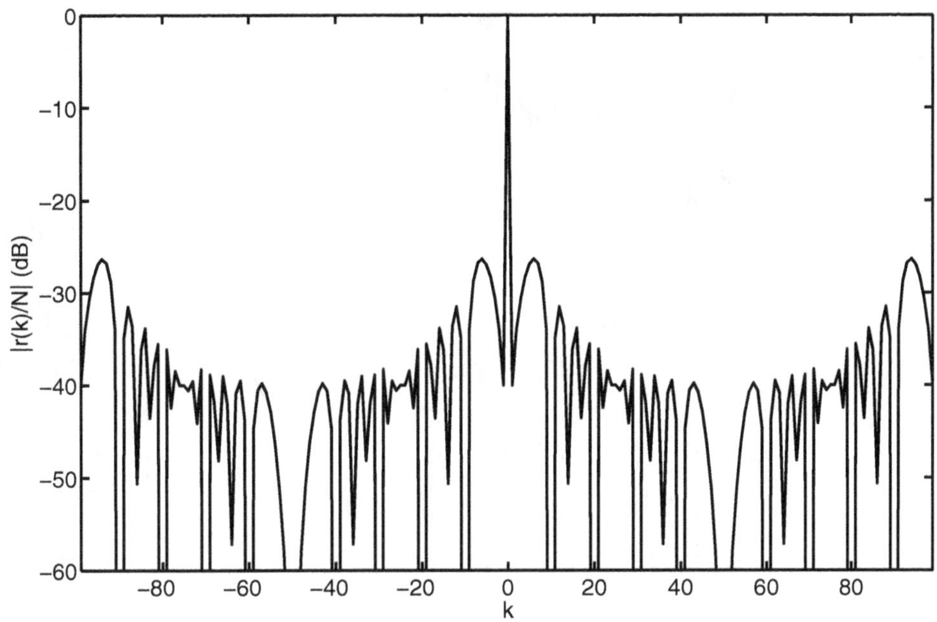

Figure 1.2 The auto-correlation function $r(k)$ of the P4 sequence, as defined in (1.24), of length $N = 100$.

sequences. An m-sequence is a type of pseudo-random binary sequence that is generated by a maximal linear feedback shift register (LFSR). Figure 1.4 shows a length-3 LFSR where the plus operator indicates "exclusive or". Each register block can store 0 or 1, so three blocks amount to eight different states. When fed with any initial binary sequence (not all zeros), such a shift register will cycle through all eight states except for the all-zero state. For instance, starting from 001, the register in Figure 1.4 will pass repeatedly through the following seven states 001, 100, 010, 101, 110, 111, 011. By taking only the output from the third block and replacing 0 with -1, we obtain a length-7 m-sequence, $\{1, -1, -1, 1, -1, 1, 1\}$. Its aperiodic as well as periodic correlations are shown in Figure 1.5.

One of the prominent features of an m-sequence is that its *periodic* correlation sidelobes are always equal to -1, as can be observed from Figure 1.5(b). Its aperiodic correlation sidelobes, though, do not have a regular pattern and can be relatively high. Also note that the LFSR can be efficiently implemented in hardware, which greatly facilitates the use of m-sequences in practice.

Figure 1.6 compares the aperiodic auto-correlation of an m-sequence of length 127 and that of a Golomb sequence (see (1.21)) of the same length. The polyphase Golomb sequence exhibits notably lower correlation sidelobes than the binary m-sequence. Although a polyphase sequence does not necessarily have lower correlation sidelobes than a binary sequence, allowing more phase values leads to more degrees of design freedom. As will be shown in Chapter 2, by allowing arbitrary phases between 0 and 2π

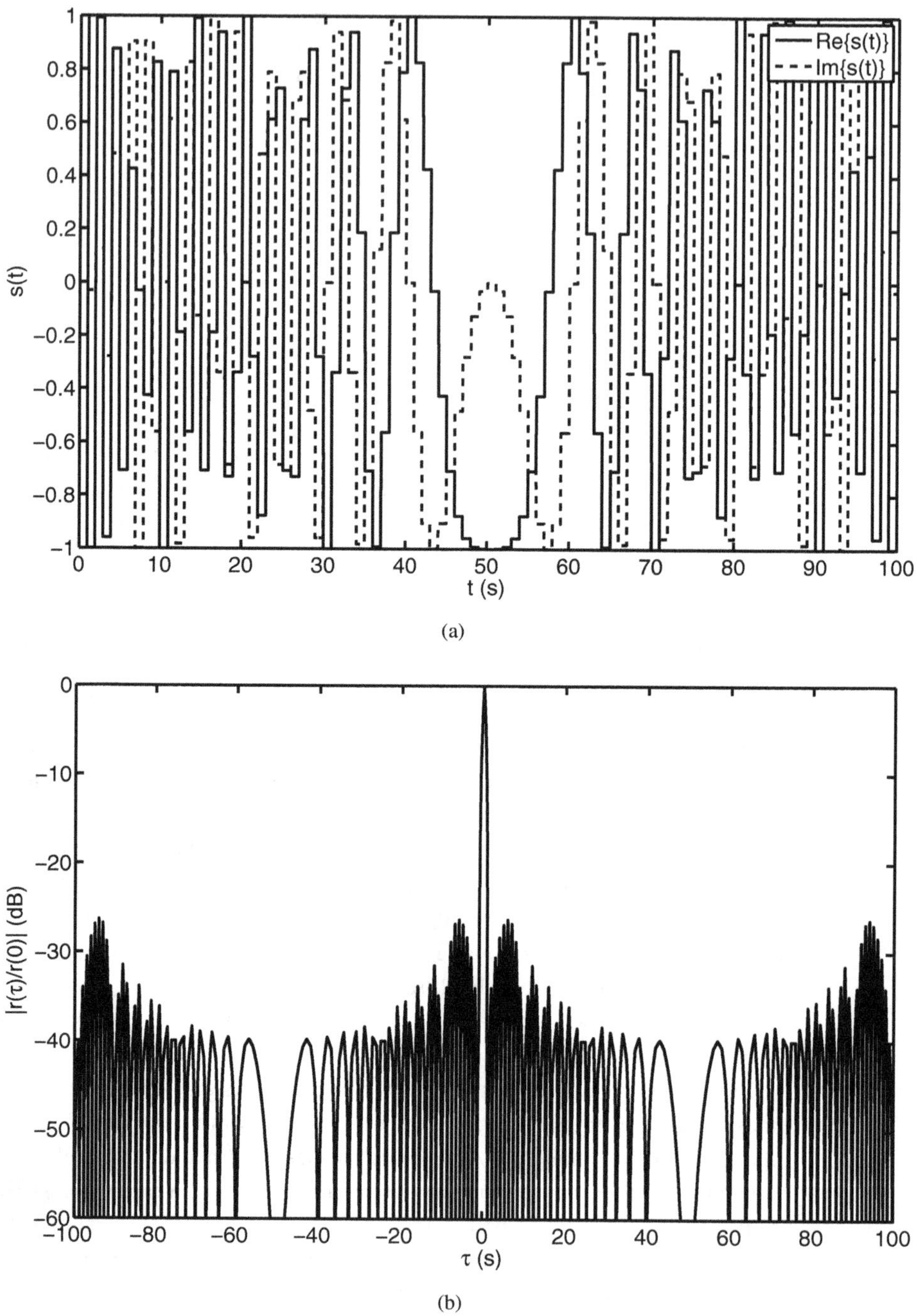

Figure 1.3 (a) The real and imaginary parts of $s(t)$ in (1.1) when a P4 sequence of length 100 and rectangular shaping pulses are used. (b) The auto-correlation function $r(\tau)$ of this $s(t)$.

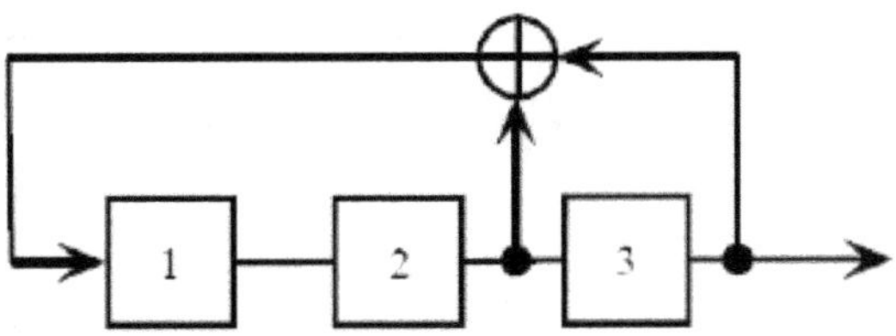

Figure 1.4 A linear feedback shift register of length 3.

instead of a fixed constellation, we can design unimodular sequences that have lower correlation sidelobes than the Golomb sequence.

The aforementioned sequences and waveforms all have closed-form construction methods. Researchers have also used gradient-descent or stochastic-optimization techniques to find sequences with lower auto-correlations [Friese & Zottmann 1994][Brenner 1998][Borwein & Ferguson 2005]. These algorithms are usually computationally expensive and work well only for small values of N, such as $N \sim 10^2$. An example is the search for polyphase Barker sequences whose auto-correlation sidelobes are all less than or equal to 1. A length-45 polyphase Barker sequence is given by [Brenner 1998]

$$x(n) = \exp\{j\phi(n)\}, \quad n = 1, \ldots, 45, \tag{1.25}$$

where

$$\begin{aligned}\{\phi(n)\} = \frac{2\pi}{90}\{&0\ 0\ 7\ 1\ 76\ 71\ 76\ 63\ 56\ 73\ 87\ 9\ 9\ 14\ 25\ 53\ 62\ 5\ 32\ 35\ 85 \\ &69\ 40\ 76\ 57\ 26\ 9\ 83\ 56\ 57\ 21\ 5\ 52\ 89\ 48\ 11\ 68\ 26\ 62\ 6\ 37\ 73\ 19\ 58\ 12\},\end{aligned} \tag{1.26}$$

and its auto-correlation is shown in Figure 1.7. Its peak sidelobe is −33.1 dB. Note that for a unimodular sequence, the lowest possible peak sidelobe is 1, since $|r(N-1)| = |x(N)x^*(1)| = 1$. In the case of $N = 45$, 1 corresponds to $20\log_{10}(1/N) = -33.1$ dB. Thus a polyphase Barker code has the best auto-correlation properties in terms of the lowest peak sidelobe. We mention in passing that the Barker code was originally defined as a *binary* sequence with correlation sidelobes not larger than 1 [Barker 1953]. The longest known Barker code has length 13 and is

$$\{x(n)\} = \left\{1\ 1\ 1\ 1\ 1\ -1\ -1\ 1\ 1\ -1\ 1\ -1\ 1\right\}; \tag{1.27}$$

its auto-correlation is plotted in Figure 1.8.

We finally point out that the design of unimodular sequences with low *aperiodic* correlations is much more difficult than designing sequences with low *periodic* correlations. The aforementioned sequences, such as Chu, P4 sequences and m-sequences, have low or zero periodic correlation sidelobes; and there are many others with the same property, such as Gold sequences [Gold 1967] and Kasami sequences [Kasami 1966]. Note the symmetric property of the periodic correlations in (1.16): $\tilde{r}(k) = \tilde{r}^*(N-k)$. It leads to the fact that, on the one hand, in order to minimize all periodic correlation sidelobes we need to consider only $\tilde{r}(1), \ldots, \tilde{r}(N/2)$ for an even N and only $\tilde{r}(1), \ldots, \tilde{r}((N-1)/2)$ for an odd N. The aperiodic correlations, on the other hand, do not have such a symmetric property. In addition, the absolute value of the maximum-

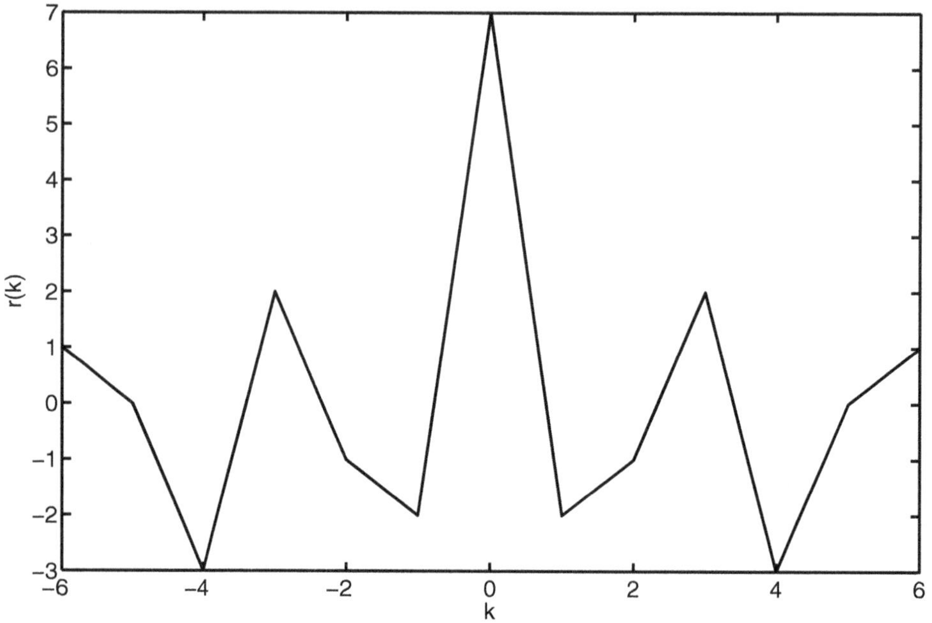

(a)

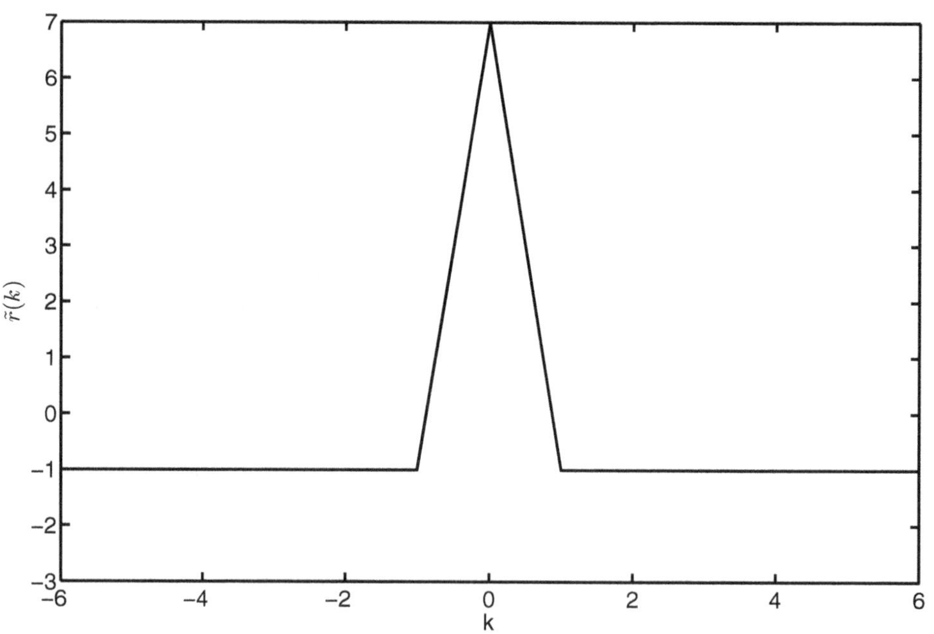

(b)

Figure 1.5 (a) The aperiodic auto-correlation and (b) the periodic correlation of the length-7 m-sequence $\{1, -1, -1, 1, -1, 1, 1\}$.

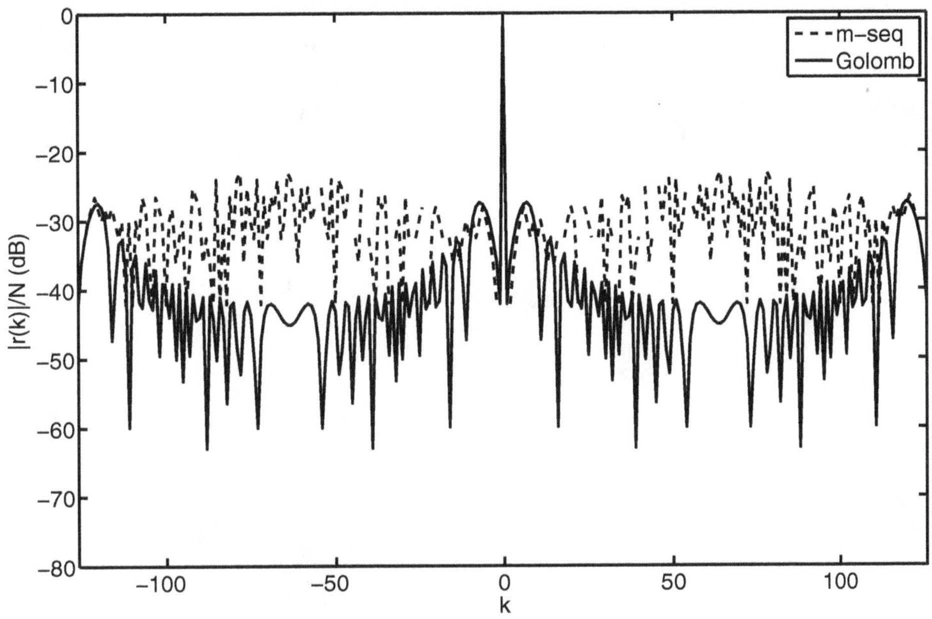

Figure 1.6 The auto-correlation function $r(k)$ of an m-sequence of length $N = 127$, and that of a Golomb sequence of the same length.

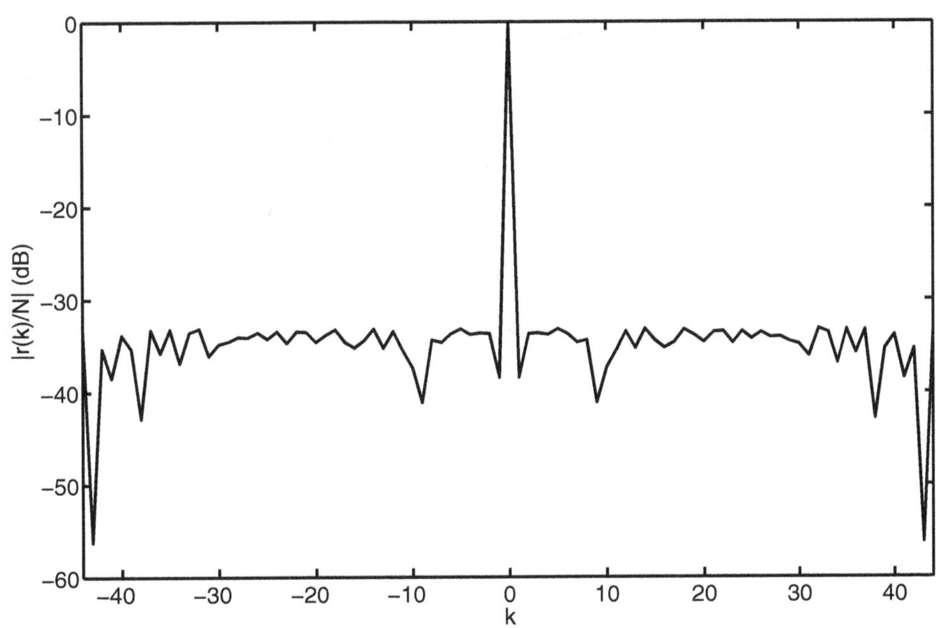

Figure 1.7 The auto-correlation function $r(k)$ of a polyphase Barker sequence of length $N = 45$, as defined in (1.25).

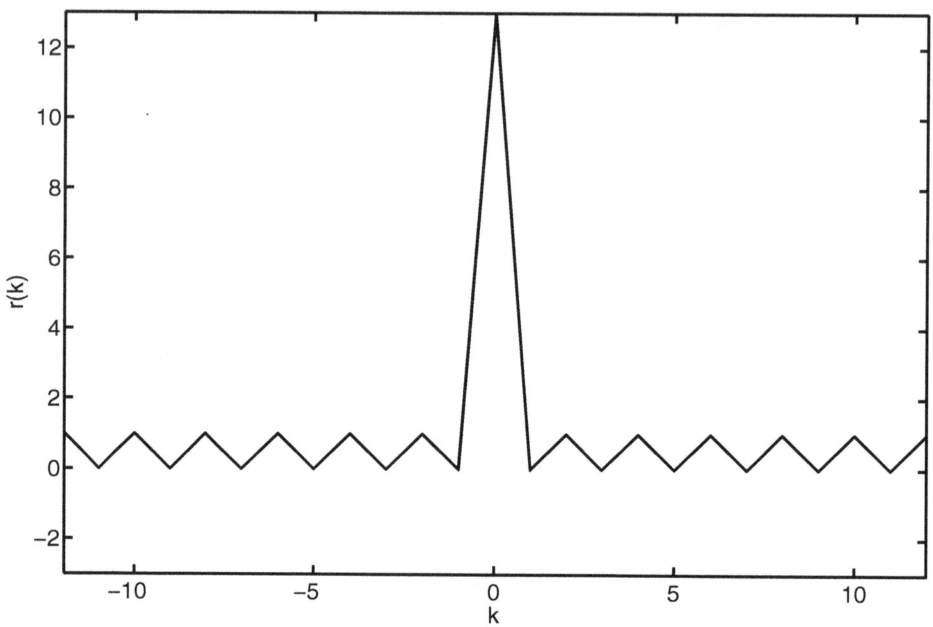

Figure 1.8 The auto-correlation function $r(k)$ of a Barker sequence of length $N = 13$, as defined in (1.27).

lag aperiodic correlation $r(N-1)$ is always equal to 1 (thus cannot be minimized): $|r(N-1)| = |x(N)x^*(1)| = 1$ because each element of the sequence has unit modulus. We can also observe from (1.18) that the case of all-zero periodic correlation sidelobes implies that $r(k) = -r^*(N-k)$ (for $k = 1, \dots, N-1$) and vice versa.

In the next chapter we start our sequence-design discussions with the more challenging aperiodic correlation case, which is relatively less mentioned in the literature. The phases of the unimodular sequences that we will design can be anywhere between 0 and 2π, which leads to increased design freedom over a fixed-phase constellation. Such phases are also hard for a foe to guess, thereby facilitating covert active sensing. Continued improvements in waveform generation hardware, indeed, allow for advanced sequence designs such as those we consider.

Part I

Aperiodic correlation synthesis

2 Single aperiodic sequence design

From the discussions given in Chapter 1, we have made it clear that the goal of aperiodic sequence design is to make $\{r(k)\}_{k\neq 0}$ as small as possible. In this chapter, we focus on the integrated sidelobe level (ISL) metric, which is defined as

$$\text{ISL} = \sum_{\substack{k=-(N-1)\\ k\neq 0}}^{N-1} |r(k)|^2 = 2\sum_{k=1}^{N-1} |r(k)|^2. \tag{2.1}$$

Our goal is to present efficient computation algorithms to minimize the ISL metric or ISL-related metrics under the constraint of synthesizing unimodular sequences. Note that the minimization of the ISL metric is equivalent to the maximization of the merit factor (MF), defined as follows:

$$\text{MF} = \frac{|r(0)|^2}{\sum_{\substack{k=-(N-1)\\ k\neq 0}}^{N-1} |r(k)|^2} = \frac{N^2}{\text{ISL}}. \tag{2.2}$$

Owing to the significant practical interest in the design of unimodular sequences with low ISL (or equivalently large MF) values, as pointed out in Chapter 1, it comes as no surprise that the literature on this topic is extensive [Borwein & Ferguson 2005][Jedwab 2005][Khan et al. 2006][Høholdt 2006][Stoica, Li & Zhu 2008].

Extensions to the ISL and MF metrics are the weighted ISL (WISL) and modified MF (MMF) metrics:

$$\text{WISL} = 2\sum_{k=1}^{N-1} w_k |r(k)|^2, \quad w_k \geq 0, \tag{2.3}$$

$$\text{MMF} = \frac{|r(0)|^2}{\text{WISL}}, \tag{2.4}$$

where $\{w_k\}_{k=1}^{N-1}$ is an arbitrary set of weights. Such weighted ISL metrics are important in applications where we want to reduce, as much as possible, the interference due to a known multipath or a known discrete clutter. For example, there are cases in which the maximum difference between the arrival times of the sequence of interest and of the interference is (much) smaller than the duration of the emitted sequence [Kretschmer Jr. & Gerlach 1991][Jakowatz Jr. et al. 1996][Khan et al. 2006][Ling, Yardibi, Su, He & Li 2009]. Consequently, in such cases the interest lies in making $\{|r(k)|\}_{k=1}^{P-1}$ small for some

$P < N$, rather than trying to make all the correlation sidelobes $\{|r(k)|\}_{k=1}^{N-1}$ small. Here the value of P is selected on the basis of the *a priori* knowledge about the application at hand (for instance, in wireless communications it is usually known that significant channel tap coefficients can occur up to a certain maximum delay, and so we can choose P as the said delay).

Because the ISL metric is highly multimodal (i.e., it has many local minima), stochastic optimization algorithms have been suggested for its minimization [Brenner 1998][Schotten & Lüke 2005][Borwein & Ferguson 2005]. However, the computational burden of these algorithms becomes prohibitive as N increases: such algorithms are hardly effective on the currently available computing machines for $N \sim 10^3$ or larger. Optimization algorithms for locally minimizing the ISL metric have also been proposed. These algorithms can be used to provide quick solutions to the problem of reducing the ISL value of a given reasonably good sequence. They can also be used as local minimization blocks of a stochastic global optimization algorithm. However, most existing local minimization algorithms for the ISL metric are descent gradient methods whose convergence problems as well as computational burdens increase significantly as N increases.

In this chapter we introduce two cyclic algorithms [Li et al. 2008][Stoica et al. 2009*a*] for the local minimization of ISL-related metrics. The first algorithm is called CAN (cyclic algorithm-new) and can be used for the local minimization of the ISL metric. CAN is based on FFT (fast Fourier transform) operations and can be used for virtually any practically relevant values of N up to $N \sim 10^6$ or even larger. We also modify CAN so that it can tackle the WISL metric. The resulting algorithm, which is called WeCAN (weighted CAN), requires N times more computations than CAN and it can be run on a PC for N up to 10^4.

2.1 Cyclic algorithm-new (CAN)

The derivation of CAN involves several steps, the first of which consists of expressing the ISL metric in the frequency domain. It is well known that, for any $\omega \in [0, 2\pi]$,

$$\left|\sum_{n=1}^{N} x(n)e^{-j\omega n}\right|^2 = \sum_{k=-(N-1)}^{N-1} r(k)e^{-j\omega k} \triangleq \Phi(\omega) \tag{2.5}$$

[Stoica & Moses 2005]. Furthermore, it can be shown that the ISL metric in (2.1) is equivalently written as follows:

$$\text{ISL} = \frac{1}{2N}\sum_{p=1}^{2N}\left[\Phi(\omega_p) - N\right]^2, \tag{2.6}$$

where $\{\omega_p\}$ are the following Fourier frequencies:

$$\omega_p = \frac{2\pi}{2N}p, \quad p = 1, \ldots, 2N. \tag{2.7}$$

(Note that (2.6) is a Parseval-type equality.) To prove (2.6), let δ_k denote the Kronecker delta and use the correlogram-based expression for $\Phi(\omega)$ in (2.5) to verify that

$$\begin{aligned}\sum_{p=1}^{2N}\left[\Phi(\omega_p)-N\right]^2 &= \sum_{p=1}^{2N}\left\{\sum_{k=-(N-1)}^{N-1}[r(k)-N\delta_k]e^{-j\omega_p k}\right\}^2\\ &= \sum_{k=-(N-1)}^{N-1}\sum_{\tilde{k}=-(N-1)}^{N-1}[r(k)-N\delta_k][r(\tilde{k})-N\delta_{\tilde{k}}]^*\left(\sum_{p=1}^{2N}e^{-j\omega_p(k-\tilde{k})}\right).\end{aligned}\tag{2.8}$$

Since, for $|k-\tilde{k}|\le 2N-2$,

$$\sum_{p=1}^{2N}e^{-j\omega_p(k-\tilde{k})} = e^{-j\frac{2\pi}{2N}(k-\tilde{k})}\frac{e^{-j2\pi(k-\tilde{k})}-1}{e^{-j\frac{2\pi}{2N}(k-\tilde{k})}-1} = 2N\delta_{(k-\tilde{k})},\tag{2.9}$$

we obtain from (2.8) the following equation:

$$\frac{1}{2N}\sum_{p=1}^{2N}\left[\Phi(\omega_p)-N\right]^2 = \sum_{k=-(N-1)}^{N-1}|r(k)-N\delta_k|^2 = 2\sum_{k=1}^{N-1}|r(k)|^2 = \text{ISL},\tag{2.10}$$

which is (2.6). Using the periodogram-based expression for $\Phi(\omega)$ (see (2.5)) in (2.6) shows that the problem of minimizing the ISL is equivalent to the minimization of the following frequency-domain metric:

$$\sum_{p=1}^{2N}\left(\left|\sum_{n=1}^{N}x(n)e^{-j\omega_p n}\right|^2-N\right)^2.\tag{2.11}$$

This equivalence result has an obvious intuitive interpretation: minimizing the ISL makes the sequence behave like white noise, and consequently its periodogram should be nearly constant in frequency.

The next point to note is that the criterion in (2.11) is a quartic function of $\{x(n)\}$. However, it can be verified that the minimization of (2.11) with respect to $\{x(n)\}$ is "almost equivalent" to the following simpler problem (whose criterion is a quadratic function of $\{x(n)\}$):

$$\min_{\{x(n)\}_{n=1}^{N};\{\psi_p\}_{p=1}^{2N}}\sum_{p=1}^{2N}\left|\sum_{n=1}^{N}x(n)e^{-j\omega_p n}-\sqrt{N}e^{j\psi_p}\right|^2.\tag{2.12}$$

Briefly speaking, if the criterion in (2.12) takes on a small value then so does (2.11), and vice versa. More specifically, (2.11) is equal to zero if and only if the criterion in (2.12) is equal to zero. Consequently, by continuity arguments, if the global minimum value of (2.11) is "sufficiently small" then the sequences minimizing (2.11) and, respectively, the criterion in (2.12) can be expected to be close to one another. See Appendix 2A for a detailed discussion on this "almost equivalence".

Let

$$\mathbf{a}_p^H = \left[e^{-j\omega_p}\ \cdots\ e^{-j2N\omega_p}\right],\tag{2.13}$$

let $\mathbf{A}^H$ be a unitary $2N \times 2N$ DFT matrix given by

$$\mathbf{A}^H = \frac{1}{\sqrt{2N}} \begin{bmatrix} \mathbf{a}_1^H \\ \vdots \\ \mathbf{a}_{2N}^H \end{bmatrix} \tag{2.14}$$

and let $\mathbf{z}$ be the sequence $\{x(n)\}_{n=1}^{N}$ padded with N zeros:

$$\mathbf{z} = \begin{bmatrix} x(1) \cdots x(N)\ 0 \cdots 0 \end{bmatrix}_{2N\times 1}^T . \tag{2.15}$$

Then the criterion in (2.12) can be rewritten in the more compact form (to within a multiplicative constant)

$$\left\|\mathbf{A}^H\mathbf{z} - \mathbf{v}\right\|^2 , \tag{2.16}$$

where

$$\mathbf{v} = \frac{1}{\sqrt{2}} \begin{bmatrix} e^{j\psi_1} \cdots e^{j\psi_{2N}} \end{bmatrix}^T . \tag{2.17}$$

For given $\{x(n)\}$, the minimization of (2.16) with respect to $\{\psi_p\}$ is immediate: let

$$\mathbf{f} = \mathbf{A}^H\mathbf{z} \tag{2.18}$$

denote the FFT of $\mathbf{z}$; then

$$\psi_p = \arg(f_p), \quad p = 1, \ldots, 2N. \tag{2.19}$$

Similarly, for a given $\mathbf{v}$, let

$$\mathbf{g} = \mathbf{A}\mathbf{v} \tag{2.20}$$

denote the inverse-FFT (IFFT) of $\mathbf{v}$. Because $||\mathbf{A}^H\mathbf{z} - \mathbf{v}||^2 = ||\mathbf{z} - \mathbf{A}\mathbf{v}||^2$, it follows that the minimizing sequence $\{x(n)\}$ is given by

$$x(n) = e^{j\arg(g_n)}, \quad n = 1, \ldots, N. \tag{2.21}$$

The CAN algorithm for the cyclic local minimization of the ISL-related metric in (2.12) is summarized in Table 2.1. A flowchart of CAN is presented in Figure 2.1. Owing to its simple (I)FFT operations, CAN can be used for very large values of N, such as $N \sim 10^6$ on an ordinary PC.

The CAN algorithm described above considers the constraint that the sequence $\{x(n)\}$ is unimodular, i.e., its peak-to-average power ratio (PAR) is equal to 1. Allowing the PAR value to be larger than 1 leads to better suppression of correlation sidelobes. See Chapter 4 for an extended CAN algorithm that deals with a relaxed PAR constraint.

In the next section we present an extended version of CAN which can cope with the WISL metric (with arbitrarily chosen weights) as defined in (2.3). The extended algorithm is called WeCAN (weighted CAN). The price paid for WeCAN's ability to deal with a general WISL metric is an increased computational burden compared with CAN. Specifically, as will be shown in the next section, each iteration of WeCAN requires N computations of $2N$-point (I)FFTs; thus the number of floating point operations required

Table 2.1. The CAN algorithm

Step 0: Set $\{x(n)\}_{n=1}^{N}$ to random initial values (e.g., $\{x(n)\}$ can be set to $\{e^{j2\pi\theta(n)}\}$ where $\{\theta(n)\}$ are independent random variables uniformly distributed in $[0, 2\pi]$), or $\{x(n)\}_{n=1}^{N}$ can be initialized by a good existing sequence such as a Golomb sequence.
Step 1: Compute the $\{\psi_p\}_{p=1}^{2N}$ that minimize the metric for $\{x(n)\}_{n=1}^{N}$ fixed at their most recent values (see (2.19)).
Step 2: Compute the sequence $\{x(n)\}_{n=1}^{N}$ that minimizes the metric, under the constraint $|x(n)| = 1$, for $\{\psi_p\}_{p=1}^{2N}$ fixed at their most recent values (see (2.21)).
Iteration: Repeat steps 1 and 2 until a pre-specified stop criterion is satisfied, e.g., $\|\mathbf{x}^{(i)} - \mathbf{x}^{(i+1)}\| < \epsilon$, where $\mathbf{x}^{(i)}$ is the sequence obtained at the ith iteration and ϵ is a predefined threshold, such as 10^{-3}.

by WeCAN is roughly N times larger than that required by CAN. Nonetheless, WeCAN can still be used for relatively large values of N, such as $N \sim 10^4$, on an ordinary PC.

2.2 Weighted cyclic algorithm-new (WeCAN)

In a similar way to that in the proof of (2.6) in Section 2.1, we can derive the following expression for the WISL metric (γ_k below is related to the weight w_k in (2.3) by $w_k = \gamma_k^2$):

$$\text{WISL} = 2\sum_{k=1}^{N-1} \gamma_k^2 |r(k)|^2 \tag{2.22}$$

$$= \frac{1}{2N}\sum_{p=1}^{2N} [\tilde{\Phi}(\omega_p) - \gamma_0 N]^2, \tag{2.23}$$

where

$$\tilde{\Phi}(\omega_p) \triangleq \sum_{k=-(N-1)}^{N-1} \gamma_k r(k) e^{-j\omega_p k}, \tag{2.24}$$

$$\omega_p = \frac{2\pi}{2N}p, \quad p = 1, \ldots, 2N,$$

and where $\{\gamma_k\}_{k=1}^{N-1}$ are real-valued (with $\gamma_k = \gamma_{-k}$). Note that, by choosing $\{\gamma_k\}_{k=1}^{N-1}$ appropriately, we can weigh the correlation lags in (2.22) in any desired way. Regarding γ_0, which does not enter into (2.22), it will be chosen to ensure that the matrix

$$\mathbf{\Gamma} = \frac{1}{\gamma_0}\begin{bmatrix} \gamma_0 & \gamma_1 & \cdots & \gamma_{N-1} \\ \gamma_1 & \gamma_0 & \ddots & \vdots \\ \vdots & \ddots & \ddots & \gamma_1 \\ \gamma_{N-1} & \cdots & \gamma_1 & \gamma_0 \end{bmatrix} \tag{2.25}$$

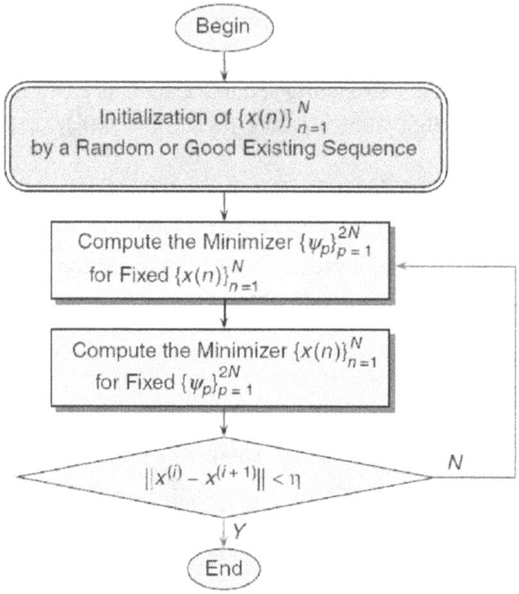

Figure 2.1 The CAN algorithm flowchart (see Table 2.1 for the detailed algorithm steps).

is positive semidefinite, which we denote by $\mathbf{\Gamma} \geq 0$. This can be done in the following simple way. Let $\tilde{\mathbf{\Gamma}}$ be the matrix $\gamma_0 \mathbf{\Gamma}$ with all diagonal elements set to 0, and let $\lambda_{\min}$ denote the minimum eigenvalue of $\tilde{\mathbf{\Gamma}}$; then $\mathbf{\Gamma} \geq 0$ if and only if $\gamma_0 + \lambda_{\min} \geq 0$, a condition that can always be satisfied by selecting γ_0.

Next we will derive a criterion which is "almost equivalent" to (2.23) and which depends quadratically on the unknowns $\{x(n)\}_{n=1}^{N}$, in a similar way to that in the previous section. To do so, we must apparently obtain a square root of $\tilde{\Phi}(\omega_p)$ in (2.24) that is linear in $\{x(n)\}_{n=1}^{N}$. Note the following DFT pairs:

$$\begin{aligned} \{r(k)\} &\longleftrightarrow \Phi(\omega) = |X(\omega)|^2, \\ \{\gamma_k r(k)\} &\longleftrightarrow \tilde{\Phi}(\omega) = \Gamma(\omega) * |X(\omega)|^2, \end{aligned} \tag{2.26}$$

where

$$X(\omega) = \sum_{n=1}^{N} x(n) e^{-jn\omega}, \quad \Gamma(\omega) = \sum_{k=-(N-1)}^{N-1} \gamma_k e^{-j\omega k} \tag{2.27}$$

and $*$ is the convolution operator. Thus $\tilde{\Phi}(\omega_p)$ can be expressed as follows:

$$\begin{aligned}\tilde{\Phi}(\omega_p) &= \frac{1}{2\pi}\int_{-\pi}^{\pi}\Gamma(\omega_p-\psi)|X(\psi)|^2\,d\psi \\ &= \frac{1}{2\pi}\int_{-\pi}^{\pi}\sum_{k=-(N-1)}^{N-1}\gamma_k e^{-jk(\omega_p-\psi)}\sum_{n=1}^{N}x(n)e^{-jn\psi}\sum_{\tilde{n}=1}^{N}x^*(\tilde{n})e^{j\tilde{n}\psi}\,d\psi \\ &= \sum_{k=-(N-1)}^{N-1}\sum_{n=1}^{N}\sum_{\tilde{n}=1}^{N}\gamma_k x(n)x^*(\tilde{n})\left(\frac{1}{2\pi}\int_{-\pi}^{\pi}e^{j\psi(k-n+\tilde{n})}\,d\psi\right)e^{-j\omega_p k}.\end{aligned} \tag{2.28}$$

It is easy to verify that

$$\frac{1}{2\pi}\int_{-\pi}^{\pi}e^{j\psi(k-n+\tilde{n})}\,d\psi = \delta_{k-(n-\tilde{n})}. \tag{2.29}$$

Thus

$$\tilde{\Phi}(\omega_p) = \sum_{n=1}^{N}\sum_{\tilde{n}=1}^{N}\gamma_{n-\tilde{n}}x(n)x^*(\tilde{n})e^{-j\omega_p(n-\tilde{n})} = \tilde{\mathbf{x}}_p^H(\gamma_0\mathbf{\Gamma})\tilde{\mathbf{x}}_p, \tag{2.30}$$

where

$$\tilde{\mathbf{x}}_p = \left[x(1)e^{-j\omega_p}\ x(2)e^{-j2\omega_p}\ \cdots\ x(N)e^{-jN\omega_p}\right]^T \tag{2.31}$$

and $\mathbf{\Gamma}$ is defined in (2.25). Therefore the WISL metric in (2.23) can be written as

$$\text{WISL} = \frac{\gamma_0^2}{2N}\sum_{p=1}^{2N}\left(\tilde{\mathbf{x}}_p^H\mathbf{\Gamma}\tilde{\mathbf{x}}_p - N\right)^2. \tag{2.32}$$

This expression suggests that the following problem can be expected to be "almost equivalent" to the minimization of the WISL metric:

$$\begin{aligned}\min_{\{x(n)\}_{n=1}^{N},\{\boldsymbol{\alpha}_p\}_{p=1}^{2N}} &\sum_{p=1}^{2N}\left\|\mathbf{C}\tilde{\mathbf{x}}_p-\boldsymbol{\alpha}_p\right\|^2 \\ \text{s.t. } &\|\boldsymbol{\alpha}_p\|^2 = N, \quad p=1,\ldots,2N, \\ &|x(n)| = 1, \quad n=1,\ldots,N,\end{aligned} \tag{2.33}$$

where "s.t." stands for "subject to" and the $N\times N$ matrix $\mathbf{C}$ is a square root of $\mathbf{\Gamma}$, i.e., $\mathbf{\Gamma} = \mathbf{C}^T\mathbf{C}$.

A cyclic algorithm for (2.33), which we will call WeCAN, can be derived as follows. For given $\{x(n)\}_{n=1}^{N}$, (2.33) decouples into $2N$ independent problems each of which has the following form:

$$\begin{aligned}\min_{\boldsymbol{\alpha}_p} &\ \|\mathbf{f}_p-\boldsymbol{\alpha}_p\|^2 \\ \text{s.t. } &\|\boldsymbol{\alpha}_p\|^2 = N,\end{aligned} \tag{2.34}$$

where the $N \times 1$ vector $\mathbf{f}_p = \mathbf{C}\tilde{\mathbf{x}}_p$ is given. Note that under the constraint $\|\boldsymbol{\alpha}_p\|^2 = N$ we have

$$\begin{aligned} \|\mathbf{f}_p - \boldsymbol{\alpha}_p\|^2 &= \text{const} - 2\,\text{Re}\{\mathbf{f}_p^H \boldsymbol{\alpha}_p\} \\ &\geq \text{const} - 2\|\mathbf{f}_p\|\,\|\boldsymbol{\alpha}_p\| = \text{const} - 2N\|\mathbf{f}_p\|, \end{aligned} \tag{2.35}$$

where equality is achieved if and only if

$$\boldsymbol{\alpha}_p = \sqrt{N}\frac{\mathbf{f}_p}{\|\mathbf{f}_p\|}. \tag{2.36}$$

This is therefore the solution to the minimization problem in (2.33) for given $\{x(n)\}_{n=1}^N$. Note that $\{\mathbf{f}_p\}_{p=1}^{2N}$ can be computed using an FFT. Indeed, let c_{kn} denote the (k,n)th element of $\mathbf{C}$ and define

$$\mathbf{z}_k = \left[c_{k1}x(1) \cdots c_{kN}x(N)\; 0 \cdots 0\right]^T_{(2N\times 1)} \tag{2.37}$$

and

$$\mathbf{F} = \sqrt{2N}\mathbf{A}^H\left[\mathbf{z}_1 \cdots \mathbf{z}_N\right]_{2N\times N}, \tag{2.38}$$

where the unitary $2N \times 2N$ DFT matrix $\mathbf{A}^H$ was defined in (2.14). Then it is not difficult to see that the transpose of the vector $\mathbf{f}_p$ is given by the pth row of $\mathbf{F}$.

Next we show that, for given $\{\boldsymbol{\alpha}_p\}_{p=1}^{2N}$, the minimization problem in (2.33) with respect to $\{x(n)\}_{n=1}^N$ also has a closed-form solution. Let α_{pk} denote the kth element of $\boldsymbol{\alpha}_p$ and let $\mathbf{a}_p^H$ be given by (2.13). Using this notation, the criterion in (2.33) can be written as

$$\begin{aligned} \sum_{p=1}^{2N} \left\|\mathbf{C}\tilde{\mathbf{x}}_p - \boldsymbol{\alpha}_p\right\|^2 &= \sum_{k=1}^{N}\sum_{p=1}^{2N} \left|\mathbf{a}_p^H \mathbf{z}_k - \alpha_{pk}\right|^2 \\ &= \sum_{k=1}^{N} \left\|\mathbf{A}^H\mathbf{z}_k - \boldsymbol{\beta}_k\right\|^2 = \sum_{k=1}^{N} \left\|\mathbf{z}_k - \mathbf{A}\boldsymbol{\beta}_k\right\|^2, \end{aligned} \tag{2.39}$$

where

$$\boldsymbol{\beta}_k = \frac{1}{\sqrt{2N}}\left[\alpha_{1k} \cdots \alpha_{2N,k}\right]^T, \quad k = 1, \ldots, N. \tag{2.40}$$

For a generic element of $\{x(n)\}_{n=1}^N$, denoted as x, (2.39) becomes

$$\sum_{k=1}^{N} |\mu_k x - \nu_k|^2 = \text{const} - 2\,\text{Re}\left[\left(\sum_{k=1}^{N}\mu_k^*\nu_k\right)x^*\right], \tag{2.41}$$

where μ_k and ν_k are given by the corresponding elements in $\mathbf{z}_k$ and $\mathbf{A}\boldsymbol{\beta}_k$, respectively. Under the unit-modulus constraint (see (1.4)), the minimizer x of the criterion in (2.41) is given by

$$x = e^{j\phi}, \quad \phi = \arg\left(\sum_{k=1}^{N}\mu_k^*\nu_k\right). \tag{2.42}$$

This observation concludes the derivation of the main steps of the WeCAN algorithm, whose summary is given in Table 2.2.

Table 2.2. The WeCAN algorithm

Step 0: Set the $\{x(n)\}_{n=1}^{N}$ to some initial values and select the desired weights $\{\gamma_k\}_{k=1}^{N-1}$; also choose γ_0 such that the matrix $\mathbf{\Gamma}$ in (2.25) is positive semidefinite.
Step 1: Compute the $\{\alpha_p\}_{p=1}^{2N}$ that minimize the criterion in (2.33) for $\{x(n)\}_{n=1}^{N}$ fixed at their most recent values (see (2.36)).
Step 2: Compute the sequence $\{x(n)\}_{n=1}^{N}$ that minimizes the criterion in (2.33) for $\{\alpha_p\}_{p=1}^{2N}$ fixed at their most recent values (see (2.42)).
Iteration: Repeat steps 1 and 2 until a prespecified stop criterion is satisfied (see the CAN algorithm in Table 2.1).

2.3 Numerical examples

2.3.1 Integrated sidelobe level (ISL) design

We compare the merit factors of the Golomb sequence (see (1.21) for the definition), the CAN sequence initialized by the Golomb sequence, the m-sequence (see Figure 1.4 and the associated discussion), and the random-phase sequence (as in (1.4) with $\{\phi(n)\}$ being independent random variables uniformly distributed in $[0, 2\pi]$). Note that the length of the m-sequence is restricted to $2^n - 1$ with integer n whereas the lengths of the other three are not restricted. We computed the merit factors of the above four types of sequences, denoted as Golomb, CAN(G), m-seq and random-phase seq, respectively, for the lengths N given by $N + 1 = 2^5, 2^6, 2^7, 2^8, 2^9, 2^{10}, 2^{11}, 2^{12}, 2^{13}$. The results are shown in Figure 2.2 using a log–log scale. For all sequence lengths considered, the CAN(G) sequence gives the largest merit factor. When $N = 2^{13} - 1$, the merit factor of the CAN(G) sequence is 1431.0, which is more than ten times larger than that of the Golomb sequence (which is 142.2). Figure 2.3 gives for comparison the correlation levels of the CAN(G), m- and random-phase sequences of length $N = 127$. The correlation level is defined as follows:

$$\text{correlation level} = 20\log_{10}\left|\frac{r(k)}{r(0)}\right|, \quad k = 1, \ldots, N-1. \tag{2.43}$$

It is easy to observe that the CAN sequence offers lower correlation sidelobes than either the m-sequence or the random-phase sequence. We also show the correlation levels of the Golomb and CAN(G) sequences of lengths $N = 10^2$ and 10^3, in Figures 2.4 and 2.5, respectively. We note that the correlation sidelobes of the Golomb sequence are comparatively large for k close to 0 and $N - 1$, while the CAN(G) sequence has relatively more uniform correlation sidelobes as k increases from 0 to $N - 1$.

2.3.2 Weighted integrated sidelobe level (WISL) design

Consider the design of a sequence of length $N = 100$. Suppose that we are interested in suppressing the correlations $r_1, \ldots, r_{25}$ and $r_{70}, \ldots, r_{79}$. We apply the WeCAN algorithm

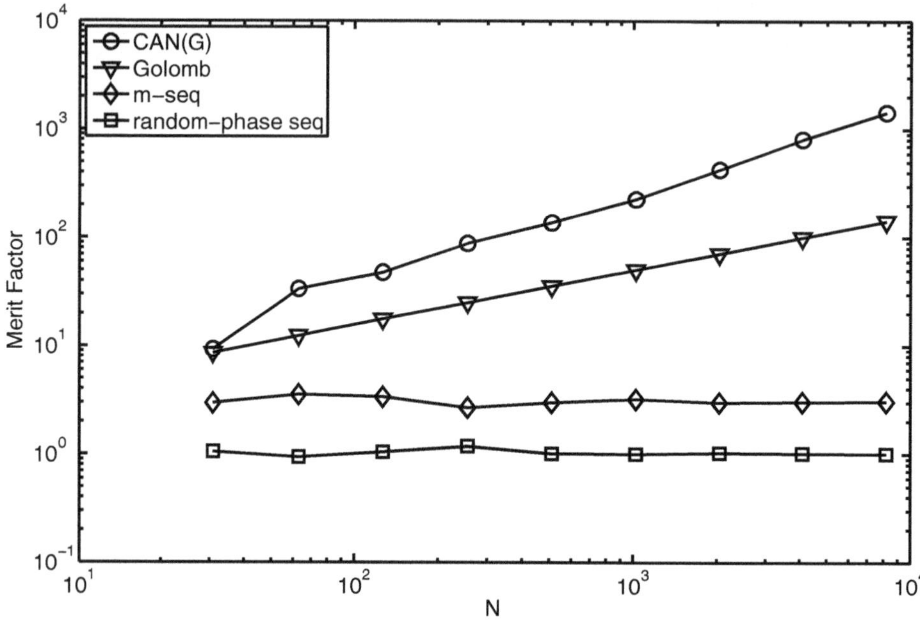

Figure 2.2 The merit factors of the Golomb, CAN(G), m- and random-phase sequences with lengths from $2^5 - 1$ up to $2^{13} - 1$.

Table 2.3. MMF values for the weights in (2.45), $N = 100$

	Golomb	CAN(G)	WeCAN
MMF	32.55	142.64	1.06×10^{21}

using the following weights in the matrix $\mathbf{\Gamma}$ in (2.25):

$$\gamma_k = \begin{cases} 1, \ k \in [1, 25] \cup [70, 79], \\ 0, \ k \in [26, 69] \cup [80, 99] \end{cases} \tag{2.44}$$

(γ_0 is chosen to ensure the positive semidefiniteness of $\mathbf{\Gamma}$; more exactly we choose $\gamma_0 = 12.05$ following the discussion after (2.25).) In this scenario, the MMF is as defined in (2.4) with

$$w_k = \gamma_k^2 = \begin{cases} 1, \ k \in [1, 25] \cup [70, 79], \\ 0, \ k \in [26, 69] \cup [80, 99]. \end{cases} \tag{2.45}$$

A randomly generated sequence is used to initialize the WeCAN algorithm. The correlation level of the designed sequence is shown in Figure 2.6. The WeCAN sequence has correlation sidelobes that are practically zero at the required lags and are much smaller than the sidelobes of the Golomb or CAN(G) sequences discussed in the last subsection (see Figures 2.4(a) and 2.4(b)). Table 2.3 presents the corresponding MMF values. The MMF of the WeCAN sequence (which is practically infinite) is significantly larger than the other MMF values in the table.

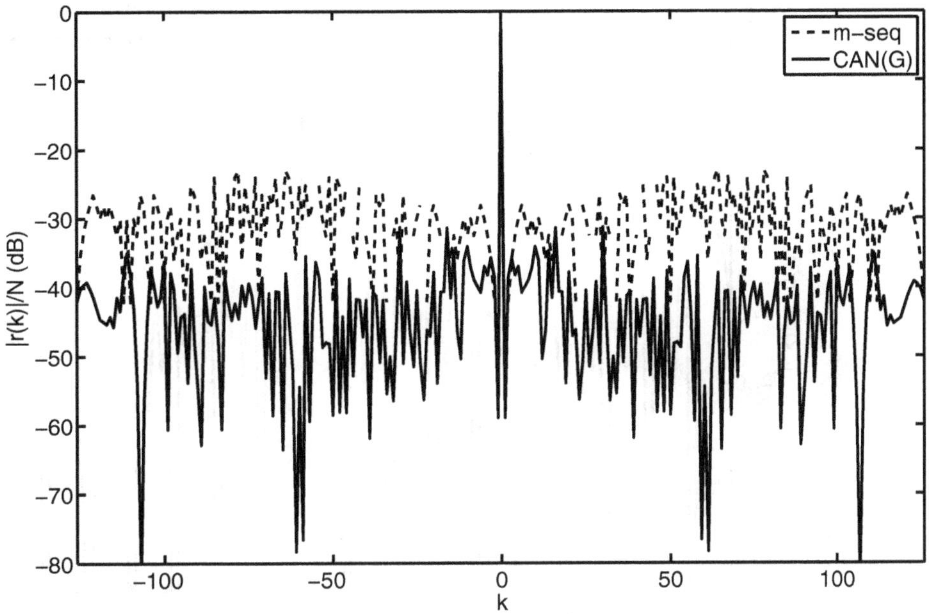

(a)

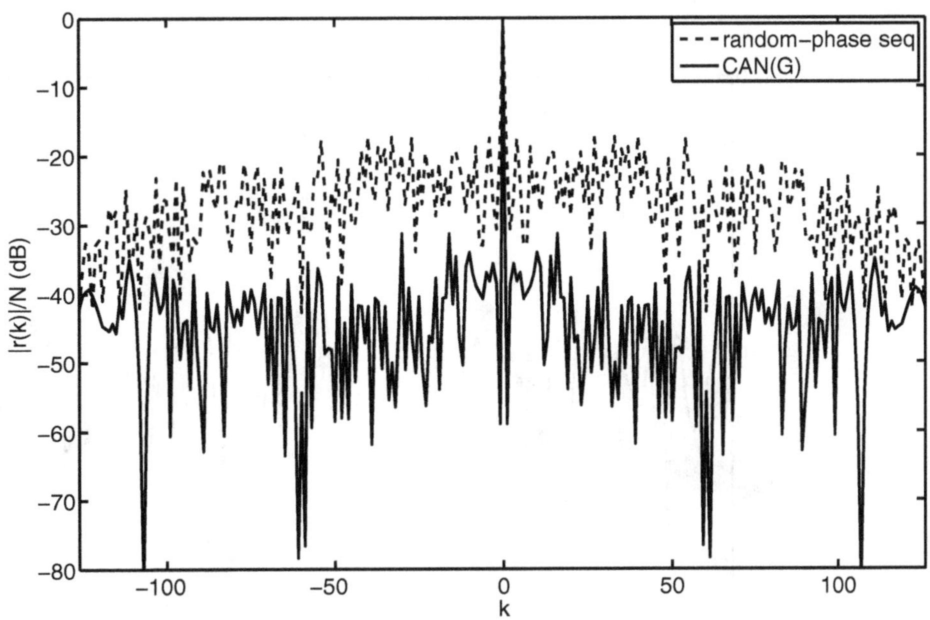

(b)

Figure 2.3 Correlation levels of the CAN(G), m- and random-phase sequences of length $N = 127$, designed under the ISL metric: (a) the CAN(G) and m- sequences and (b) the CAN(G) and random-phase sequences.

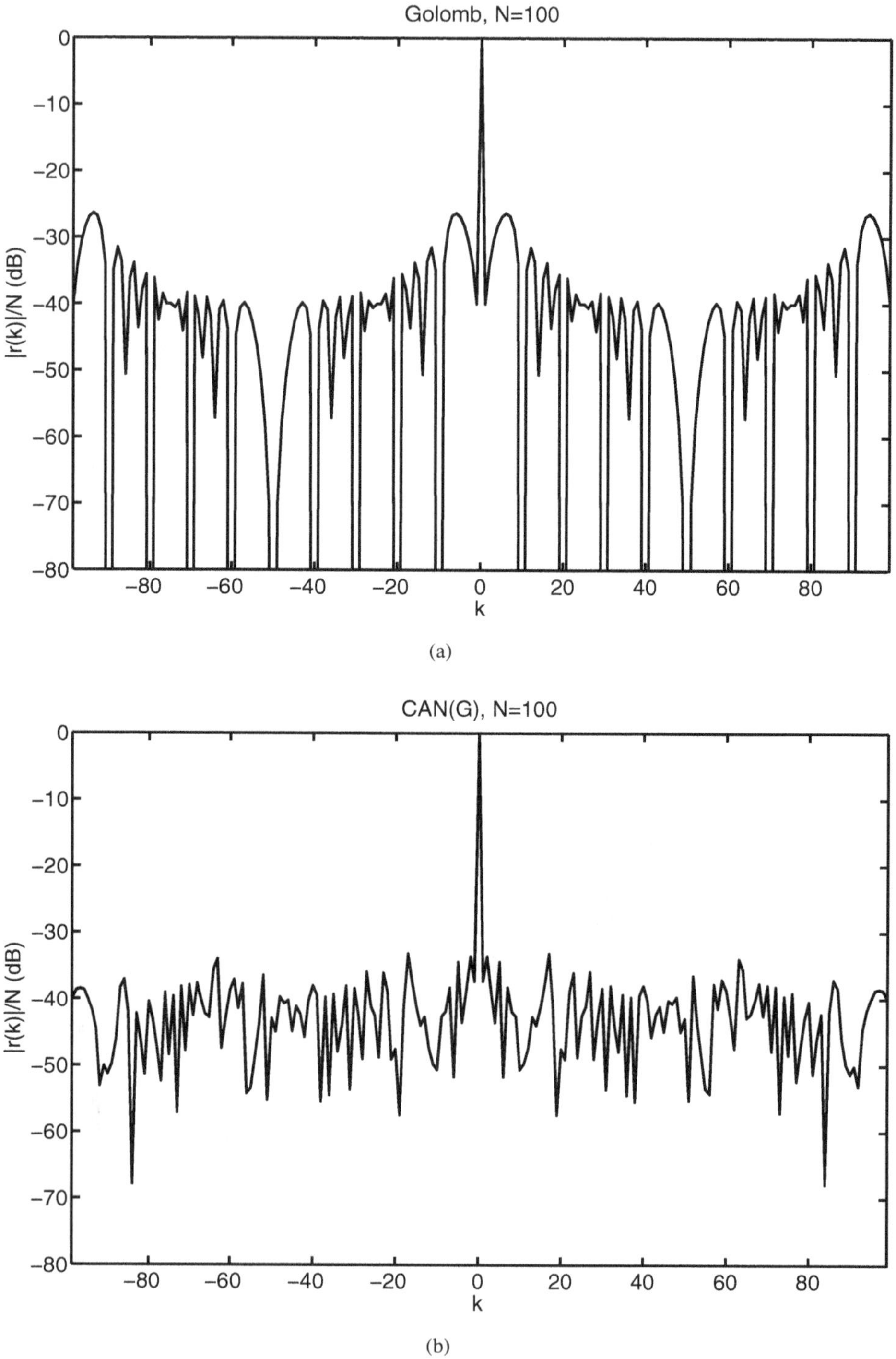

(a)

(b)

Figure 2.4 Correlation levels of the Golomb and CAN sequences of length $N = 10^2$ designed under the ISL metric: (a) the Golomb sequence and (b) the CAN(G) sequence.

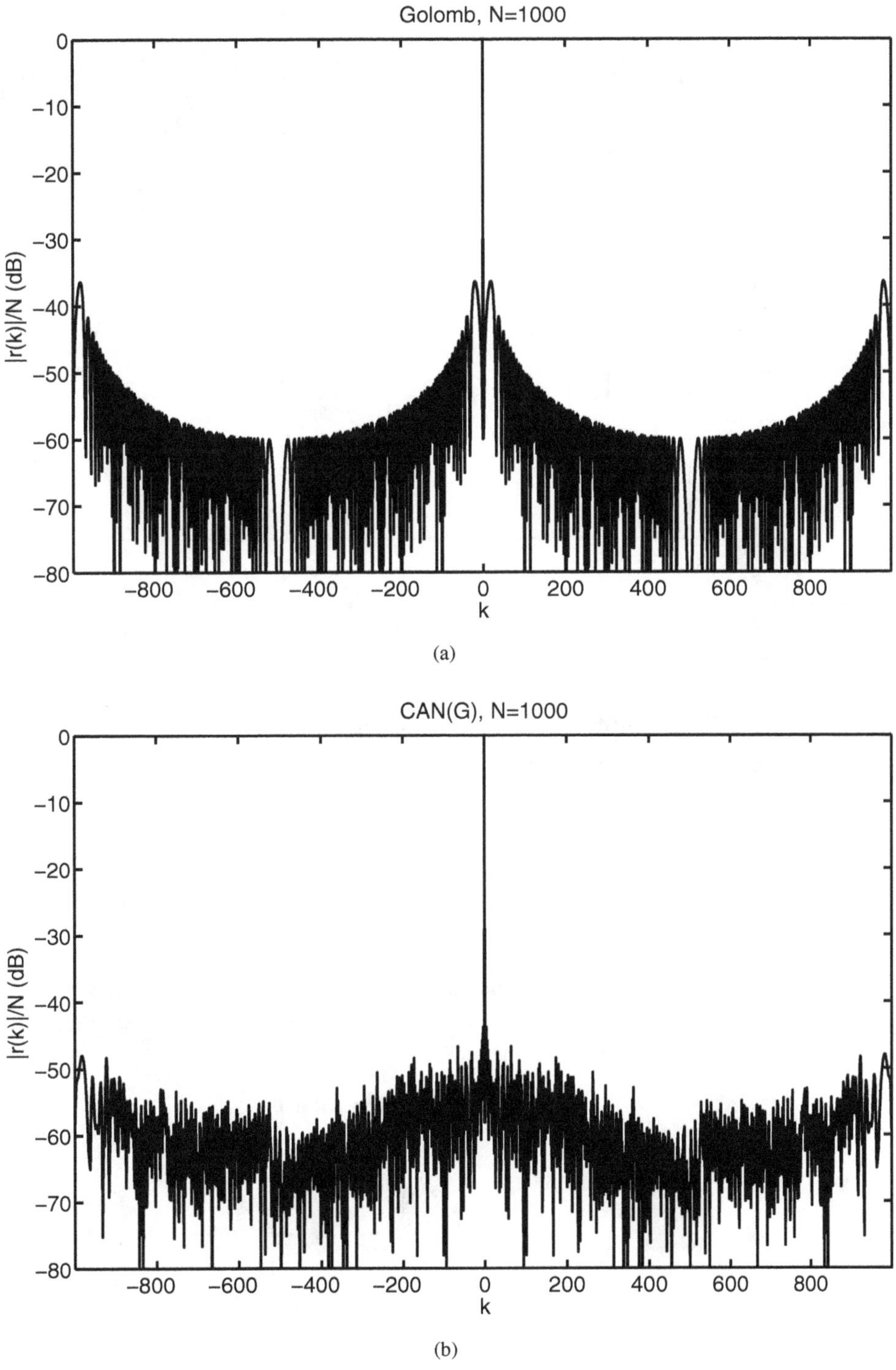

Figure 2.5 Correlation levels of the Golomb and CAN sequences of length $N = 10^3$ designed under the ISL metric: (a) the Golomb sequence and (b) the CAN(G) sequence.

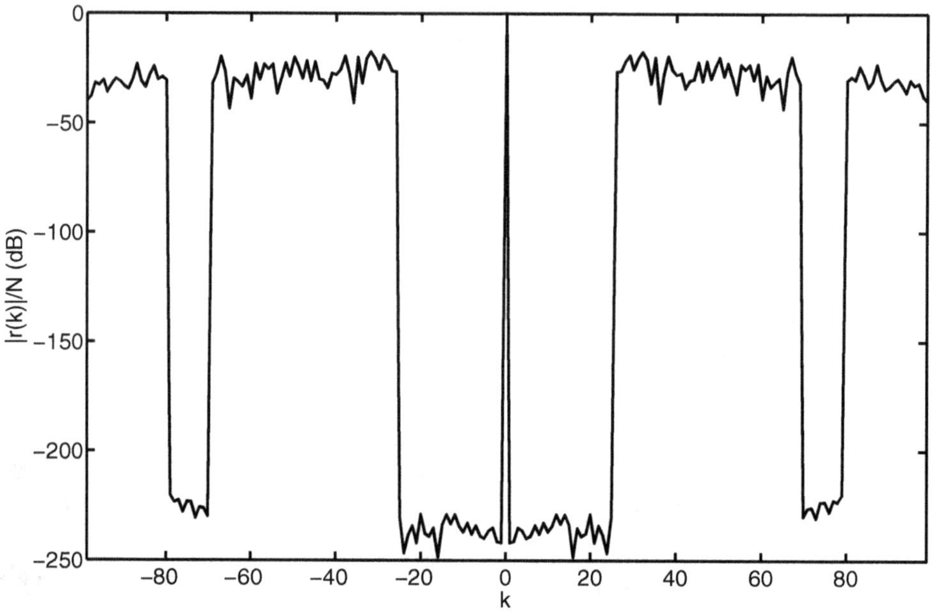

Figure 2.6 The correlation level of a WeCAN sequence designed under the WISL metric with the weights given in (2.45).

2.3.3 Channel estimation in communications

Consider an FIR (finite impulse response) channel impulse response $\{h_p\}_{p=0}^{P-1}$, whose estimation is our main goal (the number of channel taps P is assumed to be known). Suppose that we transmit a probing sequence $\{x(n)\}_{n=1}^{N}$ and obtain a received signal

$$y_n = \sum_{p=0}^{P-1} h_p x(n-p) + e_n, \quad n = 1, \dots, N+P-1, \tag{2.46}$$

where $\{e_n\}_{n=1}^{N+P-1}$ is an independent and identically distributed (i.i.d.) complex Gaussian white noise sequence with zero mean and variance σ^2. Equation (2.46) can be written in the following more compact form:

$$\mathbf{y} = \bar{\mathbf{X}}\mathbf{h} + \mathbf{e}, \tag{2.47}$$

where $\bar{\mathbf{X}}$ is defined as

$$\bar{\mathbf{X}} = \begin{bmatrix} x(1) & & \mathbf{0} \\ \vdots & \ddots & \\ \vdots & & x(1) \\ x(N) & & \vdots \\ & \ddots & \vdots \\ \mathbf{0} & & x(N) \end{bmatrix}_{(N+P-1)\times P} \tag{2.48}$$

and

$$\mathbf{y} = \begin{bmatrix} y_1 & \cdots & y_{N+P-1} \end{bmatrix}^T, \ \mathbf{h} = \begin{bmatrix} h_0 & \cdots & h_{P-1} \end{bmatrix}^T,$$
$$\mathbf{e} = \begin{bmatrix} e_1 & \cdots & e_{N+P-1} \end{bmatrix}^T. \tag{2.49}$$

Let $\bar{\mathbf{x}}_p$ denote the pth column of the matrix $\bar{\mathbf{X}}$. We use $\bar{\mathbf{x}}_p$ as a "matched filter" to determine h_p from $\mathbf{y}$, which leads to the following estimate of h_p:

$$\hat{h}_p = \frac{1}{N}\bar{\mathbf{x}}_p^*\mathbf{y}. \tag{2.50}$$

Let the number of channel taps P be 40. Figure 2.7(a) shows the magnitude of the simulated channel impulse response $\{|h_p|\}_{p=0}^{P-1}$. We performed two experiments to compare the Golomb sequence and the WeCAN sequence. The WeCAN sequence was generated with the following weights:

$$\gamma_k = \begin{cases} 1, \ k \in [1, 39], \\ 0, \ k \in [40, N-1], \end{cases}$$
$$w_k = \gamma_k^2, \quad k = 1, \ldots, N-1 \tag{2.51}$$

(as usual, γ_0 is selected such that $\mathbf{\Gamma} \geq 0$) and its auto-correlation level is shown in Figure 2.7(b) for $N = 100$. In one experiment the noise power σ^2 was fixed at 10^{-4} and the sequence length N was varied from 100 to 500; in the other experiment N was fixed at 200 and σ^2 was varied from 10^{-6} to 1. For each pair (N, σ^2), 500 Monte-Carlo trials were run (in which the noise sequence $\mathbf{e}$ was varied) and the mean-squared error (MSE) of $\hat{\mathbf{h}}$ was recorded. Figure 2.8 shows the MSE of $\hat{\mathbf{h}}$ in the two situations. Owing to its better auto-correlation properties, the WeCAN sequence generates consistently smaller MSE values than the Golomb sequence. In particular, it is interesting to observe from Figure 2.8(b) that, as σ^2 decreases, the MSE of $\hat{\mathbf{h}}$ corresponding to the WeCAN sequence decreases linearly (and becomes 0 as σ^2 goes to 0), while the performance of the Golomb sequence is limited to a certain level because of its nonzero correlation sidelobes, which induce an estimation bias.

2.3.4 Quantization effects

We have assumed that the phases of the designed sequences can take on any values from 0 to 2π. In practice it might be required that the phases are drawn from a discrete constellation. Thus we briefly examine here the performance of the designed sequences under quantization.

Let $\{x(n)\}_{n=1}^N$ denote a sequence obtained from one of the algorithms discussed in this chapter. Suppose that the quantization level is 2^q where $q \geq 1$ is an integer. Then the quantized polyphase sequence can be expressed as

$$\hat{x}(n) = \exp\left(j\left\lfloor\frac{\arg\{x(n)\}}{2\pi/2^q}\right\rfloor\frac{2\pi}{2^q}\right), \quad n = 1, \ldots, N. \tag{2.52}$$

We quantize the phases of the CAN sequence used in Figure 2.2 into 32 levels (i.e., $q = 5$) and make the same comparisons with the Golomb, m- and random-phase

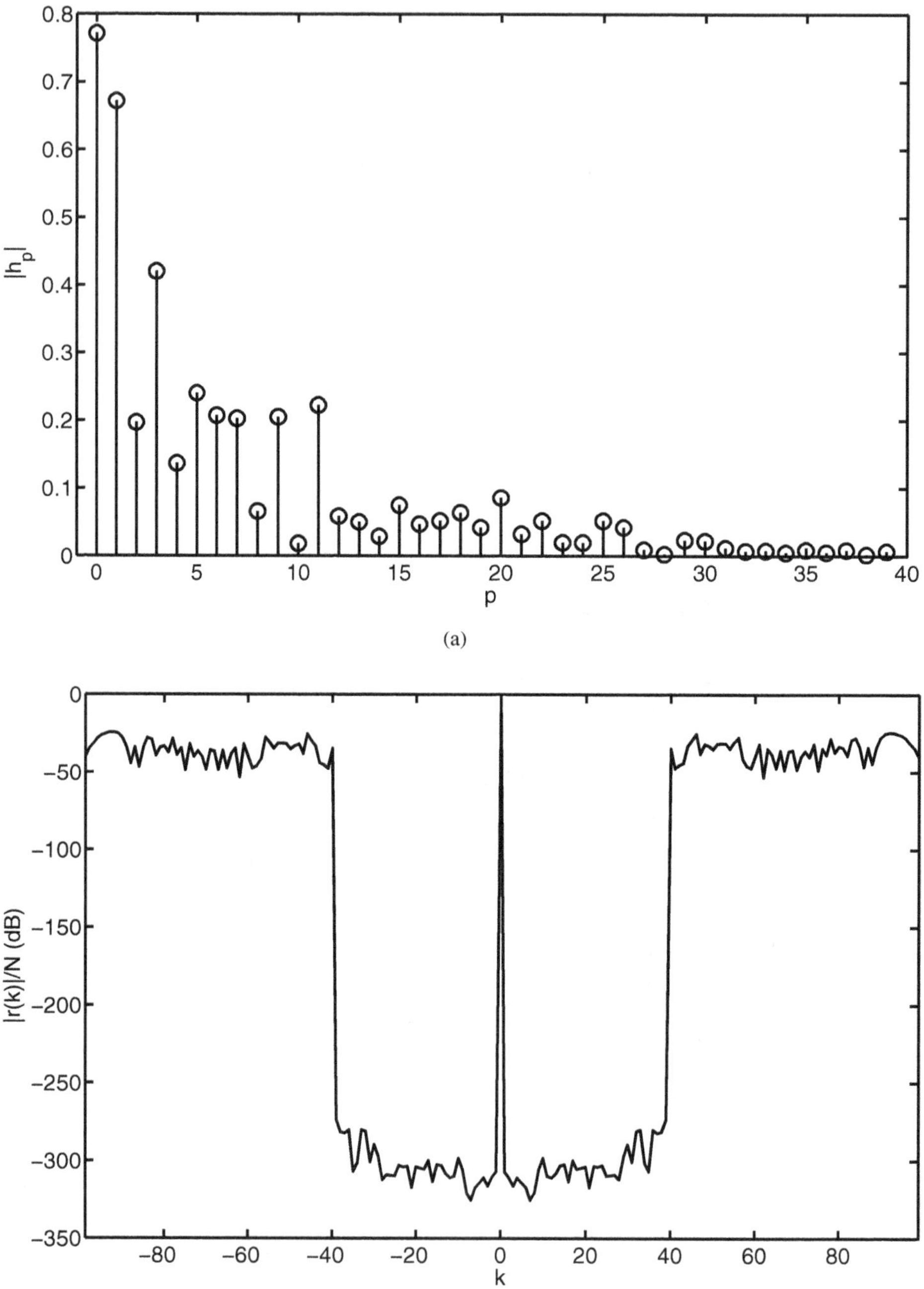

(a)

(b)

Figure 2.7 (a) The magnitude of the simulated channel impulse response $\mathbf{h}$. (b) The auto-correlation level of the WeCAN sequence designed under the WISL metric with weights in (2.51).

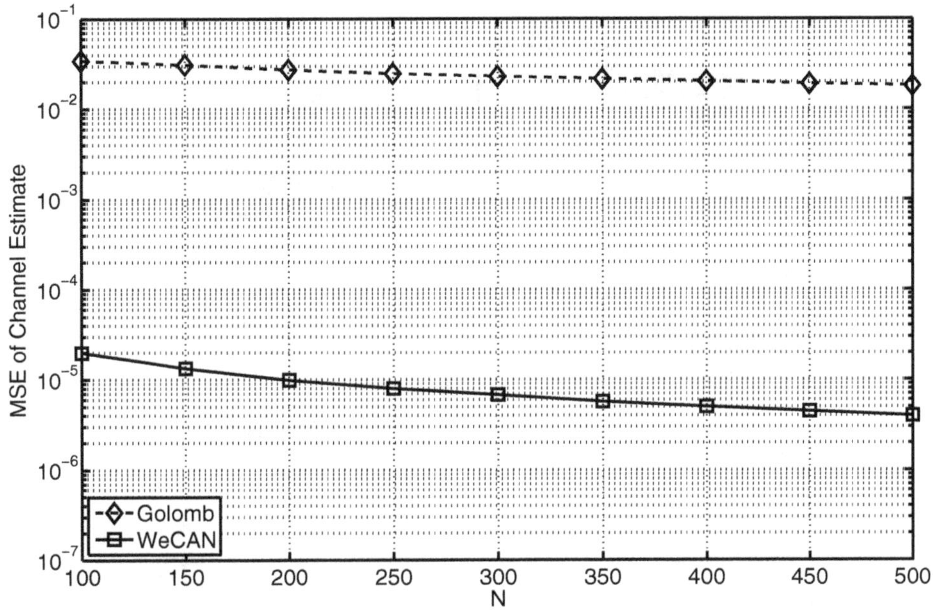

(a)

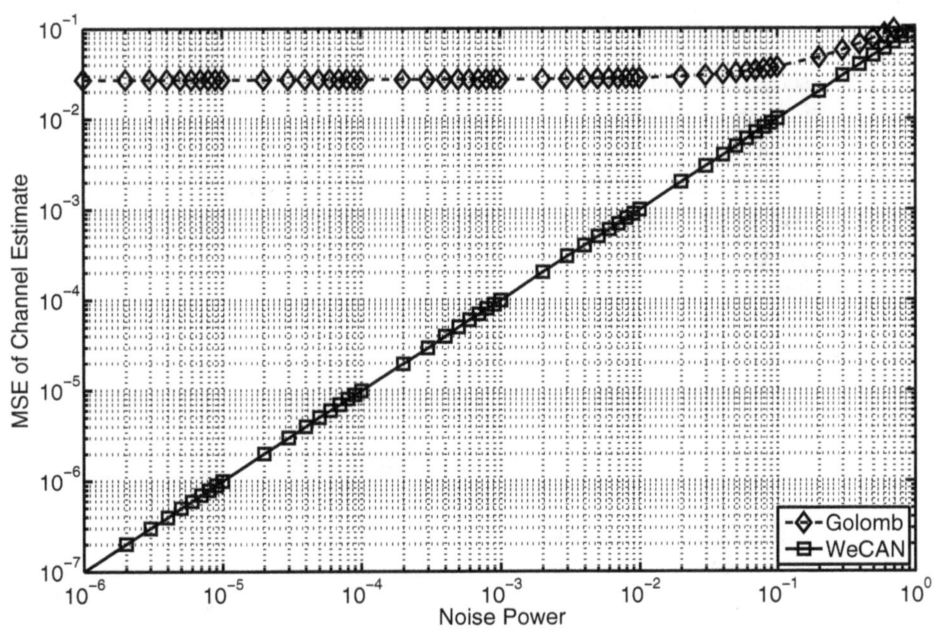

(b)

Figure 2.8 The MSE of the estimated $\hat{\mathbf{h}}$ using two training sequences, the Golomb sequence and the WeCAN sequence. (a) The noise power σ^2 was fixed at 10^{-4} and the sequence length N was varied from 100 to 500; (b) N was fixed at 200 and σ^2 was varied from 10^{-6} to 1.

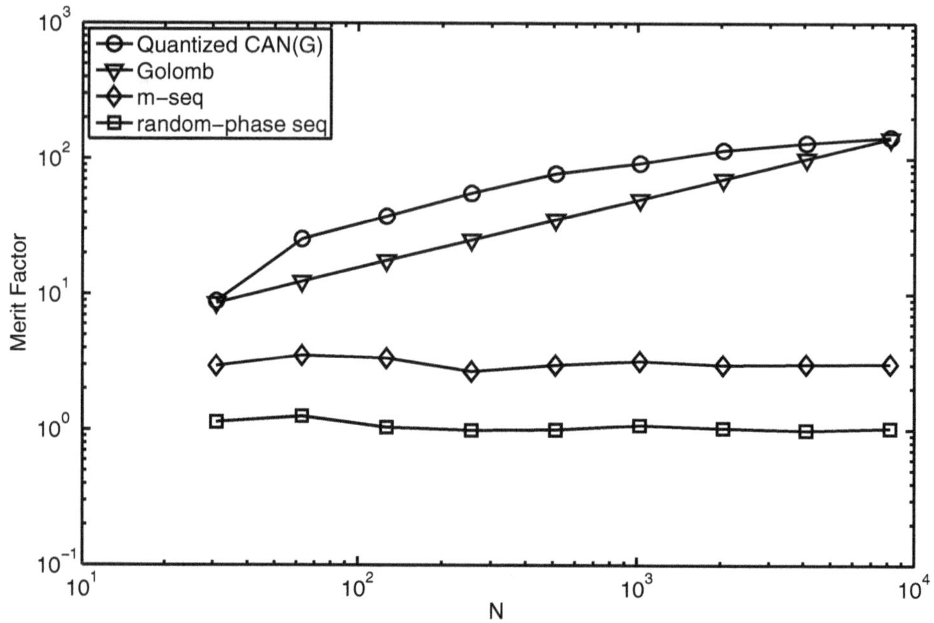

Figure 2.9 The same as Figure 2.2 except that the phases of the CAN sequence used here are quantized into 32 levels.

sequences. The results are shown in Figure 2.9. We observe that, owing to quantization, the merit factor of the CAN sequence is reduced (i.e., the correlation sidelobes are increased), but it remains the largest among the four sequences being compared. We also performed a 32-level quantization for the WeCAN sequence used in Figure 2.6; the obtained correlations are shown in Figure 2.10. For this case, the correlations in the region of interest are raised significantly, albeit they are still much lower than the correlation sidelobes outside the region of interest.

2.4 Conclusions

We have presented two cyclic algorithms, namely CAN and WeCAN, which can be used to design unimodular sequences that have good auto-correlation properties. The CAN algorithm can be used to design very long sequences (of length N up to 10^6) using an ordinary PC, a design problem that can hardly be handled by other algorithms proposed in the previous literature. CAN deals with the ISL metric, i.e., it considers all unweighted correlation lags from $r(1)$ up to $r(N-1)$, whereas WeCAN aims to minimize weighted-ISL metrics. We have shown that, in particular, the latter algorithm can be used to design sequences that have virtually zero auto-correlation sidelobes in a specified lag interval. WeCAN can be used to design sequences of lengths $N \sim 10^4$ or larger using an ordinary PC, depending on how many lags are considered. A number of numer-

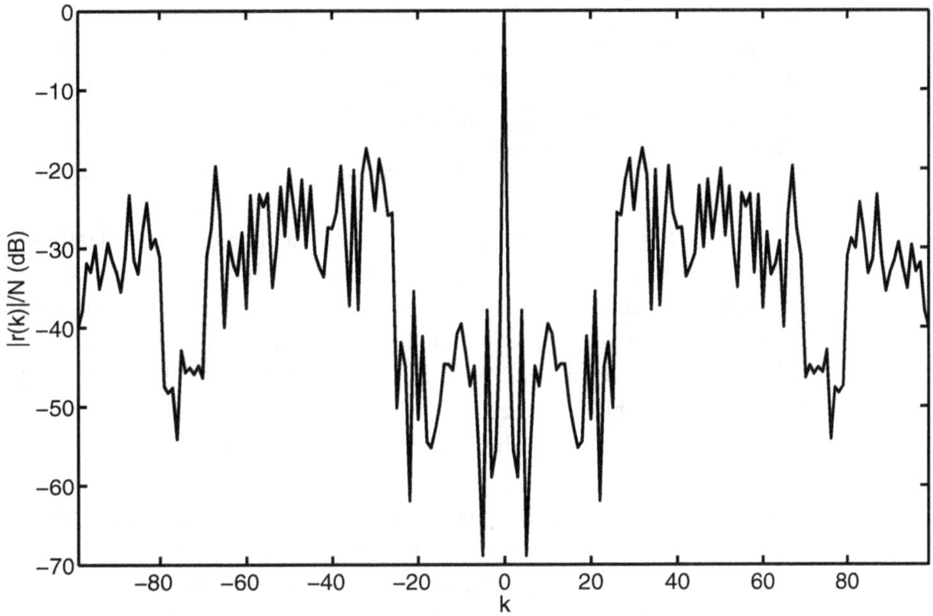

Figure 2.10 The same as Figure 2.6 except that the phases of the WeCAN sequence used here are quantized into 32 levels.

ical examples have been provided to demonstrate the good auto-correlation properties of the unimodular sequences designed using the proposed algorithms.

Appendix 2A Connections with a phase-retrieval algorithm

The CAN algorithm introduced in this chapter for designing code sequences with impulse-like correlations is related to the Gerchberg–Saxton algorithm (GSA) introduced for phase retrieval in the optics literature [Gerchberg & Saxton 1972] some 40 years ago. The technique used in GSA in fact appeared earlier in [Sussman 1962], so GSA might be better named the Sussman–Gerchberg–Saxton algorithm. However, to be consistent with most other literature, we will use the name GSA here.

In this appendix we attempt to describe and clarify the relationship between CAN and GSA. We also present some facts on GSA that appear to be useful in their own right.

Gerchberg–Saxton algorithm (GSA)

Let $\mathbf{x}$ be an $N \times 1$ vector and consider the problem of minimizing w.r.t. (with respect to) $\mathbf{x}$ the criterion

$$C(\mathbf{x}) = \sum_{k=1}^{K} \left(|\mathbf{a}_k^H \mathbf{x}| - d_k \right)^2 \tag{2.53}$$

Table 2.4. The Gerchberg–Saxton algorithm

Step 0: Given initial values $\{\phi_k^0\}_{k=1}^K$ ($\{\phi_k\}$ are auxiliary variables; see the main text for details), iterate steps 1 and 2 below for $i = 0, 1, \ldots$ until convergence.
Step 1: $\mathbf{x}^i = \arg\min_{\mathbf{x}\in S} \sum_{k=1}^K |\mathbf{a}_k^H \mathbf{x} - d_k e^{j\phi_k^i}|^2$.
Step 2: $\phi_k^{i+1} = \arg(\mathbf{a}_k^H \mathbf{x}^i)$ and $i \leftarrow i + 1$.

where $d_k \in \mathbb{R}^+$ and $\mathbf{a}_k \in \mathbb{C}^{N\times 1}$ are given and K is an integer that typically satisfies $K \geq N$. In some applications the vector $\mathbf{x}$ is free to vary in $\mathbb{C}^{N\times 1}$ [Weiss & Picard 2008]. In others, $\mathbf{x}$ is constrained to a certain subset of $\mathbb{C}^{N\times 1}$, for example to the set of vectors with unimodular elements (i.e., $|x_k| = 1$). To take this into account, we let $\mathbf{x} \in S \subseteq \mathbb{C}^{N\times 1}$.

The GSA was introduced in [Gerchberg & Saxton 1972][Sussman 1962] for tackling recovery problems typically involving a sequence and its Fourier transform. When used for problems that can be formulated as in (2.53), GSA has the form described in Table 2.4.

Note that [Gerchberg & Saxton 1972] proposed the above algorithm on heuristic grounds, without any reference to the minimization of $C(\mathbf{x})$ in (2.53). However it was realized later on in [Fienup 1982] that GSA is a minimization algorithm for (2.53) which has the appealing property of monotonically decreasing the criterion as the iteration proceeds. A simple proof of this fact is as follows:

$$
\begin{aligned}
C(\mathbf{x}^i) &= \sum_{k=1}^K \left(|\mathbf{a}_k^H \mathbf{x}^i| - d_k\right)^2 = \sum_{k=1}^K \left|\mathbf{a}_k^H \mathbf{x}^i - d_k e^{j\phi_k^{i+1}}\right|^2 \\
&\geq \sum_{k=1}^K \left|\mathbf{a}_k^H \mathbf{x}^{i+1} - d_k e^{j\phi_k^{i+1}}\right|^2 \geq \sum_{k=1}^K \left|\mathbf{a}_k^H \mathbf{x}^{i+1} - d_k e^{j\phi_k^{i+2}}\right|^2 = C(\mathbf{x}^{i+1})
\end{aligned} \tag{2.54}
$$

where the first inequality is due to step 1 of the algorithm and the second inequality is due to step 2 (these inequalities are strict if the solutions computed in steps 1 and 2 are unique, which is usually the case in applications).

The calculation in (2.54) provides a way to *motivate* GSA as a minimization algorithm for $C(\mathbf{x})$. In the following we outline a way to *derive* GSA as a minimizing procedure for $C(\mathbf{x})$.

Let $\boldsymbol{\phi}$ denote a $K \times 1$ vector of auxiliary variables and let $D(\mathbf{x}, \boldsymbol{\phi})$ be a function which has the property that

$$
\min_{\boldsymbol{\phi}} D(\mathbf{x}, \boldsymbol{\phi}) = C(\mathbf{x}). \tag{2.55}
$$

Then, under rather general conditions, the $\mathbf{x}$ that minimizes $C(\mathbf{x})$ is the same as the $\mathbf{x}$ obtained from the minimization of $D(\mathbf{x}, \boldsymbol{\phi})$ w.r.t. both $\mathbf{x}$ and $\boldsymbol{\phi}$. Evidently, for this approach to be useful the minimization of $D(\mathbf{x}, \boldsymbol{\phi})$ should be easier to handle than that of $C(\mathbf{x})$. To use the above idea in the present case of (2.53), we let

$$
D(\mathbf{x}, \boldsymbol{\phi}) = \sum_{k=1}^K |\mathbf{a}_k^H \mathbf{x} - d_k e^{j\phi_k}|^2 \tag{2.56}
$$

(where $\boldsymbol{\phi}$ is the vector made from $\{\phi_k\}_{k=1}^K$) and note that the above function has the required property:

$$\min_{\boldsymbol{\phi}} D(\mathbf{x}, \boldsymbol{\phi}) = \min_{\boldsymbol{\phi}} \sum_{k=1}^{K} \left\{ \left|\mathbf{a}_k^H \mathbf{x}\right|^2 + d_k^2 - 2\left|\mathbf{a}_k^H \mathbf{x}\right| d_k \cos\left[\arg(\mathbf{a}_k^H \mathbf{x}) - \phi_k\right] \right\}$$

$$= \sum_{k=1}^{K} \left(|\mathbf{a}_k^H \mathbf{x}| - d_k\right)^2 = C(\mathbf{x}). \tag{2.57}$$

Clearly the minimization of $D(\mathbf{x}, \boldsymbol{\phi})$ w.r.t. $\mathbf{x}$ (unconstrained as in [Weiss & Picard 2008] or constrained as in this chapter) for fixed $\boldsymbol{\phi}$ or w.r.t. $\boldsymbol{\phi}$ for fixed $\mathbf{x}$ has simple closed-form solutions. Consequently $D(\mathbf{x}, \boldsymbol{\phi})$, and hence $C(\mathbf{x})$, can be minimized conveniently via a cyclic algorithm in which $\boldsymbol{\phi}$ is fixed to its most recent value and $D(\mathbf{x}, \boldsymbol{\phi})$ is minimized w.r.t. $\mathbf{x}$, and vice versa. The algorithm thus obtained is nothing other than the GSA in Table 2.4 and its property in (2.54) follows immediately from (2.55) and the fact that the cyclic minimization of $D(\mathbf{x}, \boldsymbol{\phi})$ yields a monotonically decreasing sequence of criterion values: $C(\mathbf{x}^i) = D(\mathbf{x}^i, \boldsymbol{\phi}^{i+1}) \geq D(\mathbf{x}^{i+1}, \boldsymbol{\phi}^{i+2}) = C(\mathbf{x}^{i+1})$.

To conclude this discussion we note that the general approach based on (2.56) can be applied to other problems for which it can lead to solution algorithms (e.g., in Chapter 5) that have little, if anything, in common with GSA.

CAN and GSA

The central problem dealt with in Chapters 2 and 9 is the design of code sequences with impulse-like aperiodic or periodic correlations. A *main result* proved in these chapters is that the said problem can be reduced to that of minimizing a criterion of the form

$$\tilde{C}(\mathbf{x}) = \sum_{k=1}^{K} \left(\left|\mathbf{a}_k^H \mathbf{x}\right|^2 - d_k^2\right)^2 \tag{2.58}$$

for a certain K and certain $\{\mathbf{a}_k\}$ and $\{d_k\}$ (whose exact definitions are not of importance for the present discussion).

The criterion in (2.58) might seem rather similar to the $C(\mathbf{x})$ in (2.53), but in fact there are important differences between these two criteria. The *first* difference is that (2.55)–(2.57) obviously do not hold for $\tilde{C}(\mathbf{x})$. Consequently one cannot derive a GSA-type algorithm for (2.58) by following the approach based on (2.55) and (2.56). Of course, we could use a $\tilde{D}(\mathbf{x}, \boldsymbol{\phi})$ defined by

$$\tilde{D}(\mathbf{x}, \boldsymbol{\phi}) = \sum_{k=1}^{K} \left|\left(\mathbf{a}_k^H \mathbf{x}\right)^2 - d_k^2 e^{j\phi_k}\right|^2 \tag{2.59}$$

for which it holds that $\min_{\boldsymbol{\phi}} \tilde{D}(\mathbf{x}, \boldsymbol{\phi}) = \tilde{C}(\mathbf{x})$, as required. However, minimizing $\tilde{D}(\mathbf{x}, \boldsymbol{\phi})$ is no easier than minimizing $\tilde{C}(\mathbf{x})$.

To get around the above problem, a principal observation made in Chapters 2 and 9 is that, under certain conditions, the minimization of (2.58) is almost equivalent to that of $D(\mathbf{x}, \boldsymbol{\phi})$ in (2.56). Using this observation and the minimization approach outlined in the paragraph following (2.57), two cyclic algorithms termed CAN and PeCAN (see Chapter 9) are introduced for minimizing $D(\mathbf{x}, \boldsymbol{\phi})$. These algorithms have the same form as the GSA in Table 2.4. However, note that now the minimization of $D(\mathbf{x}, \boldsymbol{\phi})$ does not necessarily provide a solution to the problem of minimizing $\tilde{C}(\mathbf{x})$. In particular, a *second* difference between the criteria $C(\mathbf{x})$ and $\tilde{C}(\mathbf{x})$ is that the proposed algorithms do not guarantee that the criterion $\tilde{C}(\mathbf{x})$ monotonically decreases as the iteration proceeds (only $D(\mathbf{x}, \phi)$ is monotonically decreased by each iteration).

Finally, we remark on the fact that the WeCAN (weighted CAN) and multi-sequence CAN algorithms (see Chapter 3), although related to GSA in their basic principles, have a weaker connection to GSA than CAN and PeCAN. These algorithms, which have been obtained by means of the "almost equivalent" minimization approach mentioned in the previous paragraph, can be viewed as extensions of GSA to problems that have more involved forms than (2.53).

3 Aperiodic sequence set design

Chapter 2 dealt with designing a single sequence with good auto-correlation properties. In a similar way, in many applications a set of sequences that have good correlation properties is desired, for example in MIMO (multi-input and multi-output) radar and CDMA (code division multiple access) systems. When transmitting orthogonal waveforms, a MIMO radar system can achieve a greatly increased virtual aperture compared with its phased array counterpart. This increased virtual aperture enables many of the MIMO radar advantages, such as better detection performance [Fishler et al. 2006], improved parameter identifiability [Li, Stoica, Xu & Roberts 2007], enhanced resolution [Bliss & Forsythe 2003] and direct applicability of adaptive array techniques [Xu et al. 2008].

In the case of waveform-set design, which is also referred to as multi-waveform design, both auto- and cross-correlations are involved. Good auto-correlation means that a transmitted waveform is nearly uncorrelated with its own time-shifted versions, while good cross-correlation indicates that any transmitted waveform is nearly uncorrelated with other time-shifted transmitted waveforms. Good correlation properties in the above sense reduce the risk that the received signal of interest is found in correlated multipath or clutter interference.

There is an extensive literature about multi-waveform design. In [Deng 2004] and [Khan et al. 2006], orthogonal waveforms are designed that have good auto- and cross-correlation properties, a topic that is directly tied to this chapter. More related to Chapters 13 and 14, [Fuhrmann & San Antonio 2008] and [Stoica et al. 2007] focus on optimizing the covariance matrix of the transmitted waveforms so as to achieve a given transmit beampattern and in [Stoica, Li & Zhu 2008] the waveforms are designed to approximate a given covariance matrix. Other works include [Yang & Blum 2007*a*][Friedlander 2007] and [Yang & Blum 2007*b*], where some prior information is assumed (e.g., the target impulse response) and the waveforms are designed to optimize a statistical criterion (e.g., the mutual information between the target impulse response and the reflected signals). We also note that, in the area of multiple access wireless communications, spreading sequence design basically addresses the same problem of synthesizing waveforms with good auto- and cross-correlation properties [Oppermann & Vucetic 1997].

Let $\{x_m(n)\}$ ($m = 1, \ldots, M$ and $n = 1, \ldots, N$) denote a set of M sequences, each of which is of length N. The (aperiodic) cross-correlation of $\{x_{m_1}(k)\}_{k=1}^{N}$ and $\{x_{m_2}(k)\}_{k=1}^{N}$ at lag n is defined as

$$r_{m_1 m_2}(n) = \sum_{k=n+1}^{N} x_{m_1}(k) x_{m_2}^*(k-n) = r_{m_2 m_1}^*(-n), \tag{3.1}$$
$$m_1, m_2 = 1, \ldots, M, \quad n = 0, \ldots, N-1.$$

When $m_1 = m_2$, (3.1) becomes the auto-correlation of $\{x_{m_1}(k)\}_{k=1}^N$. Extending the approaches in Chapter 2, we present in this chapter cyclic algorithms for multi-sequence design. The first algorithm is called Multi-CAN and is designed to minimize all correlation sidelobes. The second algorithm is called Multi-WeCAN and is designed to minimize the correlation sidelobes in a certain time lag interval.

3.1 The Multi-CAN algorithm

The Multi-CAN algorithm aims to minimize the following metric

$$\mathcal{E} = \sum_{m=1}^{M} \sum_{n=-N+1, n\neq 0}^{N-1} |r_{mm}(n)|^2 + \sum_{m_1=1}^{M} \sum_{m_2=1, m_2\neq m_1}^{M} \sum_{n=-N+1}^{N-1} |r_{m_1 m_2}(n)|^2. \tag{3.2}$$

To facilitate the discussion, we denote the matrix of the transmitted waveforms by

$$\mathbf{X} = \begin{bmatrix} \mathbf{x}_1 \; \mathbf{x}_2 \; \cdots \; \mathbf{x}_M \end{bmatrix}_{N\times M} \tag{3.3}$$

where

$$\mathbf{x}_m = \begin{bmatrix} x_m(1) \; x_m(2) \; \cdots \; x_m(N) \end{bmatrix}^T \tag{3.4}$$

is the mth waveform. The waveform covariance matrices for different time lags are given by

$$\mathbf{R}_n = \begin{bmatrix} r_{11}(n) & r_{12}(n) & \cdots & r_{1M}(n) \\ r_{21}(n) & r_{22}(n) & \cdots & r_{2M}(n) \\ \vdots & & \ddots & \vdots \\ r_{M1}(n) & \cdots & \cdots & r_{MM}(n) \end{bmatrix}, \quad n = -N+1, \ldots, 0, \ldots, N-1. \tag{3.5}$$

Using the following shift matrix,

$$\mathbf{J}_n = \begin{bmatrix} \overbrace{\mathbf{0} \quad 1}^{n+1} & & \mathbf{0} \\ & \ddots & \\ & & 1 \\ & \mathbf{0} & \end{bmatrix}_{N\times N} = \mathbf{J}_{-n}^T, \quad n = 0, \ldots, N-1, \tag{3.6}$$

the $\mathbf{R}_n$ in (3.5) can be rewritten as

$$\mathbf{R}_n = (\mathbf{X}^H \mathbf{J}_n \mathbf{X})^T = \mathbf{R}_{-n}^H, \quad n = 0, \ldots, N-1. \tag{3.7}$$

With the above notation, the criterion (3.2) can be written more compactly as

$$\mathcal{E} = \|\mathbf{R}_0 - N\mathbf{I}_M\|^2 + 2\sum_{n=1}^{N-1}\|\mathbf{R}_n\|^2 = \sum_{n=-(N-1)}^{N-1}\|\mathbf{R}_n - N\mathbf{I}_M\delta_n\|^2. \tag{3.8}$$

The following Parseval-type equality holds true (the proof is similar to that for the case $M = 1$ in Chapter 2):

$$\sum_{n=-(N-1)}^{N-1}\|\mathbf{R}_n - N\mathbf{I}_M\delta_n\|^2 = \frac{1}{2N}\sum_{p=1}^{2N}\left\|\mathbf{\Phi}(\omega_p) - N\mathbf{I}_M\right\|^2, \tag{3.9}$$

where

$$\mathbf{\Phi}(\omega) \triangleq \sum_{n=-N+1}^{N-1}\mathbf{R}_n e^{-j\omega n} \tag{3.10}$$

is the spectral density matrix of the vector sequence $\left[x_1(n) \cdots x_M(n)\right]^T$ and

$$\omega_p = \frac{2\pi}{2N}p, \quad p = 1, \ldots, 2N. \tag{3.11}$$

The function $\mathbf{\Phi}(\omega)$ defined in (3.10) can be written in the following "periodogram-like" form [Stoica & Moses 2005]:

$$\mathbf{\Phi}(\omega) = \tilde{\mathbf{y}}(\omega)\tilde{\mathbf{y}}^H(\omega), \tag{3.12}$$

where

$$\tilde{\mathbf{y}}(\omega) = \sum_{n=1}^{N}\mathbf{y}(n)e^{-j\omega n}, \quad \mathbf{y}(n) = \left[x_1(n)\ x_2(n) \cdots x_M(n)\right]^T. \tag{3.13}$$

It follows from (3.9) and (3.12) that (3.8) can be rewritten as

$$\mathcal{E} = \frac{1}{2N}\sum_{p=1}^{2N}\left\|\tilde{\mathbf{y}}_p\tilde{\mathbf{y}}_p^H - N\mathbf{I}_M\right\|^2 \quad \left(\tilde{\mathbf{y}}_p \triangleq \tilde{\mathbf{y}}(\omega_p)\right). \tag{3.14}$$

Remark: The criterion $\mathcal{E}$ in (3.14) cannot be made very small, even without the unit-modulus constraint on the elements of $\mathbf{X}$, because the rank 1 matrix $\tilde{\mathbf{y}}_p\tilde{\mathbf{y}}_p^H$ cannot approximate well a full-rank matrix $N\mathbf{I}$. □

Equation (3.14) is a quartic (i.e., fourth-order) function of the unknowns $\{x_m(n)\}_{m=1,n=1}^{M,N}$. To get a simpler quadratic criterion function of $\{x_m(n)\}$, note that

$$\begin{aligned}\mathcal{E} &= \frac{1}{2N}\sum_{p=1}^{2N}\left\|\tilde{\mathbf{y}}_p\tilde{\mathbf{y}}_p^H - N\mathbf{I}\right\|^2 = \frac{1}{2N}\sum_{p=1}^{2N}\operatorname{tr}\left[(\tilde{\mathbf{y}}_p\tilde{\mathbf{y}}_p^H - N\mathbf{I})(\tilde{\mathbf{y}}_p\tilde{\mathbf{y}}_p^H - N\mathbf{I})^H\right] \\ &= \frac{1}{2N}\sum_{p=1}^{2N}(\|\tilde{\mathbf{y}}_p\|^4 - 2N\|\tilde{\mathbf{y}}_p\|^2 + N^2M) \\ &= 2N\sum_{p=1}^{2N}\left(\left\|\frac{\tilde{\mathbf{y}}_p}{\sqrt{2N}}\right\|^2 - \frac{1}{2}\right)^2 + N^2(M-1).\end{aligned} \tag{3.15}$$

Instead of minimizing (3.15) with respect to $\mathbf{X}$, we consider the following minimization problem:

$$\min_{\mathbf{X}, \{\boldsymbol{\alpha}_p\}_{p=1}^{2N}} \sum_{p=1}^{2N} \left\| \frac{1}{\sqrt{2N}} \tilde{\mathbf{y}}_p - \boldsymbol{\alpha}_p \right\|^2 \tag{3.16}$$
$$\text{s.t. } |x_m(n)| = 1, \quad m = 1, \dots, M \text{ and } n = 1, \dots, N$$
$$\|\boldsymbol{\alpha}_p\|^2 = \frac{1}{2}, \quad p = 1, \dots, 2N \quad (\boldsymbol{\alpha}_p \text{ is } M \times 1)$$

where s.t. stands for subject to, and $\{\boldsymbol{\alpha}_p\}$ are auxiliary variables. Evidently, if (3.15), without the constant term $N^2(M-1)$, can be made equal to zero (or "small") by choosing $\mathbf{X}$ then so can (3.16), and vice versa. To solve the minimization problem (3.16), define

$$\mathbf{a}_p^H = \left[e^{-j\omega_p} \cdots e^{-j2N\omega_p} \right], \quad \mathbf{A} = \frac{1}{\sqrt{2N}} \left[\mathbf{a}_1 \cdots \mathbf{a}_{2N} \right], \tag{3.17}$$
$$\tilde{\mathbf{X}} = \begin{bmatrix} \mathbf{X} \\ \mathbf{0} \end{bmatrix}_{2N \times M}, \quad \mathbf{V} = \left[\boldsymbol{\alpha}_1 \cdots \boldsymbol{\alpha}_{2N} \right]^T.$$

Then it is not difficult to observe that

$$\sum_{p=1}^{2N} \left\| \frac{1}{\sqrt{2N}} \tilde{\mathbf{y}}_p - \boldsymbol{\alpha}_p \right\|^2 = \|\mathbf{A}^H \tilde{\mathbf{X}} - \mathbf{V}\|^2 = \|\tilde{\mathbf{X}} - \mathbf{A}\mathbf{V}\|^2. \tag{3.18}$$

(The second equality in (3.18) follows from the fact that $\mathbf{A}$ is unitary.) The criterion (3.18) can be minimized by means of two iterative (cyclic) steps. For given $\tilde{\mathbf{X}}$ (i.e., $\mathbf{X}$ is given), the minimizer $\{\boldsymbol{\alpha}_p\}_{p=1}^{2N}$ of (3.18) is given by

$$\boldsymbol{\alpha}_p = \frac{1}{\sqrt{2}} \frac{\mathbf{c}_p}{\|\mathbf{c}_p\|}, \quad p = 1, \dots, 2N \tag{3.19}$$

where

$$\mathbf{c}_p^T \text{ is the } p\text{th row of } \mathbf{A}^H \tilde{\mathbf{X}}. \tag{3.20}$$

For given $\mathbf{V}$ (i.e., $\{\boldsymbol{\alpha}_p\}_{p=1}^{2N}$ are given), the minimizer $\{x_m(n)\}$ of (3.18) is given by

$$x_m(n) = \exp[j \arg(d_{nm})], \quad m = 1, \dots, M \quad n = 1, \dots, N \tag{3.21}$$

where

$$d_{nm} \text{ is the } (n, m)\text{th element of } \mathbf{A}\mathbf{V}. \tag{3.22}$$

The Multi-CAN (multi-sequence CAN) algorithm thus obtained is summarized in Table 3.1.

Note that the $\mathbf{A}^H \tilde{\mathbf{X}}$ in (3.20) is the FFT of each column of $\tilde{\mathbf{X}}$ and that the $\mathbf{AV}$ in (3.22) is the IFFT of each column of $\mathbf{V}$. Because of these (I)FFT-based computations, the Multi-CAN algorithm is quite fast. Indeed, it can be used to design long sequences up to $N \sim 10^3$ and $M \sim 10$ on an ordinary PC, which can hardly be handled by other algorithms suggested in the previous literature.

Table 3.1. The Multi-CAN algorithm

Step 0: Initialize $\mathbf{X}$ by a randomly generated $N \times M$ matrix or by some good existing sequences.
Step 1: Fix $\tilde{\mathbf{X}}$ and compute $\mathbf{V}$ by means of (3.19).
Step 2: Fix $\mathbf{V}$ and compute $\tilde{\mathbf{X}}$ by means of (3.21).
Iteration: Repeat steps 1 and 2 until a prespecified stop criterion is satisfied, e.g., $\|\mathbf{X}^{(i)} - \mathbf{X}^{(i+1)}\| < 10^{-3}$, where $\mathbf{X}^{(i)}$ is the waveform matrix obtained at the ith iteration.

3.2 The Multi-WeCAN algorithm

In some radar applications such as synthetic aperture radar (SAR) imaging, the transmitted pulse is relatively long (i.e., N is large) and signals backscattered from objects in the near and far range bins overlap significantly [Li et al. 2008]. In this case, only the waveform correlation properties in a certain lag interval around $n = 0$ are relevant to range resolution and a minimization criterion different from (3.8) is given by

$$\tilde{\mathcal{E}} = \|\mathbf{R}_0 - N\mathbf{I}_M\|^2 + 2\sum_{n=1}^{P-1} \|\mathbf{R}_n\|^2, \tag{3.23}$$

where $P - 1$ is the maximum lag in which we are interested. More specifically, $(P - 1)t_p$ (t_p is as mentioned in (1.1)) should be chosen to be larger than the maximum difference in round-trip delays of the signals backscattered from near and far range bins.

Remark: Recall that it is not possible to make the criterion $\mathcal{E}$ in (3.14) very small. One way to understand this problem is to examine the criterion $\tilde{\mathcal{E}}$ defined in (3.23) when only $\mathbf{R}_0, \mathbf{R}_1, \ldots, \mathbf{R}_{P-1}$ (which are complex-valued $M \times M$ matrices) are considered. The matrix $\mathbf{R}_0$ is Hermitian with all diagonal elements equal to N, so setting $\mathbf{R}_0 = N\mathbf{I}$ leads to $M(M - 1)$ (real-valued) equations. The matrices $\mathbf{R}_1, \ldots, \mathbf{R}_{P-1}$ do not have any special structure, and setting them to zero adds $2M^2$ equations for each one. Thus the total number of equations is $K = M(M - 1) + (P - 1)2M^2$. In comparison, the number of variables that we can manipulate is $M(N - 1)$ (for each of the M waveforms there are $N - 1$ free phases, as the initial phase does not matter). Therefore, a basic requirement for good performance is that $K \leq M(N - 1)$, which can be simplified to: $P \leq (N + M)/2M$. Put differently, only when $P \leq (N + M)/2M$ is it possible in principle to design unimodular waveforms $\mathbf{X}$ that make $\tilde{\mathcal{E}}$ zero; in other cases $\tilde{\mathcal{E}}$ or $\mathcal{E}$ cannot be made equal to zero. □

The Multi-WeCAN (multi-sequence weighted-CAN) algorithm is designed to minimize the following criterion:

$$\hat{\mathcal{E}} = \gamma_0^2\|\mathbf{R}_0 - N\mathbf{I}_M\|^2 + 2\sum_{n=1}^{N-1} \gamma_n^2\|\mathbf{R}_n\|^2, \tag{3.24}$$

where $\{\gamma_n\}_{n=0}^{N-1}$ are real-valued weights. For instance, if we choose $\gamma_n = 1$ for $n = 0, \ldots, P - 1$ and $\gamma_n = 0$ otherwise, then $\hat{\mathcal{E}}$ becomes the criterion $\tilde{\mathcal{E}}$ defined in (3.23).

We can show that (cf. (3.9))

$$\hat{\mathcal{E}} = \frac{1}{2N} \sum_{p=1}^{2N} \left\| \tilde{\mathbf{\Phi}}(\omega_p) - \gamma_0 N \mathbf{I}_M \right\|^2, \tag{3.25}$$

where $\{\omega_p\}_{p=1}^{2N}$ is given by (3.11),

$$\tilde{\mathbf{\Phi}}(\omega) \triangleq \sum_{n=-(N-1)}^{N-1} \gamma_n \mathbf{R}_n e^{-j\omega n}, \tag{3.26}$$

and $\gamma_n = \gamma_{-n}$ for $n = 1, \ldots, N-1$. To facilitate later developments, γ_0 is chosen such that the matrix

$$\mathbf{\Gamma} = \begin{bmatrix} \gamma_0 & \gamma_1 & \cdots & \gamma_{N-1} \\ \gamma_1 & \gamma_0 & \ddots & \vdots \\ \vdots & \ddots & \ddots & \gamma_1 \\ \gamma_{N-1} & \cdots & \gamma_1 & \gamma_0 \end{bmatrix} \tag{3.27}$$

is positive semidefinite (denoted as $\mathbf{\Gamma} \geq 0$). See the discussions following (2.25) in Chapter 2 for a way to determine γ_0. The condition $\mathbf{\Gamma} \geq 0$ is necessary because the matrix square root of $\mathbf{\Gamma}$ is needed below (see (3.31)).

Using a derivation similar to that leading to (3.12), it can be shown that (see Appendix 3A):

$$\tilde{\mathbf{\Phi}}(\omega) = \mathbf{Z}^T(\omega) \mathbf{\Gamma} \mathbf{Z}^*(\omega), \tag{3.28}$$

where

$$\mathbf{Z}^T(\omega) = \left[\mathbf{y}(1)e^{-j\omega} \; \mathbf{y}(2)e^{-j\omega 2} \; \cdots \; \mathbf{y}(N)e^{-j\omega N} \right]_{M \times N}. \tag{3.29}$$

By combining (3.25) and (3.28), the criterion becomes

$$\hat{\mathcal{E}} = \frac{1}{2N} \sum_{p=1}^{2N} \left\| \mathbf{Z}_p^H \mathbf{\Gamma} \mathbf{Z}_p - \gamma_0 N \mathbf{I}_M \right\|^2 \quad (\mathbf{Z}_p \triangleq \mathbf{Z}(\omega_p)). \tag{3.30}$$

Instead of minimizing (3.30) with respect to $\mathbf{X}$, we consider the following minimization problem (see the discussion following (3.16)):

$$\min_{\mathbf{X},\mathbf{U}} \sum_{p=1}^{2N} \left\| \mathbf{C}\mathbf{Z}_p - \sqrt{\gamma_0 N} \mathbf{U}_p \right\|^2, \tag{3.31}$$
$$\text{s.t. } |x_m(n)| = 1, \quad m = 1, \ldots, M \quad n = 1, \ldots, N,$$
$$\mathbf{U}_p^H \mathbf{U}_p = \mathbf{I}, \quad p = 1, \ldots, 2N \quad (\mathbf{U}_p \text{ is } N \times M),$$

where the $N \times N$ matrix $\mathbf{C}$ is a square root of $\mathbf{\Gamma}$ (i.e., $\mathbf{C}^H \mathbf{C} = \mathbf{\Gamma}$).

The minimization problem (3.31) can be solved in a cyclic way as follows. For given $\{\mathbf{Z}_p\}_{p=1}^{2N}$ (i.e., $\mathbf{X}$ is given), (3.31) decouples into $2N$ independent problems, each of which can be written as

$$\left\|\mathbf{C}\mathbf{Z}_p - \sqrt{\gamma_0 N}\mathbf{U}_p\right\|^2 = \text{const} - 2\,\text{Re}\left[\text{tr}\left(\sqrt{\gamma_0 N}\mathbf{U}_p\mathbf{Z}_p^H\mathbf{C}^H\right)\right], \quad p = 1, \ldots, 2N, \tag{3.32}$$

where "const" denotes a term that is independent of the variable $\mathbf{U}_p$. Let

$$\mathbf{Z}_p^H\mathbf{C}^H = \mathbf{U}_1\mathbf{\Sigma}\mathbf{U}_2^H \tag{3.33}$$

denote the "economy-size" singular value decomposition (also named the reduced SVD [Trefethen & Bau 1997]) of $\mathbf{Z}_p^H\mathbf{C}^H$, where $\mathbf{U}_1$ is $M \times M$, $\mathbf{\Sigma}$ is $M \times M$ and $\mathbf{U}_2$ is $N \times M$. Then the minimizer $\mathbf{U}_p$ of (3.32), for fixed $\mathbf{Z}_p$, is given by (see (3.46),(3.47) for the same optimization problem and see Appendix 3B for a proof):

$$\mathbf{U}_p = \mathbf{U}_2\mathbf{U}_1^H. \tag{3.34}$$

Note that the computation of $\{\mathbf{C}\mathbf{Z}_p\}_{p=1}^{2N}$ can be done by means of an FFT. To see this, let

$$\tilde{\mathbf{X}}_m = \mathbf{C}^T \odot \left[\mathbf{x}_m\ \mathbf{x}_m\ \cdots\ \mathbf{x}_m\right]_{N\times N}, \quad m = 1, \ldots, M, \tag{3.35}$$

and

$$\mathbf{F} = \sqrt{2N}\mathbf{A}^H\tilde{\mathbf{F}}, \quad \tilde{\mathbf{F}} = \begin{bmatrix} \tilde{\mathbf{X}}_1 & \tilde{\mathbf{X}}_2 & \cdots & \tilde{\mathbf{X}}_M \\ \mathbf{0}_{N\times N} & \mathbf{0}_{N\times N} & \cdots & \mathbf{0}_{N\times N} \end{bmatrix}_{2N\times NM}, \tag{3.36}$$

where $\mathbf{A}$ was defined in (3.17). Let $\mathbf{f}_p^T$ denote the pth row of $\mathbf{F}$. We divide the $NM \times 1$ vector $\mathbf{f}_p$ equally into M pieces and observe that these pieces correspond to the M columns (from left to right) of $\mathbf{C}\mathbf{Z}_p$. Thus the matrix $\mathbf{C}\mathbf{Z}_p$ can be obtained from $\mathbf{F}$, which is calculated from $\sqrt{2N}\mathbf{A}^H\tilde{\mathbf{F}}$, an FFT operation.

For given $\{\mathbf{U}_p\}_{p=1}^{2N}$, the minimization problem in (3.31) also has a closed-form solution with respect to $\mathbf{X}$. Let

$$\mathbf{G}_{2N\times NM} = \left[\mathbf{g}_1\ \mathbf{g}_2\ \cdots\ \mathbf{g}_{2N}\right]^T, \tag{3.37}$$

where $\mathbf{g}_p$ denotes the $NM \times 1$ vector given by the columns of $\sqrt{\gamma_0 N}\mathbf{U}_p$ stacked on top of each other. Then the criterion (3.31) can be written as

$$\begin{aligned} \sum_{p=1}^{2N} \left\|\mathbf{C}\mathbf{Z}_p - \sqrt{\gamma_0 N}\mathbf{U}_p\right\|^2 &= \left\|\sqrt{2N}\mathbf{A}^H\tilde{\mathbf{F}} - \mathbf{G}\right\|^2 \\ &= 2N\left\|\tilde{\mathbf{F}} - \frac{1}{\sqrt{2N}}\mathbf{A}\mathbf{G}\right\|^2. \end{aligned} \tag{3.38}$$

The above function can be minimized with respect to each element of $\{x_m(n)\}_{m=1,n=1}^{M,N}$ separately. Let x denote a generic element of $\{x_m(n)\}$. Then the problem corresponds to minimizing the following criterion with respect to x:

$$\sum_{k=1}^{N} |\mu_k x - \nu_k|^2 = \text{const} - 2\,\text{Re}\left[\left(\sum_{k=1}^{N} \mu_k^* \nu_k\right) x^*\right], \tag{3.39}$$

Table 3.2. The Multi-WeCAN algorithm

Step 0: Initialize $\mathbf{X}$ and select the desired weights $\{\gamma_n\}_{n=0}^{N-1}$ such that the matrix $\mathbf{\Gamma}$ in (3.27) is positive semidefinite.
Step 1: Fix $\{\mathbf{Z}_p\}_{p=1}^{2N}$ (i.e., $\mathbf{X}$ is given) and compute $\{\mathbf{U}_p\}_{p=1}^{2N}$ according to (3.34).
Step 2: Fix $\{\mathbf{U}_p\}_{p=1}^{2N}$ and compute $\mathbf{X}$ according to (3.40).
Iteration: Repeat steps 1 and 2 until a prespecified stop criterion is satisfied, e.g., $\|\mathbf{X}^{(i)} - \mathbf{X}^{(i+1)}\| < \epsilon$, where $\mathbf{X}^{(i)}$ is the waveform matrix obtained at the ith iteration.

where $\{\mu_k\}_{k=1}^{N}$ are given by the elements of $\tilde{\mathbf{F}}$ which contain x and $\{\nu_k\}_{k=1}^{N}$ are given by the elements of $1/\sqrt{2N}\mathbf{AG}$ whose positions are the same as those of $\{\mu_k\}_{k=1}^{N}$ in $\tilde{\mathbf{F}}$. (More specifically, for $k = 1, \ldots, N$, μ_k is given by the (k, n)th element of $\mathbf{C}$ and ν_k is given by the $(n, (m-1)N + k)$th element of $1/\sqrt{2N}\mathbf{AG}$.) Under the unit-modulus constraint, the minimizer x of the criterion (3.39) is given by

$$x = e^{j\phi}, \quad \phi = \arg\left(\sum_{k=1}^{N} \mu_k^* \nu_k\right). \tag{3.40}$$

The Multi-WeCAN algorithm follows from the above discussions and is summarized in Table 3.2.

Multi-WeCAN is not so computationally efficient as Multi-CAN but it can still be used for relatively large values of N and M, up to $N \sim 10^3$ and $M \sim 10$, on an ordinary PC.

3.3 The Multi-CA-original (Multi-CAO) algorithm

The original cyclic algorithm (CA) for waveform design proposed in [Li et al. 2008][Stoica, Li & Zhu 2008] aims at minimizing a particular form of the criterion $\hat{\mathcal{E}}$ in (3.24):

$$\hat{\mathcal{E}}_{\text{CAO}} = P\|\mathbf{R}_0 - N\mathbf{I}_M\|^2 + 2\sum_{n=1}^{P-1}(P-n)\|\mathbf{R}_n\|^2, \tag{3.41}$$

which can be obtained from (3.24) by choosing the weights $\gamma_n^2 = P - n$ for $n = 0, \ldots, P-1$ and $\gamma_n^2 = 0$ otherwise. We refer to this original CA algorithm as Multi-CAO (multi-sequence CA original) to make the naming consistent with that of Multi-CAN and Multi-WeCAN.

The above choice of $\{\gamma_n\}_{n=0}^{N-1}$ results from the following problem formulation, which is simple and direct. Consider the matrix

$$\bar{\mathbf{X}} = \left[\bar{\mathbf{X}}_1 \cdots \bar{\mathbf{X}}_M\right]_{(N+P-1)\times MP}, \tag{3.42}$$

where

$$\bar{\mathbf{X}}_m = \begin{bmatrix} x_m(1) & & \mathbf{0} \\ \vdots & \ddots & \\ \vdots & & x_m(1) \\ x_m(N) & & \vdots \\ & \ddots & \vdots \\ \mathbf{0} & & x_m(N) \end{bmatrix}_{(N+P-1)\times P}, \quad m = 1, \ldots, M. \tag{3.43}$$

Then it is easy to observe that the $\hat{\mathcal{E}}_{\text{CAO}}$ defined in (3.41) can be expressed as follows:

$$\hat{\mathcal{E}}_{\text{CAO}} = \left\|\bar{\mathbf{X}}^H\bar{\mathbf{X}} - N\mathbf{I}_{MP}\right\|^2. \tag{3.44}$$

The minimization of (3.44) can be tackled by solving the following problem

$$\begin{aligned} &\min_{\mathbf{X},\mathbf{U}} \|\bar{\mathbf{X}} - \sqrt{N}\mathbf{U}\|^2, \\ &\text{s.t. } |x_m(n)| = 1, \quad m = 1, \ldots, M, \quad n = 1, \ldots, N, \\ &\qquad \mathbf{U}^H\mathbf{U} = \mathbf{I} \quad (\mathbf{U} \text{ is an } (N+P-1) \times MP \text{ matrix}). \end{aligned} \tag{3.45}$$

Note that the problem of minimizing (3.44) and the problem (3.45) are almost equivalent, and so is the relationship between (3.15) and (3.16), as well as that between (3.30) and (3.31). See the discussion following (2.12) as well as Appendix 2A for more information.

Regarding (3.45), we note the following facts. For a given $\bar{\mathbf{X}}$, let

$$\bar{\mathbf{X}}^H = \mathbf{U}_1\mathbf{S}\mathbf{U}_2^H \tag{3.46}$$

denote the "economy-size" SVD (singular value decomposition) of $\bar{\mathbf{X}}$. Here $\mathbf{U}_1$ is an $MP \times MP$ unitary matrix, $\mathbf{U}_2$ is an $(N + P - 1) \times MP$ semi-unitary matrix and $\mathbf{S}$ is an $MP \times MP$ diagonal matrix. Then the solution $\mathbf{U}$ of (3.45), for fixed $\bar{\mathbf{X}}$, is given by (see Appendix 3B)

$$\mathbf{U} = \mathbf{U}_2\mathbf{U}_1^H. \tag{3.47}$$

Next, note that for a fixed $\mathbf{U}$ the minimization of the criterion (3.45) also has a simple closed-form solution. To see this, let x denote an arbitrary element from $\{x_m(n)\}$. Then a generic form of the minimization problem (3.45) with respect to x is given by

$$\min_x \sum_{k=1}^{P} |x - \mu_k|^2, \tag{3.48}$$

where $\{\mu_k\}_{k=1}^P$ are the elements of the matrix $\sqrt{N}\mathbf{U}$ whose positions are the same as the positions of x in $\bar{\mathbf{X}}$. More precisely, for $x = x_m(n)$ the corresponding sequence $\{\mu_k\}_{k=1}^P$ is given by the $[n - 1 + r, (m - 1)P + r]$th elements of $\sqrt{N}\mathbf{U}$ for $r = 1, \ldots, P$. Because

Table 3.3. The Multi-CAO algorithm

Step 0: Set the matrix $\bar{\mathbf{X}}$ to an initial value.
Step 1: Fix $\{x_m(n)\}$ (i.e., $\bar{\mathbf{X}}$ is given) and compute $\mathbf{U}$ according to (3.47).
Step 2: Fix $\mathbf{U}$ and compute $\{x_m(n)\}$ according to (3.50).
Iteration: Repeat steps 1 and 2 until a prespecified stop criterion is satisfied.

$|x| = 1$, the criterion in (3.48) can be rewritten as follows:

$$\begin{aligned}\sum_{k=1}^{P} |x - \mu_k|^2 &= \text{const} - 2\,\text{Re}\left(x \sum_{k=1}^{P} \mu_k^*\right) \\ &= \text{const} - 2\left|\sum_{k=1}^{P} \mu_k\right| \cos\left[\arg(x) - \arg\left(\sum_{k=1}^{P} \mu_k\right)\right]\end{aligned} \tag{3.49}$$

where const denotes a term that does not depend on x. Hence the minimizer x of the criterion (3.45) is given by

$$x = \exp\left[j \arg\left(\sum_{k=1}^{P} \mu_k\right)\right]. \tag{3.50}$$

The Multi-CAO algorithm for (3.45) follows from the above discussion and is summarized in Table 3.3.

The criterion minimized by Multi-CAO is a special case of that minimized by Multi-WeCAN. Multi-CAO and Multi-WeCAN provide similar performances in terms of correlation sidelobe suppression, although the derivation and iteration steps of Multi-CAO are relatively less involved.

3.4 Numerical examples

3.4.1 Multi-CAN

Now we consider minimizing the criterion $\mathcal{E}$ in (3.8), i.e., minimizing all correlation sidelobes: $r_{mm}(n)$ for all m and $n \neq 0$, and $r_{m_1 m_2}(n)$ for all $m_1 \neq m_2$ and n. Suppose that the number of transmit sequences is $M = 3$ and the length of each sequence is $N = 40$. We compare the Multi-CAN sequence set with the CE (cross entropy) sequence set in [Khan et al. 2006]. We use randomly generated sequences to initialize Multi-CAN (see step 0 in Table 3.1). One hundred Monte-Carlo trials are run (i.e., 100 random initializations) and the sequence set with the lowest correlation sidelobe peak is kept. The 40×3 CE sequence set is given in Table 1 of [Khan et al. 2006].

Figures 3.1, 3.2 and 3.3 show the correlations ($r_{11}, r_{12}, \ldots, r_{33}$, normalized by N) for the Multi-CAN sequence set and for the CE sequence set. The CE sequence set is slightly better than the Multi-CAN sequence set in terms of correlation sidelobe peaks. However, our goal is to minimize $\mathcal{E}$ or equivalently the following normalized fitting error:

$$\mathcal{E}_{\text{norm}} = \frac{\mathcal{E}}{MN^2} = \frac{1}{MN^2}\left(\|\mathbf{R}_0 - N\mathbf{I}\|^2 + 2\sum_{n=1}^{N-1}\|\mathbf{R}_n\|^2\right). \tag{3.51}$$

The Multi-CAN sequence set gives a fitting error of 2.00, whereas the CE sequence set has a larger fitting error equal to 2.23.

Note that although the Multi-CAN and CE sequence sets show comparable performances (which are also comparable with the performance of other sequence sets such as those in [Deng 2004]), the Multi-CAN algorithm works much faster than other existing algorithms, because Multi-CAN is based on FFT computations. For the above parameter set ($N = 40$ and $M = 3$), the Multi-CAN algorithm consumes less than one second on an ordinary PC to complete one Monte-Carlo trial. The overall computation time is still short if we run plenty of Monte-Carlo trials and pick up the best sequence set. Moreover, the computation time of Multi-CAN grows roughly as $O(MN \log N)$ so that Multi-CAN can handle very large values of N, up to $N \sim 10^5$. In contrast, the cross entropy [Khan et al. 2006] or simulated annealing based methods [Deng 2004] are relatively involved and become impractical for large values of N. In fact, we were unable to find in the literature any code set designed for good (aperiodic) correlations and, at the same time, sufficiently long to be comparable with what can be obtained using the Multi-CAN algorithm.

For relatively large values of N, we employ the Hadamard sequence set for comparison [Tse & Viswanath 2005], which is easy to generate (for virtually any length that is a power of 2) and is frequently used in wireless communications. Let $\mathbf{H}$ denote an $N \times N$ Hadamard matrix, where N is a power of 2. We construct an $N \times M$ QPSK (quadrature phase-shift keying) Hadamard sequence set:

$$\mathbf{X}_{\text{Hadamard}} = \mathbf{H}(1 : M) + j\mathbf{H}((M + 1) : 2M) \tag{3.52}$$

where $N \geq 2M$ and $\mathbf{H}(a : b)$ denotes the $N \times (b - a + 1)$ submatrix of $\mathbf{H}$ from the ath column to the bth column. We further scramble the Hadamard sequence set with a PN (pseudo-noise) sequence to lower its correlation sidelobes. More specifically, we multiplied (element-wise) each column of $\mathbf{X}_{\text{Hadamard}}$ by a PN sequence $\{p(n)\}_{n=1}^{N}$, where $\{p(n)\}_{n=1}^{N-1}$ is an m-sequence (see Section 1.3) and $p(N) = 1$. For the Multi-CAN sequence set we use random-phase sequence sets

$$x_m(n) = e^{j\phi_m(n)}, \quad \{\phi_m(n)\} \text{ i.i.d.} \sim \text{uniform}[0, 2\pi], \tag{3.53}$$

for initialization. For each N we run the Multi-CAN algorithm 100 times and keep the sequence set with the lowest correlation sidelobe peak, as well as the corresponding random-phase sequence set used for initialization. We run the simulations for $M = 3$ and $N = 2^7, \ldots, 2^{13}$. Figures 3.4 and 3.5 give the sequence sets in terms of three criteria: the auto-correlation sidelobe peak, the cross-correlation peak and the normalized fitting error (defined in (3.51)). The Multi-CAN sequence set outperforms both the random-phase and Hadamard+PN sequence sets with respect to each criterion. In fact, the advantage of the Multi-CAN algorithm lies not only in the significant length and low correlation sidelobes of the designed sequence sets, but also in the easy generation

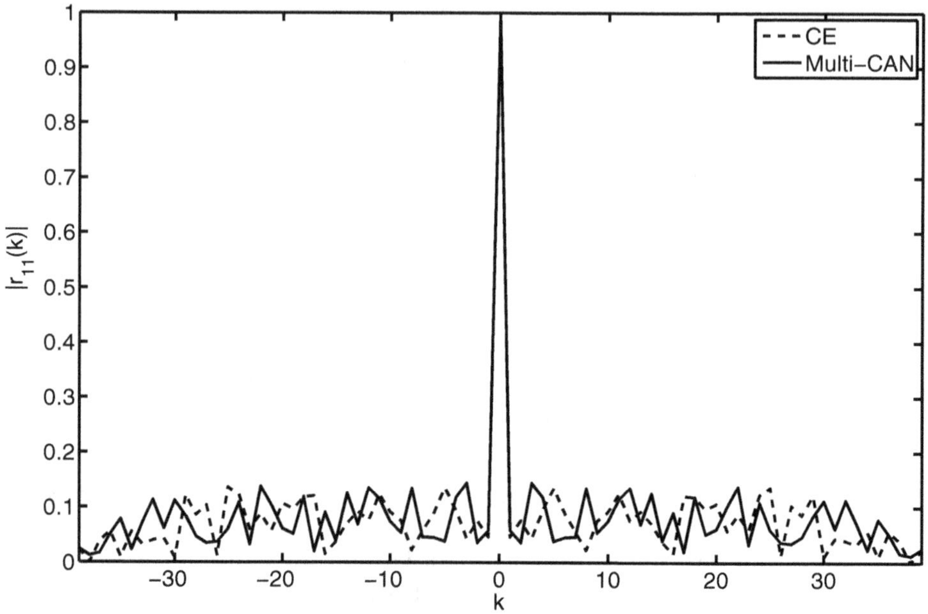

(a)

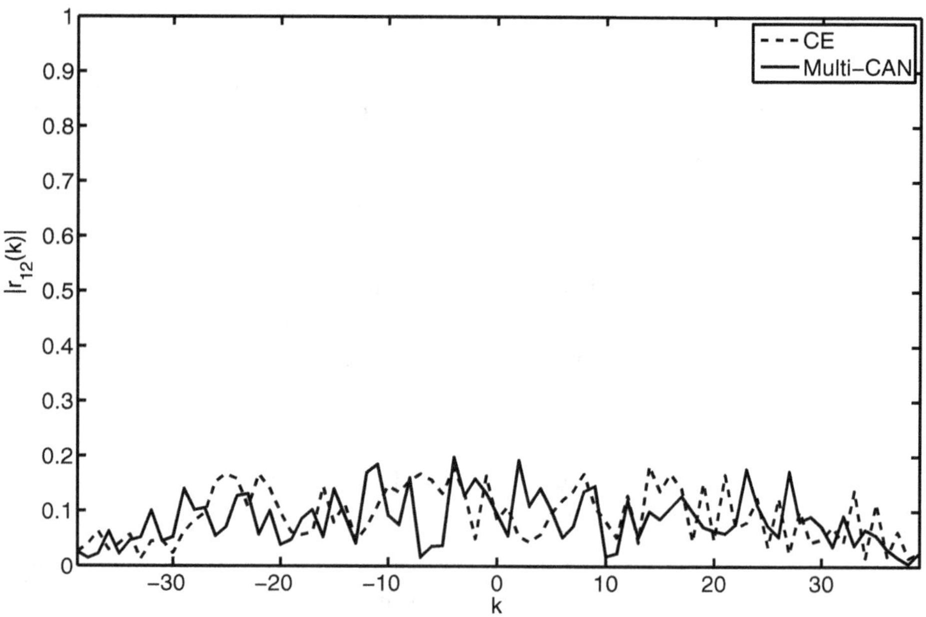

(b)

Figure 3.1 The correlations of the 40×3 CE and Multi-CAN sequence sets: (a) $r_{11}(k)$ and (b) $r_{12}(k)$.

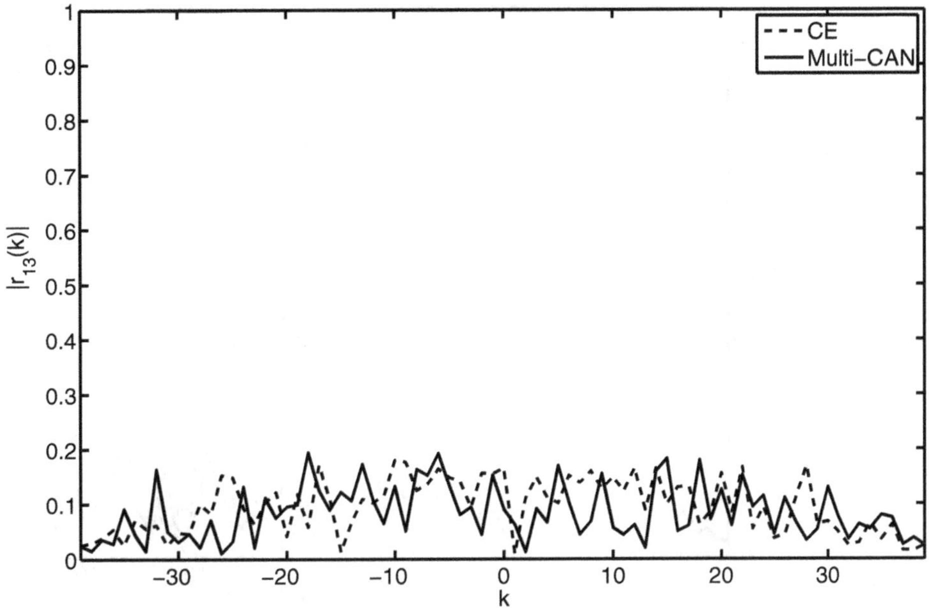

(a)

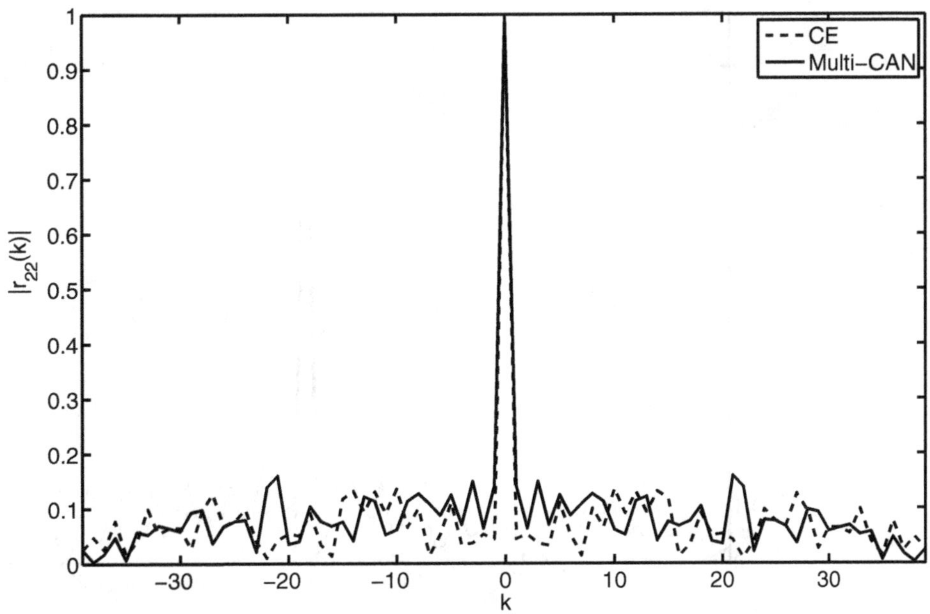

(b)

Figure 3.2 The correlations of the 40×3 CE and Multi-CAN sequence sets: (a) $r_{13}(k)$ and (b) $r_{22}(k)$.

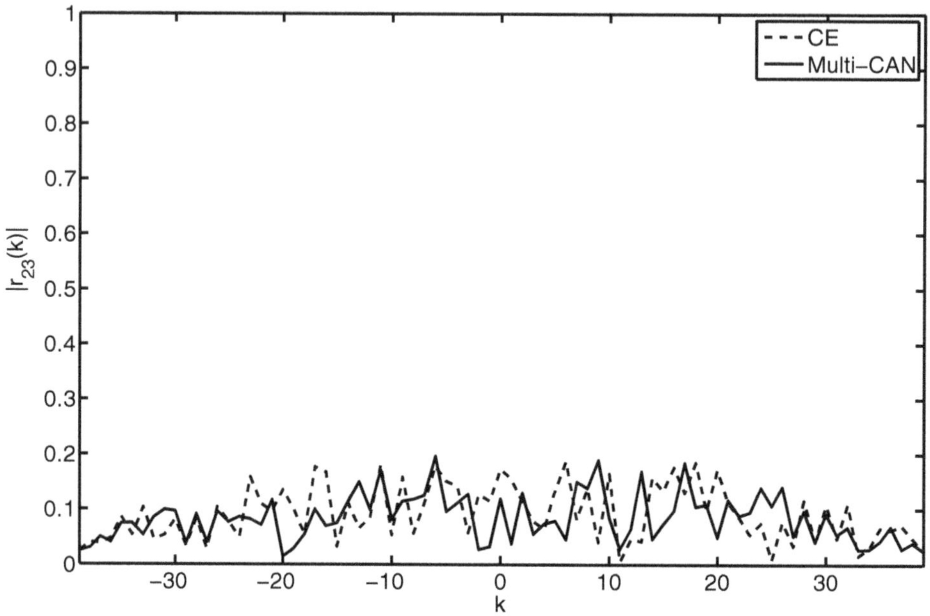

(a)

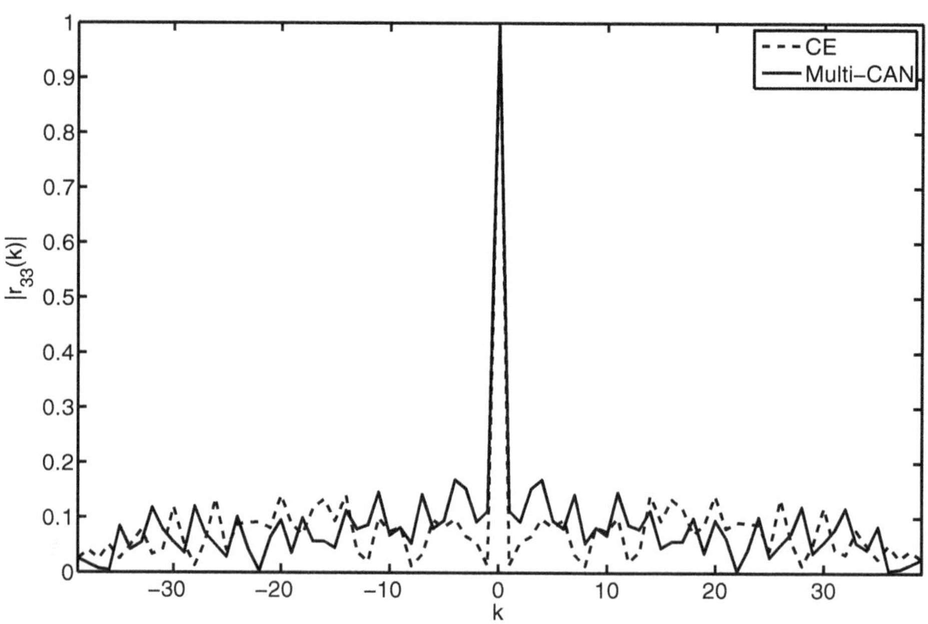

(b)

Figure 3.3 The correlations of the 40×3 CE and Multi-CAN sequence sets: (a) $r_{23}(k)$ and (b) $r_{33}(k)$.

(using different initial conditions) of many sequence sets which are of the same $N \times M$ dimension and which all have reasonably low correlation sidelobes. These randomly distributed waveform sets are useful in some application areas, such as in countering coherent repeater jamming in radar systems [Skolnik 2008][Deng 2004].

3.4.2 Multi-WeCAN

Consider minimizing the criterion $\tilde{\mathcal{E}}$ in (3.23), i.e., minimizing the correlation sidelobes for lags not larger than $P-1$; i.e., $r_{mm}(n)$ for all m and $1 \leq n \leq P-1$ and $r_{m_1m_2}(n)$ for all $m_1 \neq m_2$ and $0 \leq n \leq P-1$. Suppose that the number of transmit sequences is $M=4$, the length of each sequence is $N=256$ and the number of correlation lags we want to consider is $P=50$. The normalized fitting error for this scenario is defined similarly to (3.51) as

$$\tilde{\mathcal{E}}_{\text{norm}} = \frac{1}{MN^2}\frac{\tilde{\mathcal{E}}}{MN^2} = \left(\|\mathbf{R}_0 - N\mathbf{I}\|^2 + 2\sum_{n=1}^{P-1} \|\mathbf{R}_n\|^2 \right). \tag{3.54}$$

We also define the correlation level as follows:

$$\text{correlation level} = 20\log_{10}\frac{\|\mathbf{R}_n - N\mathbf{I}\delta_n\|}{\sqrt{MN^2}}, \quad n = -N+1,\ldots,0,\ldots,N-1. \tag{3.55}$$

It measures the correlation sidelobes for each time lag.

We compare the Multi-WeCAN algorithm and the Multi-CAO algorithm. We use a randomly generated unimodular sequence set to initialize both Multi-WeCAN and Multi-CAO. For Multi-WeCAN, we choose

$$\gamma_n^2 = \begin{cases} 1, & n \in [1, P-1], \\ 0, & n \in [P, N-1], \end{cases} \tag{3.56}$$

and γ_0 is chosen to ensure that $\mathbf{\Gamma} > 0$ (more exactly we choose $\gamma_0 = 25.5$).

Table 3.4 compares the Multi-CAO sequence set and the Multi-WeCAN sequence set in terms of the auto-correlation sidelobe peak (in the considered lag interval), the cross-correlation peak (in the considered lag interval) and the $\tilde{\mathcal{E}}_{\text{norm}}$ defined in (3.54). The 256×4 Multi-CAN sequence set is also added in Table 3.4 for comparison. The Multi-WeCAN sequence set gives the lowest correlation sidelobe peak and fitting error. Figure 3.6 shows the correlation levels of the Multi-CAO and Multi-WeCAN sequence sets. We observe from Figure 3.6 that the Multi-WeCAN sequence set provides a "uniformly low" correlation level in the required lag interval $[1, P-1]$, while the correlation level of the Multi-CAO sequence set increases as the lag increases from 1 to $P-1$. This behavior is attributed to the fact that Multi-WeCAN makes use of uniform weights $\{\gamma_n = 1\}_{n=1}^{P-1}$ in (3.56) whereas Multi-CAO implicitly assumes uneven weights $\{\gamma_n = P-n\}_{n=1}^{P-1}$ (see (3.41)), so the bigger the lag, the smaller the weight. We also note that the correlation level at $n=0$ for the Multi-WeCAN sequence set is very low (around -85 dB). The reason is that we chose $\gamma_0 = 25.5$, which is much larger than the other weights (see (3.56)) and thus the zero-lag correlation fitting error $\|\mathbf{R}_0 - N\mathbf{I}\|$ is emphasized the most in the criterion $\hat{\mathcal{E}}$ in (3.24).

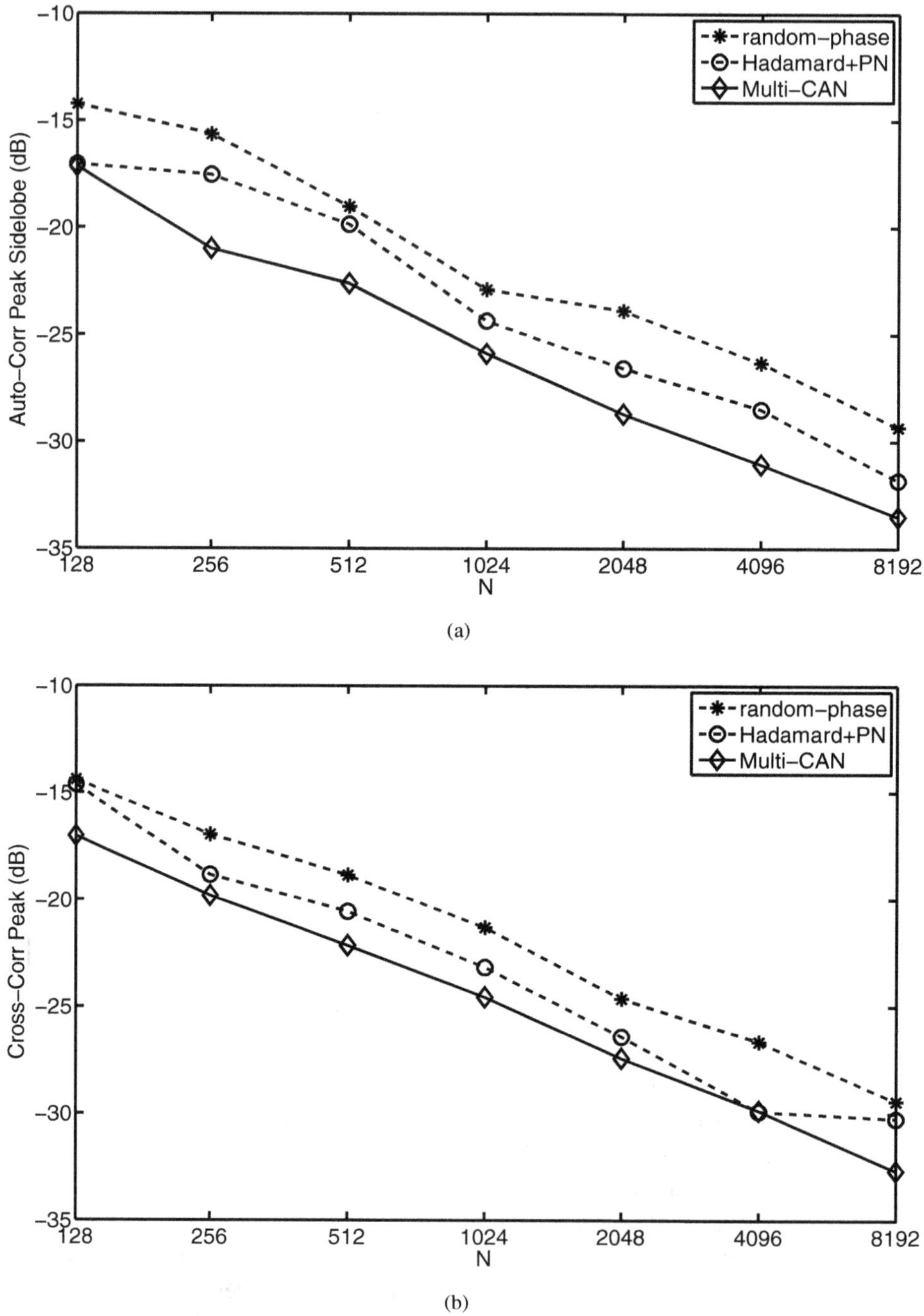

Figure 3.4 Comparison between the Multi-CAN sequence set, the Hadamard+PN sequence set and the random-phase sequence set with $M = 3$ and $N = 2^7, \ldots, 2^{13}$ in terms of (a) the auto-correlation sidelobe peak and (b) the cross-correlation peak.

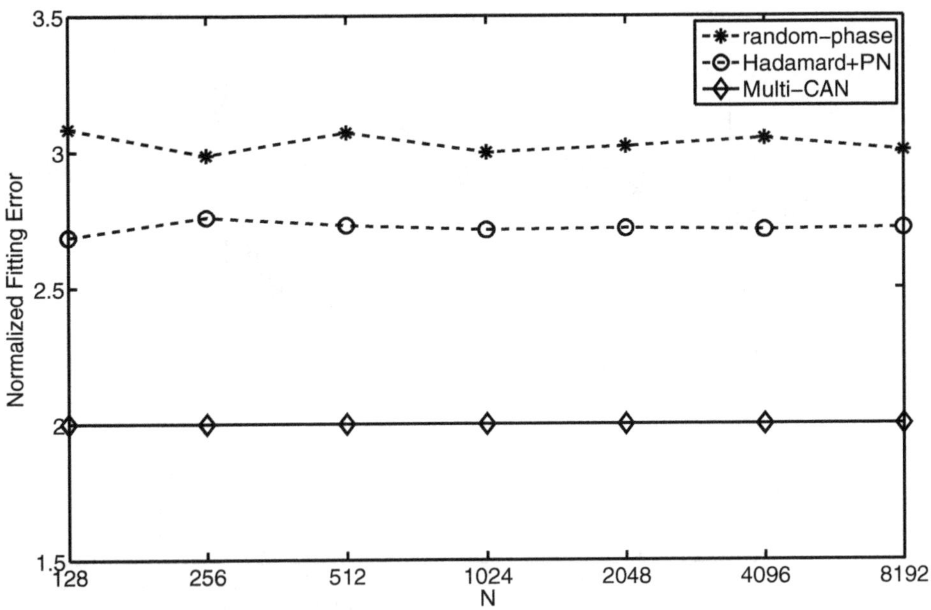

Figure 3.5 Comparison between the Multi-CAN sequence set, the Hadamard+PN sequence set and the random-phase sequence set with $M = 3$ and $N = 2^7, \ldots, 2^{13}$ in terms of the normalized fitting error as defined in (3.51).

Table 3.4. Comparison between Multi-CAN, Multi-CAO and Multi-WeCAN sequence sets under $\tilde{\mathcal{E}}$ ($N = 256, M = 4, P = 50$)

	Auto-corr sidelobe peak (dB)	Cross-corr peak (dB)	$\tilde{\mathcal{E}}_{\text{norm}}$
Multi-CAN	−20.54	−18.19	0.91
Multi-CAO	−21.08	−20.77	0.088
Multi-WeCAN	−31.10	−29.09	0.072

3.4.3 Multi-WeCAN continued

Consider using the Multi-WeCAN algorithm to minimize the criterion $\hat{\mathcal{E}}$ in (3.24) with $N = 256, M = 4$ and the following weights:

$$\gamma_n^2 = \begin{cases} 1, & n \in [1, 19] \cup [236, 255], \\ 0, & n \in [20, 235] \end{cases} \tag{3.57}$$

(as before, γ_0 is chosen to ensure the positive semidefiniteness of $\boldsymbol{\Gamma}$ in (3.27)). We still use a randomly generated sequence to initialize Multi-WeCAN. In this scenario, the normalized fitting error is defined as $\hat{\mathcal{E}}_{\text{norm}} = \hat{\mathcal{E}}/(MN^2)$.

Table 3.5 compares the Multi-WeCAN sequence and the 256×4 Multi-CAN sequence. The Multi-WeCAN sequence provides much lower correlation sidelobe peaks and a much smaller fitting error. Figure 3.7 shows the corresponding correlation levels of the Multi-CAN and Multi-WeCAN sequences, from which we see that Multi-WeCAN

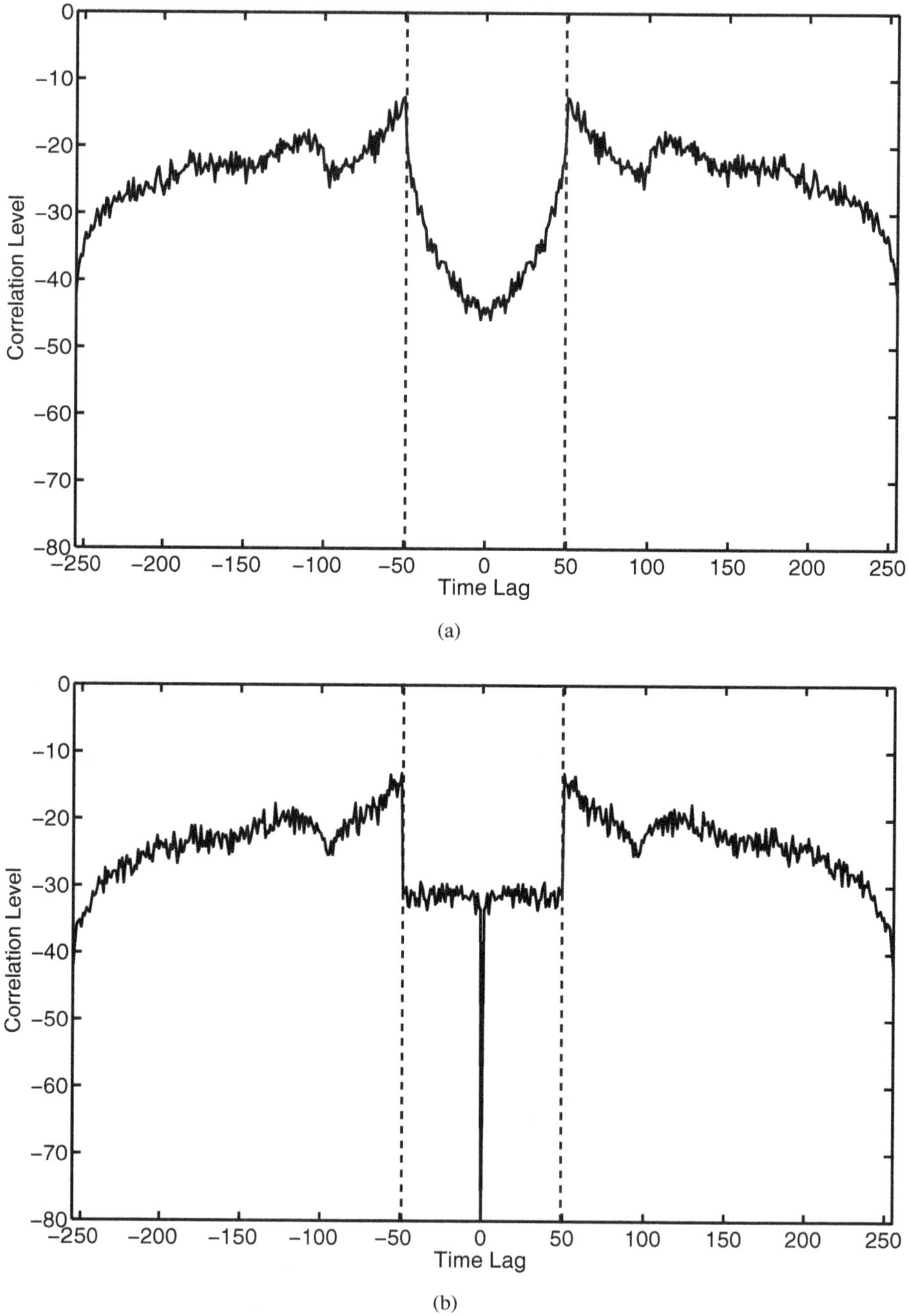

Figure 3.6 Correlation levels of (a) the Multi-CAO sequence set and (b) the Multi-WeCAN sequence set with $N = 256$, $M = 4$ and $P = 50$. (The dashed vertical lines signify the boundary of the time lag interval under consideration.)

Table 3.5. Comparison between Multi-CAN and Multi-WeCAN sequence sets under $\hat{\mathcal{E}}$ ($N = 256, M = 4$)

	Auto-corr sidelobe peak (dB)	Cross-corr peak (dB)	$\hat{\mathcal{E}}_{\text{norm}}$
Multi-CAN	−20.53	−17.68	0.40
Multi-WeCAN	−45.17	−45.81	0.001

succeeds much better in suppressing the correlations at the required lags. Note that because $|r_{m_1 m_2}(N-1)| = 1$ for all m_1 and m_2, the correlation level corresponding to the maximum lag $N-1$ is always equal to $20\log_{10}(\sqrt{M^2}/\sqrt{MN^2})$, which is −42.14 dB in this case (see the end points in both Figures 3.7(a) and 3.7(b)).

3.4.4 Quantization effects

As in Section 2.3.4, let $\{x_m(n)\}_{m=1,n=1}^{M,N}$ denote a sequence set and let 2^q denote its quantization level. Then the quantized sequence set can be expressed as

$$\hat{x}_m(n) = \exp\left\{ j \left\lfloor \frac{\arg[x_m(n)]}{2\pi/2^q} \right\rfloor \frac{2\pi}{2^q} \right\}, \quad m = 1,\ldots,M, \quad n = 1,\ldots,N. \tag{3.58}$$

We quantize the Multi-CAN sequence set used in Figures 3.4 and 3.5 into 32 levels (i.e., $q = 5$) and make the same comparisons with the random-phase and Hadamard+PN sequence sets. The results are shown in Figures 3.8 and 3.9. We observe that the curves representing the Multi-CAN sequence set move up a little but they are still below the corresponding curves of the Hadamard+PN sequence set for most of the illustrated sequence lengths. We also observe that the fitting error of the Multi-CAN sequence set hardly changes after this 32-level quantization.

Similar situations occur if we quantize sequences generated from the other algorithms discussed in this chapter. In our tests, the performance degradation (i.e., the correlation sidelobe increase) was quite limited, provided that the quantization level was not too small (e.g., $q \geq 6$).

3.4.5 Synthetic aperture radar (SAR) imaging

Consider a MIMO radar angle–range imaging example (intrapulse Doppler effects are assumed to be negligible) using uniform linear arrays with colocated $M = 4$ transmit and $M_r = 4$ receive antennas. The interelement spacings of the transmit and receive antennas are equal to 2 and 0.5 wavelengths, respectively. Suppose that all possible targets are in a far field consisting of $P = 30$ range bins (which means that the maximum round-trip delay difference within the illuminated scene is not larger than 29 subpulses) and a scanning angle area of (−30, 30) degrees. The length of the probing waveform for each transmit antenna is $N = 256$.

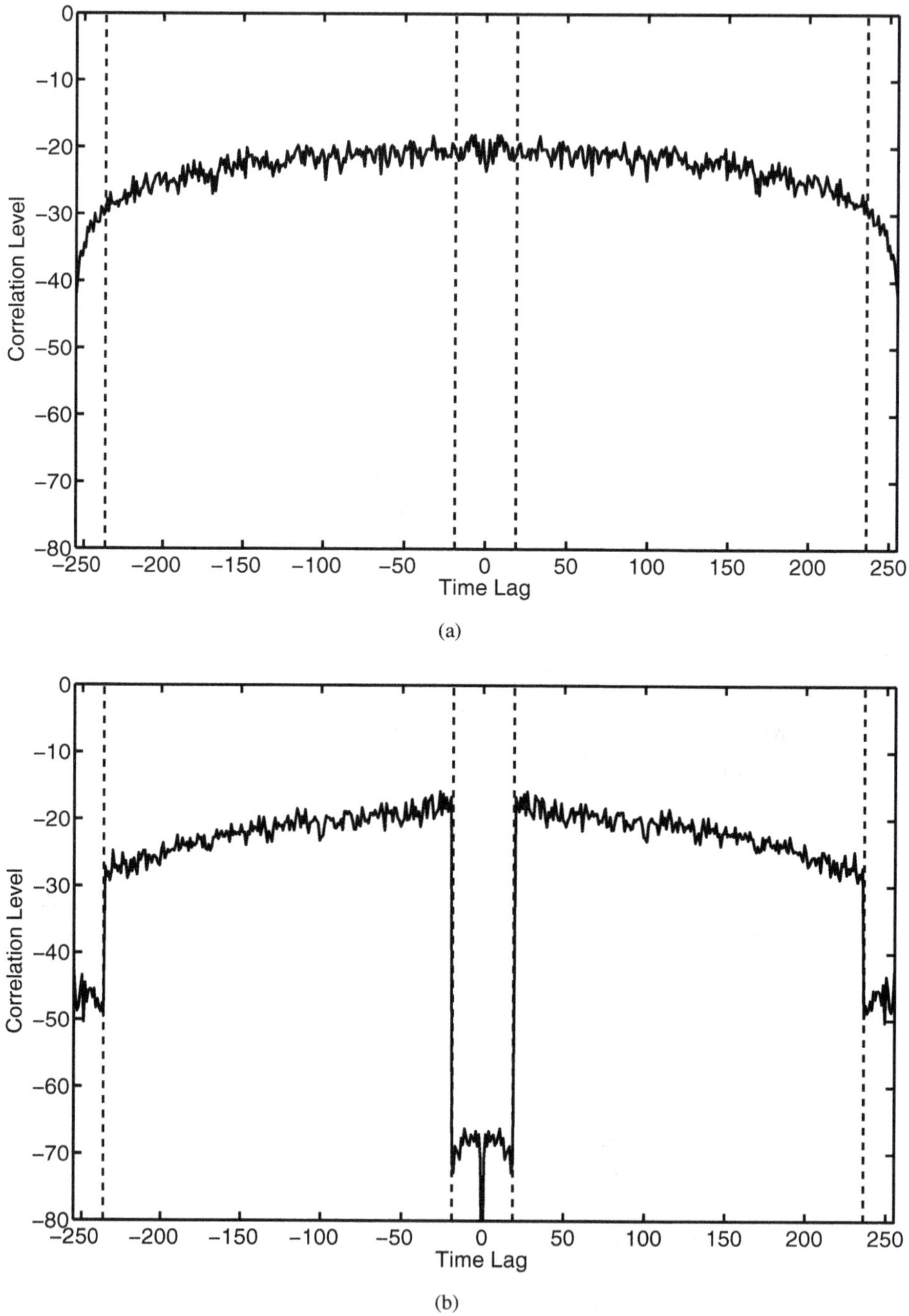

Figure 3.7 Correlation levels of (a) the Multi-CAN sequence set and (b) the Multi-WeCAN sequence set with $N = 256$, $M = 4$ and weights $\{\gamma_n\}_{n=0}^{N-1}$ as specified in (3.57). (The dashed vertical lines signify the boundaries of the time lag intervals under consideration.)

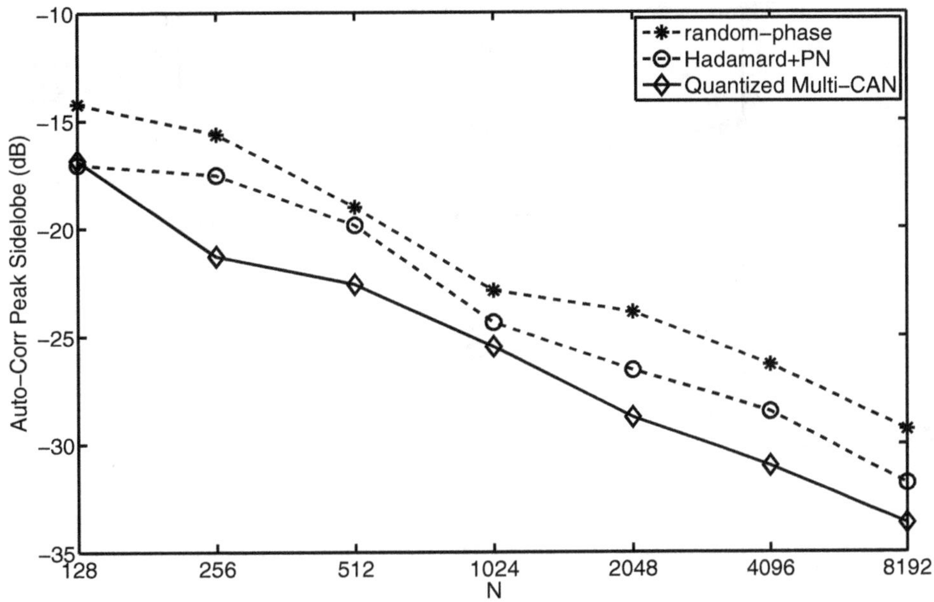

(a)

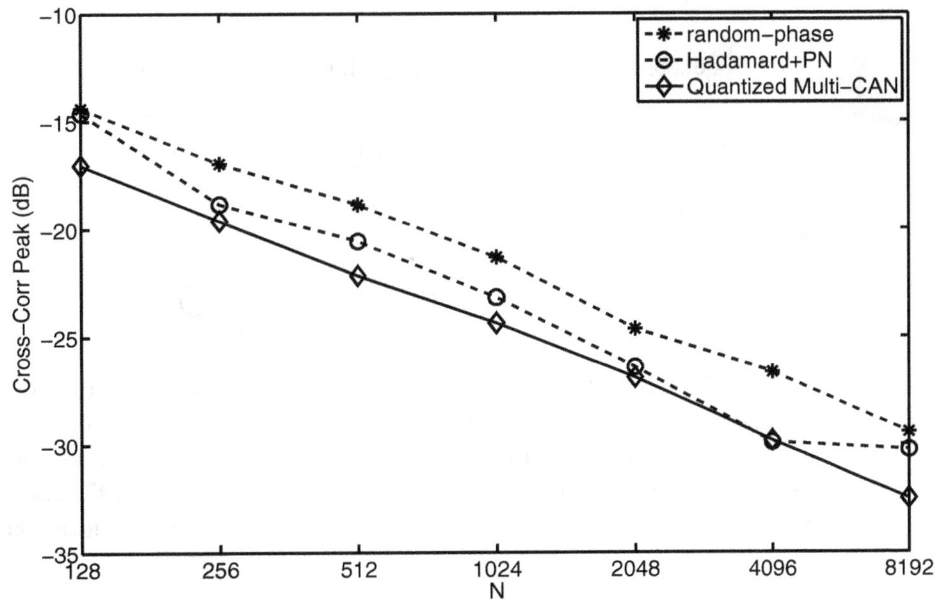

(b)

Figure 3.8 The same as Figure 3.4 except that the phases of the Multi-CAN sequence set used here are quantized into 32 levels.

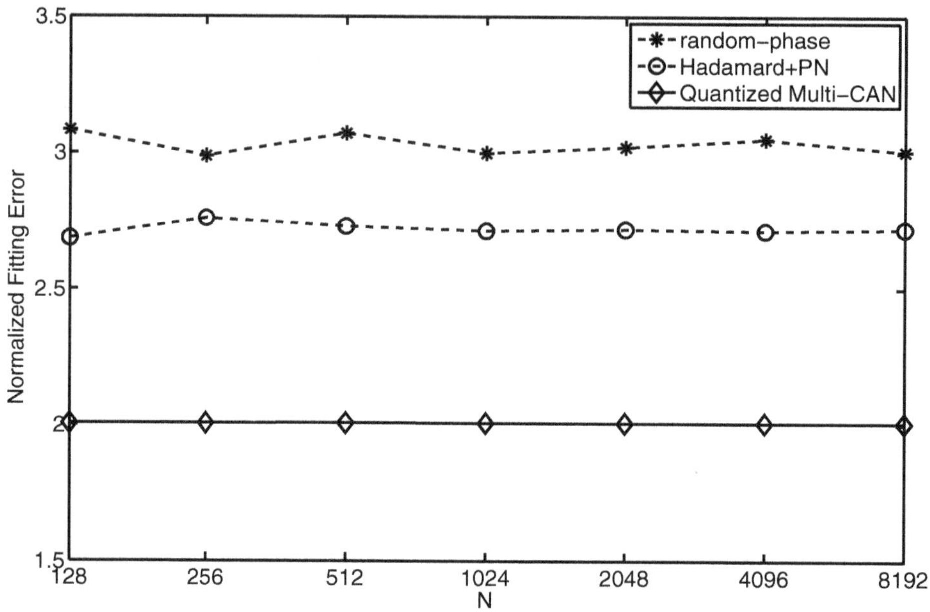

Figure 3.9 The same as Figure 3.5 except that the phases of the Multi-CAN sequence set used here are quantized into 32 levels.

Let $\mathbf{X}$ denote the $N \times M$ transmitted probing waveform matrix (see (3.3)), and let

$$\tilde{\mathbf{X}} = \begin{bmatrix} \mathbf{X} \\ \mathbf{0} \end{bmatrix}_{(N+P-1)\times M}, \tag{3.59}$$

where $\mathbf{0}$ is a $(P-1) \times M$ matrix of zeros. Then the $M_r \times (N+P-1)$ received data matrix can be written as [Xu et al. 2008]

$$\mathbf{D}^H = \sum_{p=0}^{P-1} \sum_{k=1}^{K} \alpha_{pk} \mathbf{a}_k \mathbf{b}_k^T \tilde{\mathbf{X}}^H \mathbf{J}_p + \mathbf{E}^H, \tag{3.60}$$

where $\mathbf{J}_p$ is an $(N+P-1) \times (N+P-1)$ shift matrix as defined in (3.6) (with the same structure but different dimensions), $\mathbf{E}^H$ is the noise matrix, whose columns are i.i.d. random vectors with mean zero and covariance matrix $\mathbf{Q}$, $\{\alpha_{pk}\}_{p=0,k=1}^{P-1,K}$ are complex amplitudes which are proportional to the radar cross sections (RCS) of the scatterers, and $\mathbf{a}_k$ and $\mathbf{b}_k$ are the receive and transmit steering vectors, respectively, and are given by

$$\mathbf{a}_k = \begin{bmatrix} 1 & e^{-j\pi \sin(\theta_k)} & \dots & e^{-j\pi(M_r-1)\sin(\theta_k)} \end{bmatrix}^T \tag{3.61}$$

and

$$\mathbf{b}_k = \begin{bmatrix} 1 & e^{-j\pi M_r \sin(\theta_k)} & \dots & e^{-j\pi(M-1)M_r \sin(\theta_k)} \end{bmatrix}^T, \tag{3.62}$$

where $\{\theta_k\}_{k=1}^{K}$ are the scanning angles. Our goal is to estimate $\{\alpha_{pk}\}_{p=0,k=1}^{P-1,K}$ from the collected data $\mathbf{D}^H$.

First we apply the following matched filter to the data matrix $\mathbf{D}^H$:

$$\tilde{\mathbf{X}}_p^{\text{MF}} \triangleq \mathbf{J}_p^H \tilde{\mathbf{X}} (\tilde{\mathbf{X}}^H \tilde{\mathbf{X}})^{-1}, \quad (N+P-1) \times M \tag{3.63}$$

(note that we assume $N + P - 1 \geq M$ and thus $\tilde{\mathbf{X}}^H \mathbf{J}_p \tilde{\mathbf{X}}_p^{\text{MF}} = \mathbf{I}_M$) to perform range compression for the pth range bin, i.e.,

$$\begin{aligned} \tilde{\mathbf{D}}_p^H \triangleq \mathbf{D}^H \tilde{\mathbf{X}}_p^{\text{MF}} &= \left(\sum_{q=0}^{P-1} \sum_{k=1}^{K} \alpha_{qk} \mathbf{a}_k \mathbf{b}_k^T \tilde{\mathbf{X}}^H \mathbf{J}_q \right) \tilde{\mathbf{X}}_p^{\text{MF}} + \mathbf{E}^H \tilde{\mathbf{X}}_p^{\text{MF}} \\ &= \sum_{k=1}^{K} \alpha_{pk} \mathbf{a}_k \mathbf{b}_k^T \tilde{\mathbf{X}}^H \mathbf{J}_p \tilde{\mathbf{X}}_p^{\text{MF}} + \mathbf{Z}_p \\ &= \sum_{k=1}^{K} \alpha_{pk} \mathbf{a}_k \tilde{\mathbf{b}}_k^H + \mathbf{Z}_p \quad (\tilde{\mathbf{b}}_k^H \triangleq \mathbf{b}_k^T), \end{aligned} \tag{3.64}$$

where

$$\mathbf{Z}_p = \left(\sum_{\substack{q=0 \\ q \neq p}}^{P-1} \sum_{k=1}^{K} \alpha_{qk} \mathbf{a}_k \mathbf{b}_k^T \tilde{\mathbf{X}}^H \mathbf{J}_q \right) \tilde{\mathbf{X}}_p^{\text{MF}} + \mathbf{E}^H \tilde{\mathbf{X}}_p^{\text{MF}}. \tag{3.65}$$

The filtered data in (3.64) leads naturally to the following least squares (LS) estimate of α_{pk}:

$$\hat{\alpha}_{pk}^{\text{LS}} = \frac{\mathbf{a}_k^H \tilde{\mathbf{D}}_p^H \tilde{\mathbf{b}}_k}{||\mathbf{a}_k||^2 ||\tilde{\mathbf{b}}_k||^2}, \quad k = 1, \ldots, K \quad p = 0, 1, \ldots, P-1, \tag{3.66}$$

as well as to the following Capon estimate:

$$\hat{\alpha}_{pk}^{\text{Capon}} = \frac{\mathbf{a}_k^H \hat{\mathbf{R}}_p^{-1} \tilde{\mathbf{D}}_p^H \tilde{\mathbf{b}}_k}{\mathbf{a}_k^H \hat{\mathbf{R}}_p^{-1} \mathbf{a}_k ||\tilde{\mathbf{b}}_k||^2}, \quad k = 1, \ldots, K \quad p = 0, 1, \ldots, P-1, \tag{3.67}$$

where $\hat{\mathbf{R}}_p = \tilde{\mathbf{D}}_p^H \tilde{\mathbf{D}}_p$ denotes the covariance matrix of the "compressed" received data (see [Xu et al. 2008] for more details about these estimates of α_{pk}).

To obtain a larger *synthetic* aperture, we use the SAR principle and thus repeat the process of sending a probing waveform and collecting data at $\tilde{N} = 20$ different positions; the collected data matrices are denoted as $\mathbf{D}_1^H, \mathbf{D}_2^H \ldots, \mathbf{D}_{\tilde{N}}^H$, respectively. Suppose that two adjacent positions are spaced $MM_r/2$ wavelengths apart; this will induce a phase shift of $\psi_k = -\pi M M_r \sin(\theta_k)$ for both the transmit and receive steering vectors corresponding to the two adjacent positions. (As long as the "targets in the far-field" assumption holds, the distance between two adjacent positions can be chosen at will and can be different for different adjacent positions; we only need to change the phase shift ψ_k accordingly.) In this case, we let

$$\tilde{\mathbf{D}}_p^H = \left[\mathbf{D}_1^H \tilde{\mathbf{X}}_p^{\text{MF}} \ \mathbf{D}_2^H \tilde{\mathbf{X}}_p^{\text{MF}} \cdots \ \mathbf{D}_{\tilde{N}}^H \tilde{\mathbf{X}}_p^{\text{MF}} \right]_{M_r \times \tilde{N} M}, \tag{3.68}$$

and

$$\tilde{\mathbf{b}}_k^H = \left[\mathbf{b}_k^T \ \mathbf{b}_k^T e^{j2\psi_k} \cdots \ \mathbf{b}_k^T e^{j2(\tilde{N}-1)\psi_k} \right]_{1 \times \tilde{N} M}. \tag{3.69}$$

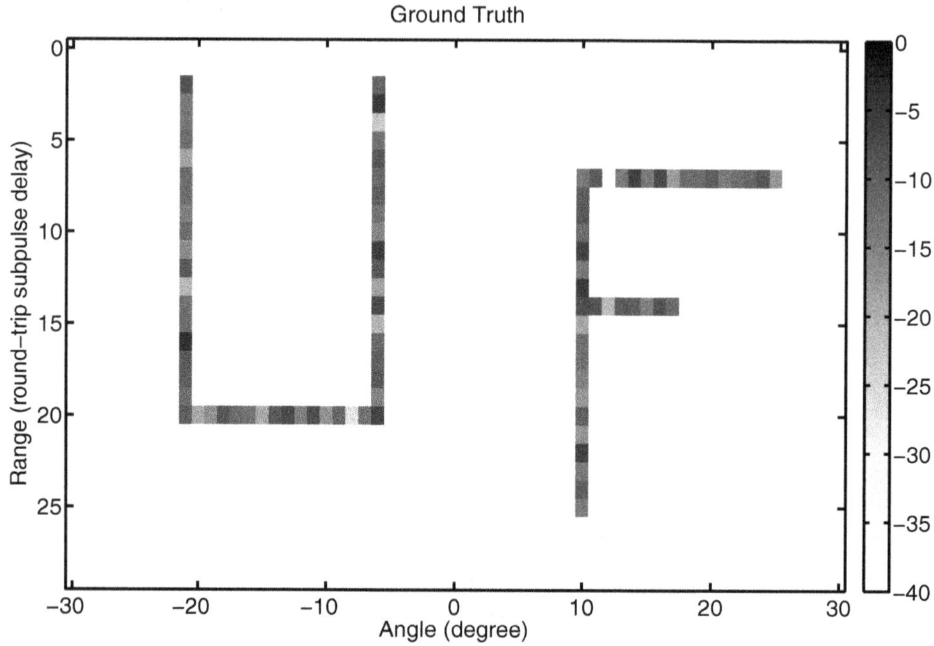

Figure 3.10 The true target image (the absolute values of $\{\alpha_{pk}\}_{p=0,k=1}^{P-1,K}$ are shown).

Using this notation, the expressions for the estimates of α_{pk} in (3.66) and (3.67) can be used *mutatis mutandis*.

In the numerical simulation, the noise covariance matrix $\mathbf{Q}$ is chosen as $\sigma^2\mathbf{I}_{M_r}$, where $\sigma^2 = 0.001$. The targets are chosen to form a "UF" shape (see Figure 3.10) and the RCS-related parameters $\{\alpha_{pk}\}_{p=0,k=1}^{P-1,K}$ are simulated as i.i.d. complex symmetric Gaussian random variables with mean 0 and variance 1 at the target locations and 0 elsewhere. The average (transmitted) signal-to-noise ratio (SNR) is given by

$$\text{SNR} = \frac{\text{tr}(\mathbf{X}^H\mathbf{X})/N}{\text{tr}(\mathbf{Q})} = \frac{M}{M_r\sigma^2} = 30 \text{ dB}. \tag{3.70}$$

We use two different probing sequence sets: a QPSK Hadamard+PN sequence set (see Section 3.4.1) and a Multi-WeCAN sequence set with $N = 256$, $M = 4$ and $P = 30$. The transmitted waveform is phase-modulated by the probing sequence set (one sequence element corresponds to one subpulse) and we assume proper sampling so that the considered discrete models are appropriate. The absolute values of the parameters $\{\alpha_{pk}\}_{p=0,k=1}^{P-1,K}$ estimated using these two waveforms are shown in Figures 3.11 and 3.12. The Multi-WeCAN waveform set gives much clearer angle–range images than the Hadamard+PN waveform set. Note from Figure 3.12(b) that the Multi-WeCAN waveform set leads to an almost perfect range compression via the matched filter (the false scatterers are due to the presence of noise) and that the Capon estimator provides a radar image with a high angle resolution.

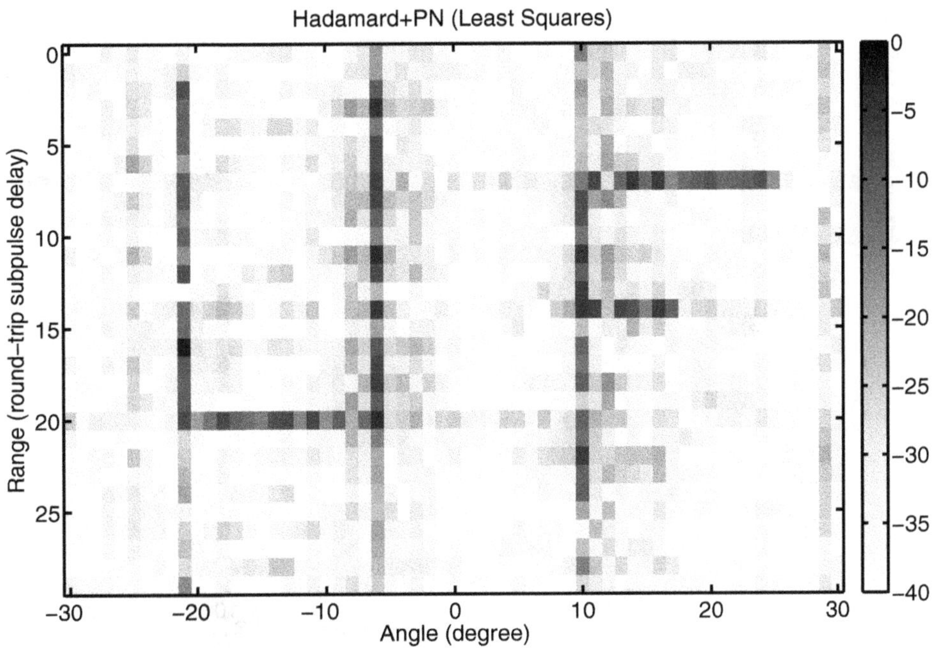

(a)

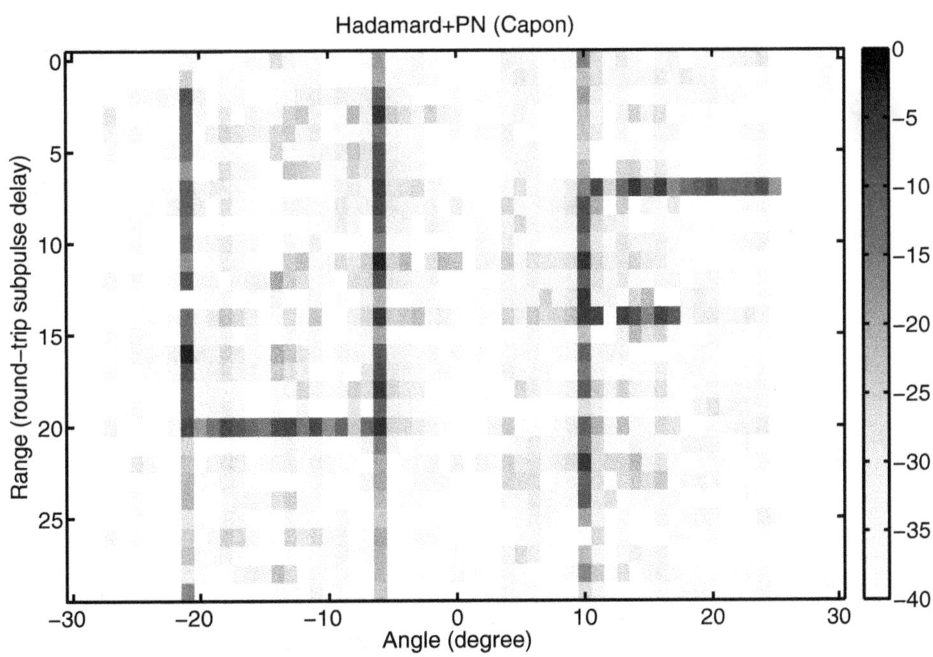

(b)

Figure 3.11 The estimated target images in terms of the RCS-related parameters $\{|\alpha_{pk}|\}_{p=0,k=1}^{P-1,K}$ using the Hadamard+PN waveform. (a) The LS estimate and (b) the Capon estimate.

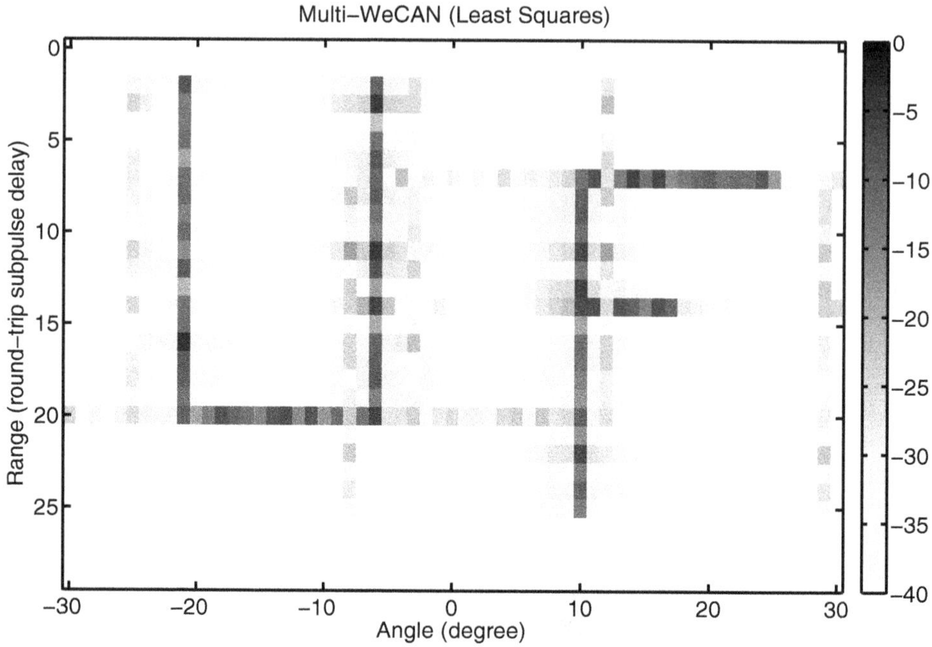

(a)

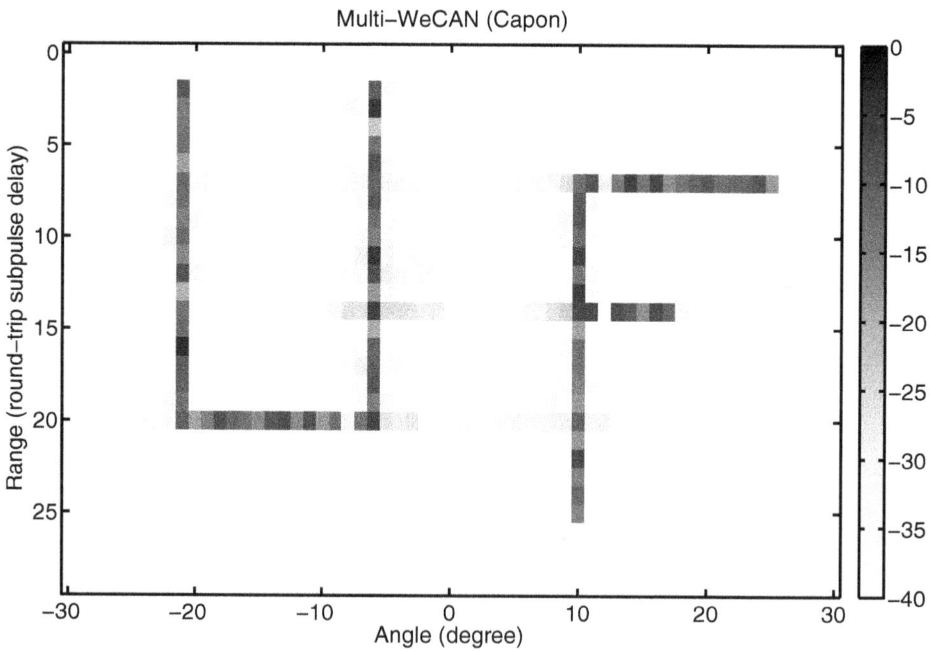

(b)

Figure 3.12 The estimated target images in terms of the RCS-related parameters $\{|\alpha_{pk}|\}_{p=0,k=1}^{P-1,K}$ using the Multi-WeCAN waveform. (a) The LS estimate and (b) the Capon estimate.

3.5 Conclusions

We have extended the CAN and WeCAN algorithms from the previous chapter, which dealt with a single sequence, to cope with a set of sequences. The obtained algorithms are called Multi-CAN and Multi-WeCAN, respectively. The Multi-CAN algorithm can be used to design very long sequences (of length up to $N \sim 10^5$), which can hardly be handled by other algorithms previously suggested in the literature. The Multi-WeCAN algorithm is useful when only a few selected correlation lags are of interest. The Multi-WeCAN algorithm can make the correlation levels almost zero if the lag interval of interest is sufficiently small.

Appendix 3A Proof of Equation (3.28)

Note the following DFT pairs:

$$\{\mathbf{R}_n\} \longleftrightarrow \mathbf{\Phi}(\omega) = \tilde{\mathbf{y}}(\omega)\tilde{\mathbf{y}}^H(\omega) \quad \text{(see (3.10) and (3.12))} \tag{3.71}$$

and

$$\{\gamma_n \mathbf{R}_n\} \longleftrightarrow \tilde{\mathbf{\Phi}}(\omega) = \tilde{\mathbf{y}}(\omega)\tilde{\mathbf{y}}^H(\omega) * \left(\sum_{k=-(N-1)}^{N-1} \gamma_k e^{-j\omega k} \right) \quad \text{(see (3.71))}. \tag{3.72}$$

Thus $\tilde{\mathbf{\Phi}}(\omega)$ can be written as follows:

$$\begin{aligned}
\tilde{\mathbf{\Phi}}(\omega) &= \frac{1}{2\pi} \int_{-\pi}^{\pi} \tilde{\mathbf{y}}(\psi)\tilde{\mathbf{y}}^H(\psi) \left(\sum_{k=-(N-1)}^{N-1} \gamma_k e^{-jk(\omega-\psi)} \right) d\psi \\
&= \frac{1}{2\pi} \int_{-\pi}^{\pi} \sum_{n=1}^{N} \mathbf{y}(n) e^{-j\psi n} \sum_{\tilde{n}=1}^{N} \mathbf{y}^H(\tilde{n}) e^{j\psi\tilde{n}} \sum_{k=-(N-1)}^{N-1} \gamma_k e^{-jk(\omega-\psi)}\, d\psi \\
&= \sum_{k=-(N-1)}^{N-1} \sum_{n=1}^{N} \sum_{\tilde{n}=1}^{N} \gamma_k \mathbf{y}(n)\mathbf{y}^H(\tilde{n}) \left(\frac{1}{2\pi} \int_{-\pi}^{\pi} e^{j\psi(\tilde{n}-n+k)}\, d\psi \right) e^{-j\omega k}.
\end{aligned} \tag{3.73}$$

It is easy to verify that $\frac{1}{2\pi} \int_{-\pi}^{\pi} e^{j\psi(k-n+\tilde{n})}\, d\psi = \delta_{k-(n-\tilde{n})}$. Therefore,

$$\begin{aligned}
\tilde{\mathbf{\Phi}}(\omega) &= \sum_{n=1}^{N} \sum_{\tilde{n}=1}^{N} \gamma_{n-\tilde{n}} \mathbf{y}(n)\mathbf{y}^H(\tilde{n}) e^{-j\omega(n-\tilde{n})} \\
&= \mathbf{Z}^T(\omega)\mathbf{\Gamma}\mathbf{Z}^*(\omega),
\end{aligned} \tag{3.74}$$

where

$$\mathbf{Z}^T(\omega) = \left[\mathbf{y}(1)e^{-j\omega} \;\; \mathbf{y}(2)e^{-j\omega 2} \;\; \cdots \;\; \mathbf{y}(N)e^{-j\omega N} \right]_{M \times N}. \tag{3.75}$$

Appendix 3B Proof of Equation (3.47)

We prove in the following that the solution of (3.45) for fixed $\mathbf{X}$ is given by (3.47).

The criterion in (3.45) can be written as

$$\left\|\bar{\mathbf{X}} - \sqrt{N}\mathbf{U}\right\|^2 = \mathrm{tr}\left[\left(\bar{\mathbf{X}}^H - \sqrt{N}\mathbf{U}^H\right)\left(\bar{\mathbf{X}} - \sqrt{N}\mathbf{U}\right)\right] \\ = \mathrm{const} - 2\sqrt{N}\,\mathrm{Re}\left[\mathrm{tr}\left(\bar{\mathbf{X}}^H\mathbf{U}\right)\right], \tag{3.76}$$

where const denotes a term that does not depend on $\mathbf{U}$ (note that $\mathbf{X}$ is assumed known and that $\mathbf{U}^H\mathbf{U} = \mathbf{I}$).

Using (3.46) we obtain

$$\mathrm{Re}\left[\mathrm{tr}\left(\bar{\mathbf{X}}^H\mathbf{U}\right)\right] = \mathrm{Re}\left[\mathrm{tr}\left(\mathbf{U}_1\mathbf{S}\mathbf{U}_2^H\mathbf{U}\right)\right] \\ = \mathrm{Re}\left[\mathrm{tr}\left(\mathbf{U}_2^H\mathbf{U}\mathbf{U}_1\mathbf{S}\right)\right] \\ = \sum_{k=1}^{MP} \mathrm{Re}\left\{\left[\mathbf{U}_2^H\mathbf{U}\mathbf{U}_1\right]_{kk}\right\} S_{kk} \tag{3.77}$$

where $[\cdot]_{kk}$ denotes the (k, k)th element of a matrix.

For notational simplicity, let $\mathbf{B} = \mathbf{U}_2^H\mathbf{U}\mathbf{U}_1$; then we have

$$|\mathrm{Re}\{B_{kk}\}|^2 \leq |B_{kk}|^2 \leq [\mathbf{B}\mathbf{B}^H]_{kk} \\ = \left[\mathbf{U}_2^H\mathbf{U}\mathbf{U}_1\mathbf{U}_1^H\mathbf{U}^H\mathbf{U}_2\right]_{kk} \\ = \left[\mathbf{U}_2^H\mathbf{U}\mathbf{U}^H\mathbf{U}_2\right]_{kk}. \tag{3.78}$$

Note that $\mathbf{U}$ is a “tall” semi-unitary matrix, which leads to the fact that $\mathbf{U}\mathbf{U}^H \leq \mathbf{I}$ and thus

$$|\mathrm{Re}\,(B_{kk})|^2 \leq \left[\mathbf{U}_2^H\mathbf{U}_2\right]_{kk} = 1. \tag{3.79}$$

It follows from (3.76) – (3.79) that

$$\left\|\bar{\mathbf{X}} - \sqrt{N}\mathbf{U}\right\|^2 = \mathrm{const} - 2\sqrt{N}\sum_{k=1}^{MP} \mathrm{Re}\left\{[\mathbf{U}_2^H\mathbf{U}\mathbf{U}_1]_{kk}\right\} S_{kk} \\ \geq \mathrm{const} - 2\sqrt{N}\sum_{k=1}^{MP} S_{kk}, \tag{3.80}$$

which is another constant independent of $\mathbf{U}$. It is not difficult to see that the equality in (3.80) holds if and only if $\mathbf{U} = \mathbf{U}_2\mathbf{U}_1^H$, which concludes the proof.

4 Lower bounds for aperiodic sequences

The previous chapters have shown the importance of designing sequences or sequence sets with low correlation sidelobes. Let $\{x_k(n)\}$ ($k = 1, \ldots, M$ and $n = 1, \ldots, N$) denote a set of M sequences of length N, which are restricted to have the same energy:

$$\sum_{n=1}^{N} |x_k(n)|^2 = N, \quad k = 1, \ldots, M. \tag{4.1}$$

Excluding the in-phase (i.e., the zero-delay) auto-correlations, all other correlations are categorized as being correlation sidelobes; correspondingly the peak sidelobe level (PSL) metric is defined as

$$\begin{aligned} \text{PSL} &= \max \{|r_{ks}(l)|\}, \\ k, s &= 1, \ldots, M, \quad l = 0, \ldots, N-1 \ (l \neq 0 \text{ if } k = s). \end{aligned} \tag{4.2}$$

The following PSL lower bound is due to Welch [Welch 1974]:

$$\text{PSL} \geq N \sqrt{\frac{M-1}{2NM - M - 1}} \triangleq B_{\text{PSL}}. \tag{4.3}$$

The integrated sidelobe level (ISL) metric, used previously in the single sequence case (see (2.1)), is defined in the present multi-sequence case as (see also (3.2)):

$$\text{ISL} = \sum_{k=1}^{M} \sum_{\substack{p=-N+1 \\ p \neq 0}}^{N-1} |r_{kk}(p)|^2 + \sum_{k=1}^{M} \sum_{\substack{s=1 \\ s \neq k}}^{M} \sum_{p=-N+1}^{N-1} |r_{ks}(p)|^2. \tag{4.4}$$

A lower bound on ISL was implicitly derived in [Sarwate 1999]. In the next section we derive both the ISL and PSL lower bounds using the Multi-CAN framework.

4.1 Bound derivation

It is shown in Chapter 3 that the ISL metric in (4.4) can be transformed to the frequency domain as follows:

$$\text{ISL} = \frac{1}{2N} \sum_{p=1}^{2N} \left(\|\mathbf{y}_p\|^2 - N \right)^2 + (M-1)N^2, \tag{4.5}$$

where

$$\mathbf{y}_p = \begin{bmatrix} y_1(p) \\ \vdots \\ y_M(p) \end{bmatrix}, \quad y_k(p) = \sum_{n=1}^{N} x_k(n) e^{-j\frac{2\pi}{2N}(p-1)(n-1)}, \quad k = 1, \ldots, M. \tag{4.6}$$

Note that $\{y_k(p)\}_{p=1}^{2N}$ is the DFT (discrete Fourier transform) of the sequence $\{x_k(n)\}_{n=1}^{N}$ padded with N zeros in the tail. In this section we start from this frequency-domain expression of ISL in order to derive a lower bound on it.

Let $z_{kp} = |y_k(p)|^2$. Then the energy constraint (4.1) is related to $\{z_{kp}\}$ by the Parseval equality:

$$\sum_{p=1}^{2N} z_{kp} = 2N \sum_{n=1}^{N} |x_k(n)|^2 = 2N^2, \quad k = 1, \ldots, M. \tag{4.7}$$

Expanding (4.5) and substituting (4.7), we obtain

$$\text{ISL} = \frac{1}{2N} \sum_{p=1}^{2N} \left(\sum_{k=1}^{M} z_{kp} \right)^2 - MN^2. \tag{4.8}$$

Making use of the Cauchy–Schwartz inequality leads to the following result:

$$\begin{aligned} \text{ISL} &= \frac{1}{(2N)^2} \left[\sum_{p=1}^{2N} 1^2 \right] \left[\sum_{p=1}^{2N} \left(\sum_{k=1}^{M} z_{kp} \right)^2 \right] - MN^2 \\ &\geq \frac{1}{4N^2} \left[\sum_{p=1}^{2N} 1 \left(\sum_{k=1}^{M} z_{kp} \right) \right]^2 - MN^2 \qquad (4.9) \\ &= M^2N^2 - MN^2, \qquad (4.10) \end{aligned}$$

where (4.7) was used to get (4.10) from (4.9).

The above result on the ISL lower bound is summarized as:

$$\text{ISL} \geq N^2 M(M-1) \triangleq B_{\text{ISL}}. \tag{4.11}$$

The PSL lower bound in (4.3) can be easily obtained from B_{ISL} as a corollary. It follows from the definition of ISL in (4.4) that

$$\text{ISL} \leq 2M(N-1)\text{PSL}^2 + M(M-1)(2N-1)\text{PSL}^2. \tag{4.12}$$

Substituting (4.11) into (4.12), we obtain B_{PSL} in (4.3).

Note that the equality (4.9) holds if and only if $\sum_{k=1}^{M} z_{kp} = c$ for all $p = 1, \ldots, 2N$, where c is a constant. Because of the energy constraint (4.7), it is easy to see that $c = NM$. In other words, a set of energy-constrained sequences $\{x_k(n)\}$ meets the ISL lower bound if and only if their $2N$-point DFT values satisfy $\|\mathbf{y}_p\|^2 = NM$ for all $p = 1, \ldots, 2N$ (see (4.6) for the definition of $\|\mathbf{y}_p\|$). An example of such a sequence set is given in (4.13) below.

Table 4.1. Values of ISL and B_{ISL} for various random-phase and Multi-CAN sequence sets

	ISL of random-phase	ISL of Multi-CAN	B_{ISL}
$M = 2, N = 200$	146351	80031	8000
$M = 2, N = 512$	1014134	524378	524288
$M = 4, N = 512$	4154326	3145746	3145728
$M = 4, N = 1000$	15620957	12000088	12000000

4.2 Approaching the bound

The question naturally arises as to whether we can generate sequence sets that achieve the correlation lower bound B_{ISL} or B_{PSL}. Here we focus on trying to meet B_{ISL}.

A trivial solution to meeting B_{ISL} is the following sequence set (recall that the energy constraint in (4.1) is always imposed):

$$x_k(n) = \begin{cases} \sqrt{N}, & n = 1, \\ 0, & n = 2, \ldots, N, \end{cases} \qquad k = 1, \ldots, M, \tag{4.13}$$

whose correlation sidelobes are all zero except for the zero-lag cross-correlation, which is N. A set of M sequences leads to $M(M-1)$ pairs of cross-correlations and thus the ISL for the above sequence set is exactly equal to the lower bound $N^2M(M-1)$. However, the sequence set in (4.13) is not practically useful because its PSL is as high as the in-phase auto-correlation. Moreover, transmiting only at one time instant while keeping silent at all other times, as evidenced by the zeros for $n = 2, \ldots, N$ in (4.13), results in a high (in fact, the maximum possible) peak-to-average power ratio (PAR), which is again undesirable in practice.

The Multi-CAN algorithm introduced in Chapter 3 aims to find unimodular sequence sets with low ISL values. The unimodular constraint refers to the fact that every sequence element has unit modulus, i.e., $|x_k(n)| = 1$. In this case the energy constraint in (4.1) is automatically satisfied. Note that unimodular sequences are often preferred in practice owing to hardware restrictions, such as the need to use an economical nonlinear amplifier which essentially works well only when the PAR is 1 or close to 1.

Although the unimodular constraint is certainly more stringent than the energy constraint, the unimodular sequence sets generated by Multi-CAN have an ISL that is fairly close to B_{ISL}, provided that there are at least two sequences in the set (the $M = 1$ situation turns out to be special and is taken care of later on). To illustrate this, we show in Table 4.1 the ISL values of sequence sets generated by Multi-CAN and the corresponding B_{ISL} values for various combinations of M and N. The ISL values of the random-phase sequence sets (i.e., $\{x_k(n) = e^{j\phi_k(n)}\}$ where each $\phi_k(n)$ is drawn independently from the uniform distribution in $[0, 2\pi]$), which are significantly larger than those of the Multi-CAN sequence sets, are also shown in Table 4.1 for comparison. Note that the Multi-CAN algorithm is run from a random initialization and that different random initializations lead to different sequence sets but with similarly low correlations. In Table 4.1 only one such realization is presented for each (M, N) pair.

The good performance of Multi-CAN synthesized unimodular sequence sets, when compared with the B_{ISL} values, can no longer be guaranteed when $M = 1$, in which case $B_{\text{ISL}} = 0$. Note that the Multi-CAN algorithm becomes the CAN algorithm when $M = 1$. Hereafter, in this section only the auto-correlation of a single sequence is considered. For a sequence $\{x(n)\}_{n=1}^{N}$ with $|x(n)| = 1$ for all n, it holds that $|r(N-1)| = |x(N)x^*(1)| = 1$ and thus ISL ≥ 1. Hence, obviously B_{ISL} cannot be reached by using unimodular sequences. Actually the ISL of a single sequence generated by CAN is much larger than 1 (e.g., on the order of 10^3 when $N = 200$), although a CAN sequence can possess much lower correlation sidelobes than many well-known unimodular sequences in the literature, such as the Golomb or Frank sequences (see Chapter 2). We consider below relaxing the unimodular constraint in the CAN algorithm so as to obtain lower correlations. More precisely, define the PAR of the sequence $\mathbf{x} = \left[x(1) \cdots x(N)\right]^T$ as follows:

$$\text{PAR}(\mathbf{x}) = \frac{\max_n |x(n)|^2}{\frac{1}{N}\sum_{n=1}^{N}|x(n)|^2} = \max_n |x(n)|^2, \tag{4.14}$$

where the second equality is due to the energy constraint. The CAN algorithm generates sequences with PAR $= 1$. Here we extend it to the more general case of PAR $\leq \rho$ where ρ can be any number between 1 and N.

Following the discussions in Chapter 2, the ISL metric (4.5) for the case of $M = 1$ can be made small by solving the following minimization problem:

$$\begin{aligned} \min_{\{x(n)\}_{n=1}^{N};\{\psi(p)\}_{p=1}^{2N}} \quad & f = \|\mathbf{A}^H\mathbf{z} - \mathbf{v}\|^2 = \|\mathbf{z} - \mathbf{A}\mathbf{v}\|^2 \\ \text{s.t.} \quad & \|\mathbf{x}\|^2 = N, \\ & \text{PAR}(\mathbf{x}) \leq \rho, \end{aligned} \tag{4.15}$$

where

$$\begin{aligned} \mathbf{z} &= \left[x(1) \cdots x(N)\ 0 \cdots 0\right]_{2N\times 1}^T, \\ \mathbf{v} &= \frac{1}{\sqrt{2}}\left[e^{j\psi(1)} \cdots e^{j\psi(2N)}\right]_{2N\times 1}^T, \end{aligned} \tag{4.16}$$

$\{\psi(p)\}$ are auxiliary variables and $\mathbf{A}^H$ is a unitary $2N \times 2N$ DFT matrix (i.e., $\mathbf{A}^H\mathbf{x}$ gives the $2N$-point DFT of any vector $\mathbf{x}$ of length $2N$). Note that (4.15) would reduce to the problem discussed in Chapter 2 (see Section 2.1) if its second constraint were replaced by PAR$(\mathbf{x}) = 1$.

The problem (4.15) can be solved in a cyclic way. We first fix $\mathbf{z}$ and compute the $\mathbf{v}$ that minimizes f:

$$\psi(p) = \arg\left(p\text{th element of } \mathbf{A}^H\mathbf{z}\right), \quad p = 1, \ldots, 2N. \tag{4.17}$$

Next we fix $\mathbf{v}$ and note that the minimization problem can be cast as

$$\begin{aligned} \min_{\mathbf{x}} \quad & \|\mathbf{x} - \mathbf{s}\|^2 \\ \text{s.t.} \quad & \|\mathbf{x}\|^2 = N, \\ & \text{PAR}(\mathbf{x}) \leq \rho, \end{aligned} \tag{4.18}$$

where $\mathbf{s}$ is an $N \times 1$ vector consisting of the first N elements of $\mathbf{Av}$.

The "nearest-vector" problem (4.18) was considered in [Tropp et al. 2005]; here we briefly outline its solution. To begin with, note that the solution to (4.18) without the PAR constraint is given by $\hat{\mathbf{x}} = \sqrt{N}\mathbf{s}/||\mathbf{s}||$. Then note that the PAR constraint is equivalent to $\max_n |x(n)| \leq \sqrt{\rho}$. Hence if the magnitudes of all elements in $\hat{\mathbf{x}}$ are below $\sqrt{\rho}$, then $\hat{\mathbf{x}}$ is a solution; if not, we resort to a recursive procedure as follows. The element in $\mathbf{x}$ corresponding to the largest element (in terms of magnitude) in $\mathbf{s}$, say s_α, is given by $\sqrt{\rho}\exp\left[j\arg(s_\alpha)\right]$. The other $N-1$ elements in $\mathbf{x}$ are obtained by solving the same problem as in (4.18), except that now $\mathbf{x}$ and $\mathbf{s}$ are $(N-1)\times 1$ and the energy constraint is $||\mathbf{x}||^2 = N-\rho$. Since the scalar case of (4.18) is trivial, such a recursive procedure is guaranteed to yield a solution. We refer the readers to [Tropp et al. 2005] for more details.

To summarize, we iterate between (4.17) and (4.18) until convergence (for instance, until the norm of the difference between the $||\mathbf{x}||$ values obtained in two consecutive iterations is less than a predefined threshold, e.g., 10^{-3}). The criterion (4.15) decreases in every iteration step, so local convergence is guaranteed (i.e., the $\mathbf{x}$ thus obtained is at least a local minimum solution to (4.15)). The iterative process can be started from a random phase initialization of $\mathbf{x}$, e.g., $\{x(n) = e^{j\phi(n)}\}_{n=1}^{N}$, where each $\phi(n)$ is drawn independently from a uniform distribution over $[0, 2\pi]$; such an initialization is used whenever we consider random initialization in what follows. Alternatively $\mathbf{x}$ can be initialized by any good existing sequence ("good" meaning that the sequence itself already has relatively low correlations), e.g., the P4 sequence. The resulting algorithm is still called CAN in view of the fact that the CAN algorithm proposed in Chapter 2 is just a special case of (4.15) (corresponding to PAR = 1) and certainly an important one; no ambiguity will be introduced by using this name since hereafter in this chapter we will specify the PAR value whenever we apply CAN.

Consider next using CAN to generate a sequence of length $N = 512$ with energy N. Figure 4.1(a) shows the auto-correlations (normalized by N and in dB) of two CAN sequences, one with PAR = 1 and the other with PAR = 4, both initialized by a randomly generated sequence. Figure 4.1(b) is for the same setting as Figure 4.1(a) except that the P4 sequence was used to initialize the CAN algorithm. Clearly ρ plays an important role: a larger ρ leads to significantly lower correlation sidelobe levels. (Note that we have not plotted, for comparison, the correlations of the P4 or other well-known sequences such as the Golomb or Frank sequences, because they have higher correlation sidelobes than the CAN sequence with PAR = 1; see Chapter 2 for examples.)

Figure 4.2 illustrates the ISL for a CAN sequence with length $N = 512$ and ρ ranging from 1 to 10. As before, we use either a randomly generated sequence or the P4 sequence to initialize CAN. The P4 initialization gives a lower ISL than random initialization. Interestingly, when ρ is relatively small the decrease of ISL caused by even a small increase of ρ is significant. Note that in the case of P4 initialization, the ISL can be decreased by more than two orders of magnitude if ρ is increased just from 1 to 1.2. However, after reaching a certain point, an increase of ρ does not push the ISL any lower. The ISL of the CAN sequence initialized by P4 when $\rho = 4$ is 5.38, a value relatively close to the ISL lower bound, $B_{\mathrm{ISL}} = 0$. A full explanation is still lacking as

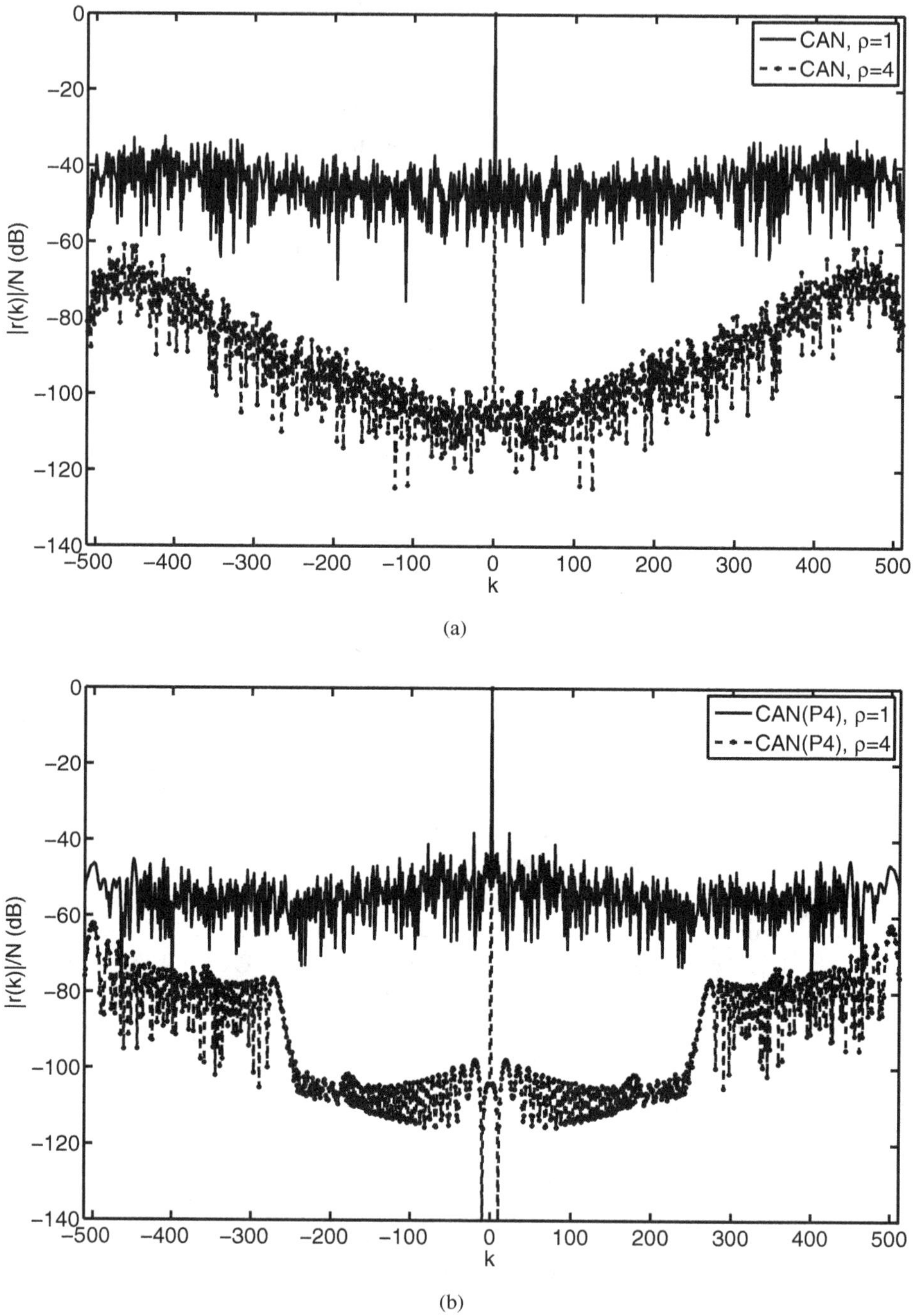

Figure 4.1 (a) The auto-correlations (normalized by N and shown in dB) of two CAN sequences of length $N = 512$, one with PAR = 1 and the other with PAR = 4, both initialized by a randomly generated sequence. (b) The same as (a) except that the P4 sequence is used to initialize the CAN algorithm.

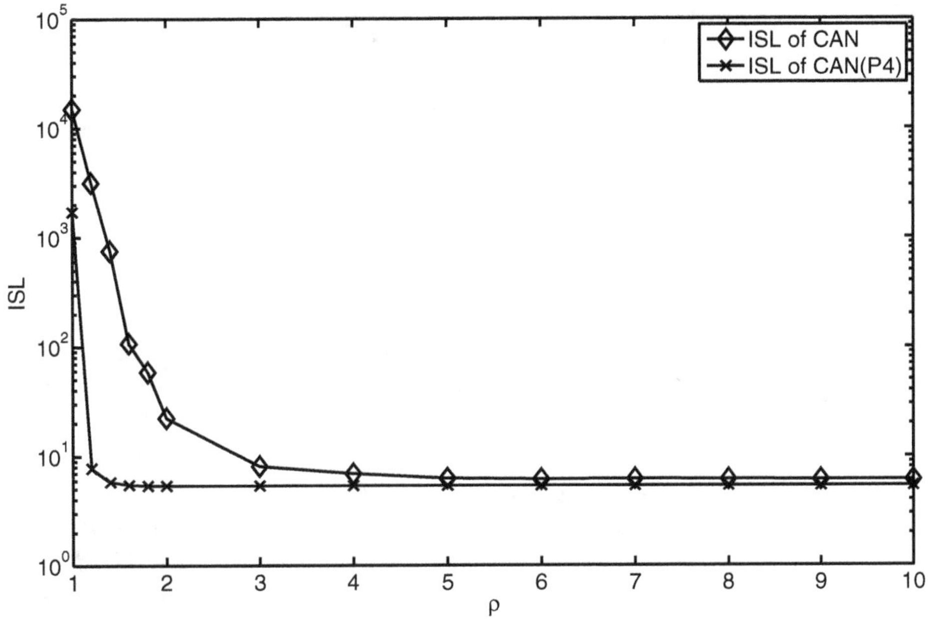

Figure 4.2 The ISL of CAN sequences (with length $N = 512$ and initialized either randomly or by P4) versus ρ.

to why the ISL of the CAN sequence does not go to zero when ρ is sufficiently large, though the possible trapping of the algorithm in local minima is a likely reason.

4.3 Conclusions

In this chapter, using a different framework from that in the literature, we have derived lower bounds on the ISL and PSL metrics of a sequence set under the total energy constraint. We have shown that if a sequence set has more than one sequence, then the ISL lower bound B_{ISL} can be nearly met by the unimodular sequence sets generated by the Multi-CAN algorithm of Chapter 3. For the more challenging single-sequence case where B_{ISL} equals zero, we have accommodated the CAN algorithm to PAR values larger than 1 so that the ISL value can be lowered further.

We finally comment on the fact that the "relaxing PAR" technique can also be utilized for zero-correlation zone (ZCZ) sequence synthesis. The Multi-WeCAN algorithm can generate sequence sets whose correlation sidelobes are almost zero within a certain time lag interval – thus such sets are called ZCZ. An increased PAR is expected to lead to a sequence with a longer ZCZ.

5 Stopband constraint case

Among the tasks associated with cognitive radar [Haykin 2006], an important one is to adapt the spectrum of transmitted waveforms to the changing environment. In particular, the transmitted signal should not use certain frequency bands that have already been reserved, such as the bands for navigation and military communications; or there could exist strong emitters whose operating frequencies should be avoided. Therefore it is required that the spectral power of transmitted waveforms be small for certain frequency bands [Lindenfeld 2004][Salzman et al. 2001][Wang & Lu 2011][Headrick & Skolnik 1974].

The main focus in this chapter is on designing a discrete sequence whose spectral power is small in certain specified frequency bands. The designed sequence can be used in active sensing systems such as radar or sonar as a probing sequence. It can also be used as a spreading sequence in spread-spectrum applications such as CDMA (code division multiple access) systems.

Besides frequency notching, we also need to take into account the correlation properties of the designed sequence. As pointed out several times in previous chapters (e.g., in Chapter 1), in radar or sonar applications low auto-correlation of the probing sequence improves target detection when range compression is applied in the receiver. Furthermore, practical hardware components such as analog-to-digital converters and power amplifiers have a maximum signal-amplitude clip. In order to maximize the transmitted power that is available in the system, unimodular sequences are desirable.

In this chapter we propose an algorithm named SCAN (stopband CAN) for unimodular transmit sequence design. SCAN is an extension of the CAN algorithm introduced in Chapter 2. CAN aims at generating unimodular sequences with low correlation sidelobes. SCAN extends CAN in such a way that both frequency stopbands and correlation sidelobes are considered. The SCAN algorithm is computationally efficient (as it is based on FFT operations) and thus it facilitates long sequence generation and, possibly, real-time waveform update. Another advantage of SCAN is that the algorithm can start from random initializations and that different initializations lead to different sequences all having similarly good properties.

The problem formulation and the SCAN algorithm are presented in Section 5.1. A variation of the SCAN algorithm named WeSCAN, which has more flexibility in controlling the correlation levels (but at the cost of increased computation), is discussed in Section 5.2. Several simulation results are given in Section 5.3.

5.1 Stopband CAN (SCAN)

We first formulate two design criteria which are related, respectively, to spectral band suppression and correlation sidelobe suppression. Without loss of generality, only normalized frequencies (from 0 to 1 Hz) are considered for notational simplicity.

Suppose that the set of frequencies which $\{x(n)\}_{n=1}^{N}$ should avoid can be expressed as

$$\Omega = \bigcup_{k=1}^{N_s} (f_{k1}, f_{k2}) \tag{5.1}$$

where (f_{k1}, f_{k2}) identifies one stopband and N_s is the number of stopbands. For the set Ω, a number $\tilde{N}$ is chosen large enough that points of the DFT (discrete Fourier transform) frequency grid $\{p/\tilde{N}\}_{p=0}^{\tilde{N}-1}$ cover Ω densely (see later for details). Let $\mathbf{F}_{\tilde{N}}$ denote the $\tilde{N} \times \tilde{N}$ DFT matrix whose (k, l)th element is given by

$$[F_{\tilde{N}}]_{kl} = \frac{1}{\sqrt{\tilde{N}}} \exp\left(j2\pi \frac{kl}{\tilde{N}}\right), \quad k, l = 0, \ldots, \tilde{N} - 1, \tag{5.2}$$

where the coefficient $1/\sqrt{\tilde{N}}$ makes $\mathbf{F}_{\tilde{N}}$ unitary. We form a matrix $\mathbf{S}$ from the columns of $\mathbf{F}_{\tilde{N}}$ corresponding to the frequencies in Ω. For example, if $\Omega = [0.2, 0.3]$ Hz and we choose $\tilde{N} = 100$, $\mathbf{S}$ will be the 100×11 submatrix of $\mathbf{F}_{\tilde{N}}$ comprising its 20th to 30th columns (indexed from 0). After constructing $\mathbf{S}$, let $\mathbf{G}$ denote the matrix comprising the remaining columns in $\mathbf{F}_{\tilde{N}}$.

It follows from the above discussion that we can suppress the spectral power of $\{x(n)\}$ in Ω by minimizing the criterion

$$||\mathbf{S}^H \tilde{\mathbf{x}}||^2 \tag{5.3}$$

where

$$\tilde{\mathbf{x}} = \left[x(1) \cdots x(N) \underbrace{0 \cdots 0}_{\tilde{N}-N} \right]^T. \tag{5.4}$$

Observe that (5.3) would be zero if $\tilde{\mathbf{x}}$ lies in the null space of $\mathbf{S}^H$. Since the null space of $\mathbf{S}^H$ is spanned by the columns of $\mathbf{G}$, the problem of minimizing (5.3) can be equivalently formulated as

$$\begin{aligned} &\min_{\mathbf{x},\boldsymbol{\alpha}} J_1(\mathbf{x}, \boldsymbol{\alpha}) = ||\tilde{\mathbf{x}} - \mathbf{G}\boldsymbol{\alpha}||^2 \\ &\text{s.t. } |x(n)| = 1, \quad n = 1, \ldots, N, \end{aligned} \tag{5.5}$$

where $\mathbf{x} = \left[x(1) \cdots x(N) \right]^T$ and $\boldsymbol{\alpha}$ is an auxiliary variable vector.

The problem of suppressing the auto-correlation sidelobes can be dealt with using the CAN algorithm formulation discussed in Chapter 2. As shown in Section 2.1, the auto-correlation sidelobes can be suppressed by solving the following problem:

$$\begin{aligned}\min_{\mathbf{x},\mathbf{v}} J_2(\mathbf{x},\mathbf{v}) &= \left\| \mathbf{F}_{2N}^H \begin{bmatrix} \mathbf{x} \\ \mathbf{0}_{N\times 1} \end{bmatrix} - \mathbf{v} \right\|^2 \\ \text{s.t. } |x(n)| &= 1, \quad n = 1,\ldots,N, \\ |v_n| &= \frac{1}{\sqrt{2}}, \quad n = 1,\ldots,2N,\end{aligned} \tag{5.6}$$

where $\mathbf{F}_{2N}$ is the $2N \times 2N$ DFT matrix, $\mathbf{x} = [x(1) \ \cdots \ x(N)]$ and $\mathbf{v}$ is an auxiliary variable.

Combining (5.5) and (5.6), we obtain the following minimization problem, which incorporates both frequency stopband and correlation sidelobe constraints:

$$\begin{aligned}\min_{\mathbf{x},\boldsymbol{\alpha},\mathbf{v}} J(\mathbf{x},\boldsymbol{\alpha},\mathbf{v}) &= \lambda\|\tilde{\mathbf{x}} - \mathbf{G}\boldsymbol{\alpha}\|^2 + (1-\lambda)\left\| \mathbf{F}_{2N}^H \begin{bmatrix} \mathbf{x} \\ \mathbf{0}_{N\times 1} \end{bmatrix} - \mathbf{v} \right\|^2 \\ \text{s.t. } |x(n)| &= 1, \quad n = 1,\ldots,N, \\ |v_n| &= \frac{1}{\sqrt{2}}, \quad n = 1,\ldots,2N,\end{aligned} \tag{5.7}$$

where $0 \le \lambda \le 1$ controls the relative weight of the two penalty functions J_1 and J_2. Note that there are three variables in the criterion $J(\mathbf{x},\boldsymbol{\alpha},\mathbf{v})$. We minimize $J(\mathbf{x},\boldsymbol{\alpha},\mathbf{v})$ with respect to only one variable at a time and then iterate. The iteration steps are summarized in Table 5.1. The resulting algorithm is named SCAN (stopband CAN), as it is an extension of the CAN algorithm discussed in Chapter 2. Note that all matrix operations involved in the updating formulae of (5.8), (5.9) and (5.13) can be done via FFT. Therefore the SCAN algorithm is computationally efficient. Indeed, it can be used to generate sequences of length N up to 10^6 on an ordinary PC.

Before proceeding to the next section, we point out that a more general constraint than unimodularity is a constraint on the peak-to-average power ratio (PAR) of the transmitted sequence (see (4.14)). If a PAR larger than 1 is allowed, the SCAN algorithm outlined in Table 5.1 can be kept as is with the exception that the minimizer $\mathbf{x}$ in (5.13) is now given by the solution to the following problem:

$$\begin{aligned}&\min_{\mathbf{x}} \|\mathbf{x} - [\lambda\mathbf{c}_1 + (1-\lambda)\mathbf{c}_2]\|^2 \\ &\text{s.t. } \mathrm{PAR}(\mathbf{x}) \le \rho,\end{aligned} \tag{5.14}$$

where $1 \le \rho \le N$ is the prescribed largest allowable PAR. The "nearest-vector" problem in (5.14) has already been discussed in Section 4.2 to which we refer for details. In the examples of Section 5.3, however, we will use mostly the unimodular constraint unless otherwise stated.

Table 5.1. The SCAN algorithm

Step 0: Initialize $\{x(n)\}_{n=1}^{N}$ with a randomly generated unimodular sequence.
Step 1: For fixed $\mathbf{x}$ and $\mathbf{v}$, $J(\mathbf{x}, \alpha, \mathbf{v})$ is a convex quadratic function of α. By setting $\partial J / \partial \alpha = 0$ and using $\mathbf{G}^H \mathbf{G} = \mathbf{I}$, we get the minimizer α:

$$\alpha = \mathbf{G}^H \tilde{\mathbf{x}}. \tag{5.8}$$

Step 2: For fixed $\mathbf{x}$ and α, it is easy to see that the minimizer $\mathbf{v}$ is given by (component-wise)

$$\mathbf{v} = \frac{1}{\sqrt{2}} \exp\left[j \arg \left(\mathbf{F}_{2N}^H \left[\mathbf{x}^T \ \mathbf{0}_{1\times N} \right]^T \right) \right]. \tag{5.9}$$

Step 3: For fixed α and $\mathbf{v}$, J can be written as

$$J = \text{const} - 2\text{Re}\left\{ \mathbf{x}^H \left[\lambda \mathbf{c}_1 + (1-\lambda)\mathbf{c}_2 \right] \right\}, \tag{5.10}$$

where

$$\mathbf{c}_1 \text{ represents the first } N \text{ elements of } \mathbf{G}\alpha \tag{5.11}$$

and

$$\mathbf{c}_2 \text{ represents the first } N \text{ elements of } \mathbf{F}_{2N}\mathbf{v}. \tag{5.12}$$

Then the minimizer $\mathbf{x}$ is given by

$$\mathbf{x} = \exp\left\{ j \arg \left[\lambda \mathbf{c}_1 + (1-\lambda)\mathbf{c}_2 \right] \right\}. \tag{5.13}$$

Iteration: Repeat (5.8), (5.9) and (5.13) until convergence (for instance, until the norm of the difference between the $\mathbf{x}$'s obtained in two consecutive iterations is less than a predefined threshold, e.g., 10^{-3}).

5.2 Weighted SCAN (WeSCAN)

In this section we present the WeSCAN (weighted SCAN) algorithm which can be viewed as an extension of both SCAN and WeCAN.

For SCAN, minimizing the function J_2 in (5.6) is a way of minimizing the integrated sidelobe level (ISL) metric (see (2.1)):

$$\text{ISL} = 2 \sum_{k=1}^{N-1} |r(k)|^2. \tag{5.15}$$

The following more general WISL (weighted ISL) metric associates a weight γ_k^2 with each correlation term $r(k)$ (see (2.3)):

$$\text{WISL} = 2 \sum_{k=1}^{N-1} \gamma_k^2 |r(k)|^2. \tag{5.16}$$

The weights $\{\gamma_k\}$ can be chosen to satisfy our needs. For example, we can set $\gamma_1 = 0, \gamma_2 = 0$ and $\gamma_k = 1$ for $k = 3, \ldots, N-1$ to trade the correlation mainlobe width for sidelobe suppression.

Define the weighting matrix

$$\mathbf{\Gamma} = \frac{1}{\gamma_0} \begin{bmatrix} \gamma_0 & \gamma_1 & \cdots & \gamma_{N-1} \\ \gamma_1 & \ddots & & \gamma_{N-2} \\ \vdots & & \ddots & \vdots \\ \gamma_{N-1} & \cdots & & \gamma_0 \end{bmatrix} \tag{5.17}$$

where $\gamma_0 > 0$ is large enough to make $\mathbf{\Gamma}$ positive semidefinite. Let $\mathbf{D}$ be a square root of $\mathbf{\Gamma}$ and let d_{kl} denote the (k, l)th element of $\mathbf{D}$. Then it is shown in Section 2.2 that the WISL metric can be minimized by solving the following problem:

$$\begin{aligned} \min_{\mathbf{x}, \{\boldsymbol{\alpha}_p\}_{p=1}^{2N}} \quad & \tilde{J}_2(\mathbf{x}, \mathbf{V}) = \|\mathbf{F}_{2N}^H \mathbf{Z} - \mathbf{V}\|^2 \\ \text{s.t.} \quad & |x(n)| = 1, \quad n = 1, \ldots, N \\ & \|\boldsymbol{\alpha}_p\|^2 = 1, \quad p = 1, \ldots, 2N \end{aligned} \tag{5.18}$$

where

$$\begin{aligned} \mathbf{Z} &= \begin{bmatrix} \mathbf{z}_1 & \cdots & \mathbf{z}_N \end{bmatrix}_{2N \times N}, \\ \mathbf{z}_k &= \begin{bmatrix} d_{k1} x(1) & \cdots & d_{kN} x(N) & 0 & \cdots & 0 \end{bmatrix}_{2N \times 1}^T, \quad k = 1, \ldots, N, \end{aligned} \tag{5.19}$$

and

$$\mathbf{V} = \frac{1}{\sqrt{2}} \begin{bmatrix} \boldsymbol{\alpha}_1 & \cdots & \boldsymbol{\alpha}_{2N} \end{bmatrix}^T \quad (2N \times N). \tag{5.20}$$

We replace the penalty function J_2 in (5.6) by the function $\tilde{J}_2$ in (5.18), follow the discussion after (5.7) in Section 5.1 and make necessary changes that are straightforward. The resulting algorithm is named WeSCAN (weighted SCAN). We will show in Section 5.3 that, compared with SCAN, WeSCAN is able to generate sequences with much better frequency stopband suppression, for instance at the cost of an increased correlation mainlobe width.

To conclude this section, we remark that the frequency stopband penalty function (e.g., (5.3)) can be formulated using continuous frequencies [Lindenfeld 2004]. Let

$$X(f) = \sum_{n=1}^{N} x(n) e^{-j2\pi f(n-1)}, \quad f \in [0, 1], \tag{5.21}$$

be the discrete-time Fourier transform of $\{x(n)\}_{n=1}^{N}$. The frequency stopband is again given by (5.1). The spectral power in the kth band is calculated as

$$\int_{f_{k1}}^{f_{k2}} |X(f)|^2 df = \int_{f_{k1}}^{f_{k2}} \left| \sum_{n=1}^{N} x(n) e^{-j2\pi f(n-1)} \right|^2 df \tag{5.22}$$

$$= \sum_{n=1}^{N} \sum_{m=1}^{N} x^*(m) \left(\int_{f_{k1}}^{f_{k2}} e^{j2\pi f(m-n)}\, df \right) x(n),$$

and the stopband criterion $\tilde{J}_1$ is defined to be the summation of (5.22) over the stopbands:

$$\tilde{J}_1 = \sum_{k=1}^{N_s} \int_{f_{k1}}^{f_{k2}} |X(f)|^2 df \tag{5.23}$$

$$= \sum_{n=1}^{N} \sum_{m=1}^{N} x^*(m) \left(\sum_{k=1}^{N_s} \int_{f_{k1}}^{f_{k2}} e^{j2\pi f(m-n)}\, df \right) x(n).$$

Letting $\mathbf{R}$ be an $N \times N$ matrix whose (m, n)th element is given by

$$R_{mn} = \sum_{k=1}^{N_s} \int_{f_{k1}}^{f_{k2}} e^{j2\pi f(m-n)} df \tag{5.24}$$

$$= \sum_{k=1}^{N_s} \begin{cases} \dfrac{e^{j2\pi f_{k2}(m-n)} - e^{j2\pi f_{k1}(m-n)}}{j2\pi(m-n)}, & m \neq n, \\ f_{k2} - f_{k1}, & m = n, \end{cases}$$

leads to the following expression for $\tilde{J}_1$:

$$\tilde{J}_1 = \mathbf{x}^H \mathbf{R} \mathbf{x}. \tag{5.25}$$

In general $\mathbf{R}$ will be rank deficient [Stoica & Moses 2005]. Suppose that rank$(\mathbf{R}) = \hat{N} < N$. Let $\mathbf{R} = \mathbf{U}\mathbf{\Sigma}\mathbf{U}^*$ denote the eigenvalue decomposition of $\mathbf{R}$ (with diagonal elements of $\mathbf{\Sigma}$ in decreasing order), and let $\mathbf{B}$ denote the $N \times (N - \hat{N})$ matrix formed by the last $N - \hat{N}$ columns of $\mathbf{U}$. Using an argument similar to that leading from (5.3) to (5.5), the weighted problem of notching the power in Ω can be formulated as

$$\min_{\mathbf{x}, \boldsymbol{\alpha}} \tilde{J}_1(\mathbf{x}, \boldsymbol{\alpha}) = \|\mathbf{x} - \mathbf{B}\boldsymbol{\alpha}\|^2 \tag{5.26}$$

$$\text{s.t. } |x(n)| = 1, \quad n = 1, \ldots, N.$$

Compared with (5.5), the formulation in (5.26) precludes the use of an FFT in the algorithm iterations and thus is not particularly appealing from a computational efficiency perspective. Another problem with (5.26) relates to the need to select $\hat{N}$ (since the rank of a matrix is not always easy to determine). The penalty function $\tilde{J}_1$ in (5.26) will be used in the examples in the next section only when it is thus specified.

5.3 Numerical examples

5.3.1 SCAN

Suppose that we want to design a unimodular sequence of length $N = 100$ and that the frequency stopband is given by $\Omega = [0.2, 0.3)$ Hz. We use the SCAN algorithm to generate such a sequence with parameters $\lambda = 0.7$ and $\tilde{N} = 1000$. Letting $\mathbf{x}$ denote the so-generated sequence, we measure the properties of $\mathbf{x}$ in terms of the peak stopband power (P_{stop}) and the peak correlation sidelobe (P_{corr}). More specifically, we define

$$P_{\text{corr}} = 20 \log_{10} \left(\max_{k=1,\ldots,N-1} \frac{|r(k)|}{N} \right). \tag{5.27}$$

To calculate P_{stop}, letting $\{y(k)\}_{k=1}^{\tilde{N}}$ denote the $\tilde{N}$-point FFT of $\{x(n)\}_{n=1}^{N}$, we normalize $\{y(k)\}_{k=1}^{\tilde{N}}$ in such a way that the average value of $|y(k)|^2$ in the passband is 1. Then P_{stop} is calculated as

$$P_{\text{stop}} = 10 \log_{10} \left(\max_{k} |y(k)|^2 \right) \quad \text{for} \quad \frac{k-1}{\tilde{N}} \in \Omega. \tag{5.28}$$

For this example the range of k in (5.28) is from 201 to 300 (note that k corresponds to the frequency $(k-1)/\tilde{N}$).

The spectral power (i.e., the normalized $\{|y(k)|^2\}$) and correlation level ($\{|r(k)|/N\}$) of the generated sequence are shown in Figures 5.1(a) and 5.1(b), respectively. Corresponding to the two figures we have $P_{\text{stop}} = -8.3$ dB and $P_{\text{corr}} = -19.2$ dB.

To illustrate how λ affects the performance, we increase λ from 0.1 to 1 and plot the values of P_{stop} and P_{corr} in Figure 5.2 (the other parameters are kept the same as before). We do not show the results for λ less than 0.1 because such a small λ leads to very little stopband suppression. From Figure 5.2 it is easy to see that a larger λ gives more weight to the stopband penalty function and thus results in a smaller P_{stop} at the cost of an increased P_{corr}.

Remark: The curve of P_{corr} in Figure 5.2 does not increase strictly monotonically as λ increases, nor does P_{stop} decrease monotonically. The reason is that the SCAN algorithm is initialized by a random sequence. Different initializations lead to different final sequences whose P_{stop} or P_{corr} varies. While generally a larger λ favors stopband suppression, it is not guaranteed that, e.g., $\lambda = 0.75$ gives a smaller P_{stop} and a larger P_{corr} than $\lambda = 0.7$.

Next we consider a situation where the allowed band is highly segmented by stopbands:

$$\begin{aligned} \Omega = [0, 0.11) \bigcup [0.13, 0.19) \bigcup [0.25, 0.36) \\ \bigcup [0.40, 0.65) \bigcup [0.8, 0.87) \bigcup [0.94, 1) \text{ Hz.} \end{aligned} \tag{5.29}$$

The sequence length is $N = 10^4$ and we choose $\tilde{N} = N$ since N is already large enough to ensure a dense DFT-frequency grid. We set $\lambda = 0.9$ to emphasize the stopband suppression. Figure 5.3 shows the spectral power and correlation level of the so-generated SCAN sequence. Here $P_{\text{stop}} = -15.1$ dB and $P_{\text{corr}} = -7.3$ dB. Note that the peak correla-

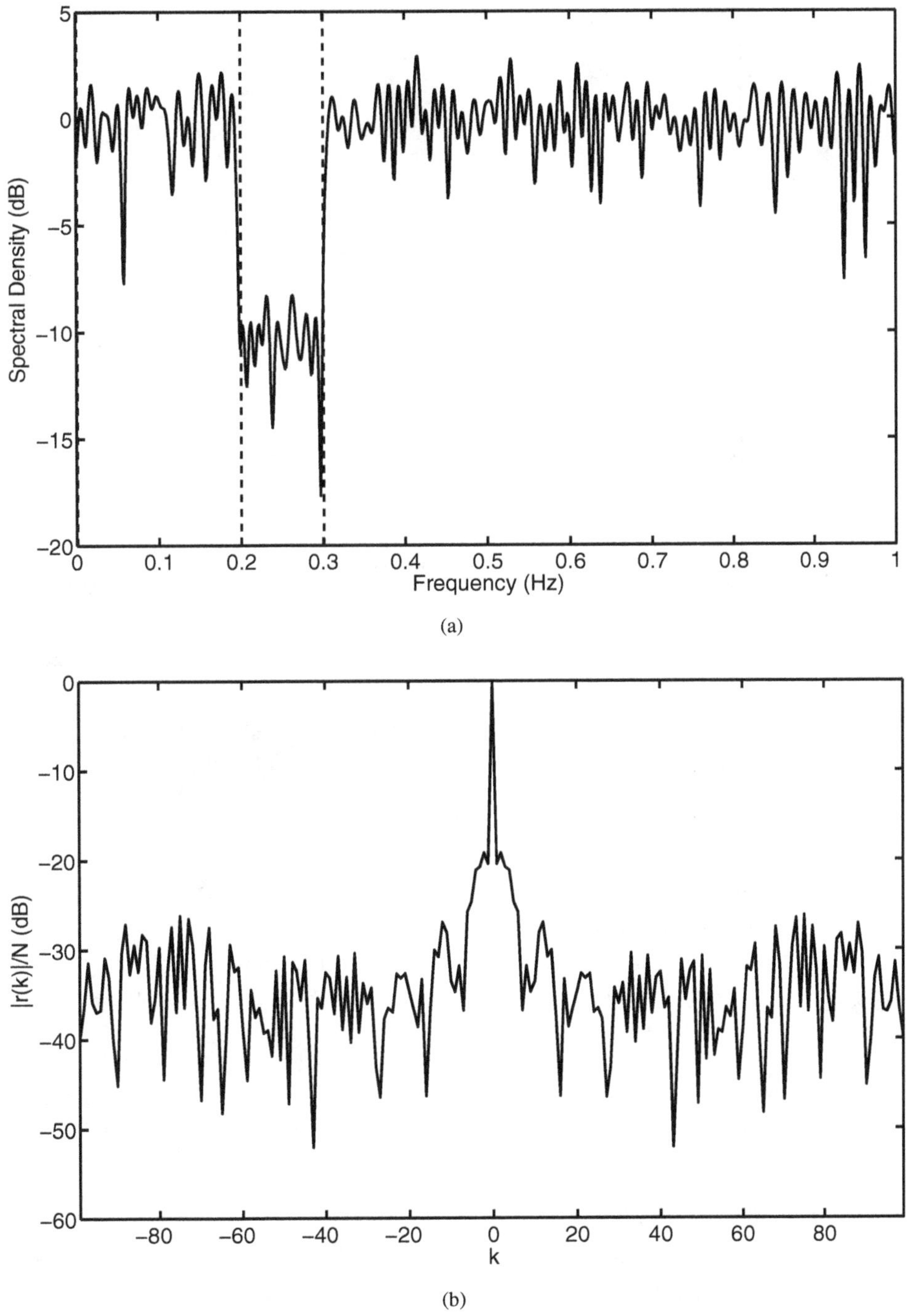

Figure 5.1 The spectral power and correlation level of a SCAN sequence generated with parameters $N = 100$, $\tilde{N} = 1000$, $\lambda = 0.7$ and $\Omega = [0.2, 0.3)$ Hz. (a) The spectral power (the dashed vertical lines signify the stopband borders) and (b) the correlation level.

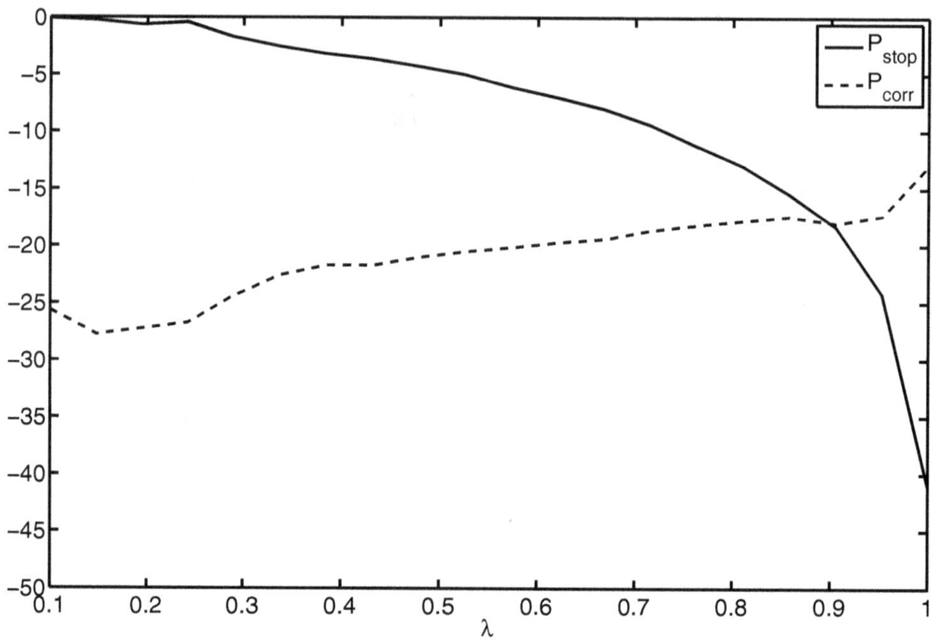

Figure 5.2 P_{stop} and P_{corr} vs. λ (the other settings are the same as those in Figure 5.1).

tion sidelobe occurs close to the origin (i.e., the in-phase correlation point) and that the correlation sidelobes far from the origin are much lower than P_{corr}.

5.3.2 WeSCAN

In this subsection we show that by using the WeSCAN algorithm, we can trade off correlation properties for an improvement in P_{stop}.

We use the same setting as in Figure 5.1 except that we apply the WeSCAN algorithm. Assuming that a relatively wide correlation mainlobe is acceptable, we choose the correlation weights (see (5.16)) as $\gamma_1 = 0$, $\gamma_2 = 0$ and $\gamma_k = 1$ for $k = 3, \ldots, N-1$. We show the spectral power and correlation level of the so-generated WeSCAN sequence in Figure 5.4. In comparison with Figure 5.1, the stopband power ($P_{\text{stop}} = -34.9$ dB) is smaller by more than 20 dB, but at the cost of an enlarged correlation mainlobe.

5.3.3 Relaxed amplitude constraint

In this example we illustrate the effect of a relaxed peak-to-average ratio (i.e., PAR > 1). The same setting as in Figure 5.1 is used except that (5.14) replaces (5.13) in the SCAN algorithm. We choose $\rho = 2$ which constrains the PAR of the designed sequence to be less than or equal to 2. The spectral power and correlation level of the so-generated SCAN sequence are shown in Figure 5.5, where $P_{\text{stop}} = -9.0$ dB and $P_{\text{corr}} = -19.3$ dB. In terms of P_{stop} and P_{corr}, this relaxed PAR does not lead to a significantly better per-

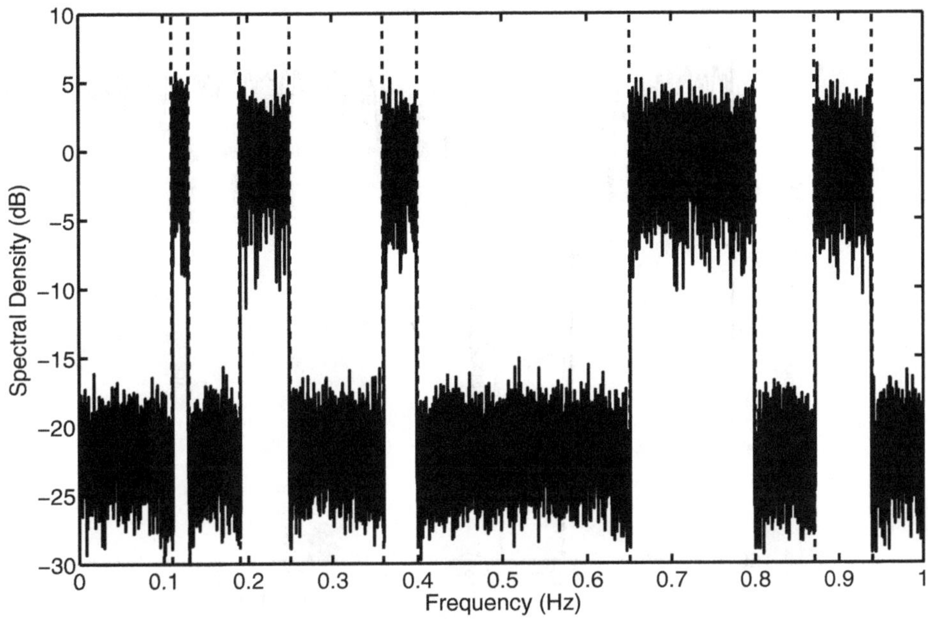

(a)

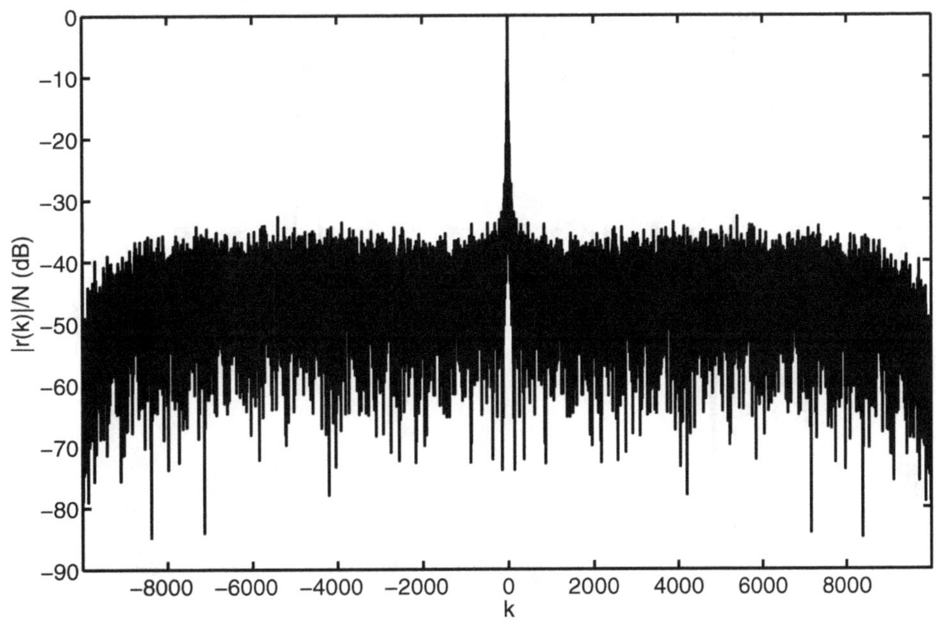

(b)

Figure 5.3 The spectral power and correlation level of a SCAN sequence generated with parameters $N = 10^5$, $\tilde{N} = 10^5$, $\lambda = 0.9$ and $\Omega = [0, 0.11) \bigcup [0.13, 0.19) \bigcup [0.25, 0.36) \bigcup [0.40, 0.65) \bigcup [0.8, 0.87) \bigcup [0.94, 1)$ Hz. (a) The spectral power and (b) the correlation level.

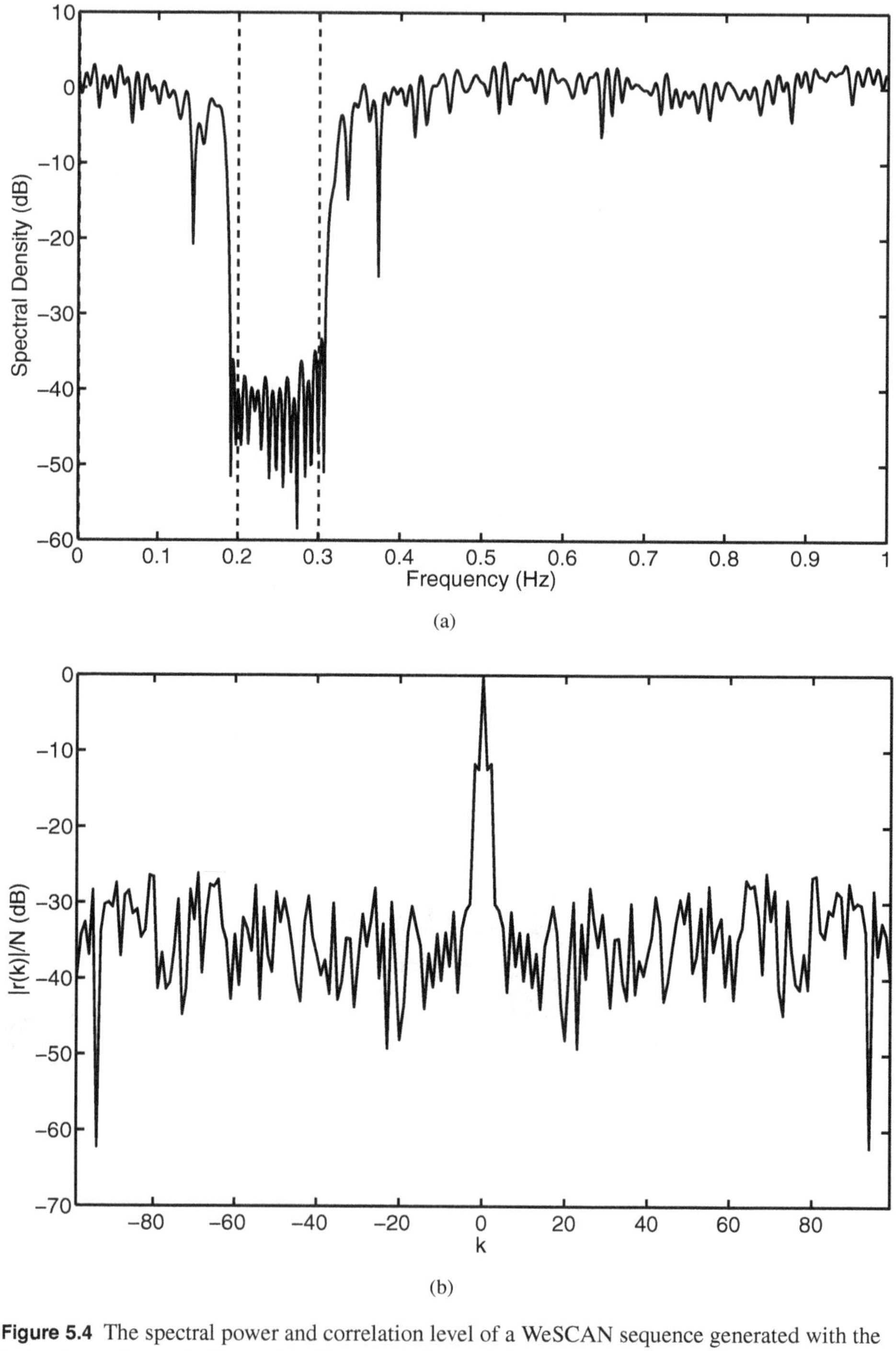

Figure 5.4 The spectral power and correlation level of a WeSCAN sequence generated with the same parameters as in Figure 5.1. The correlation weights are chosen as $\gamma_1 = 0$, $\gamma_2 = 0$ and $\gamma_k = 1$ for $k = 3, \ldots, N-1$. (a) The spectral power and (b) the correlation level.

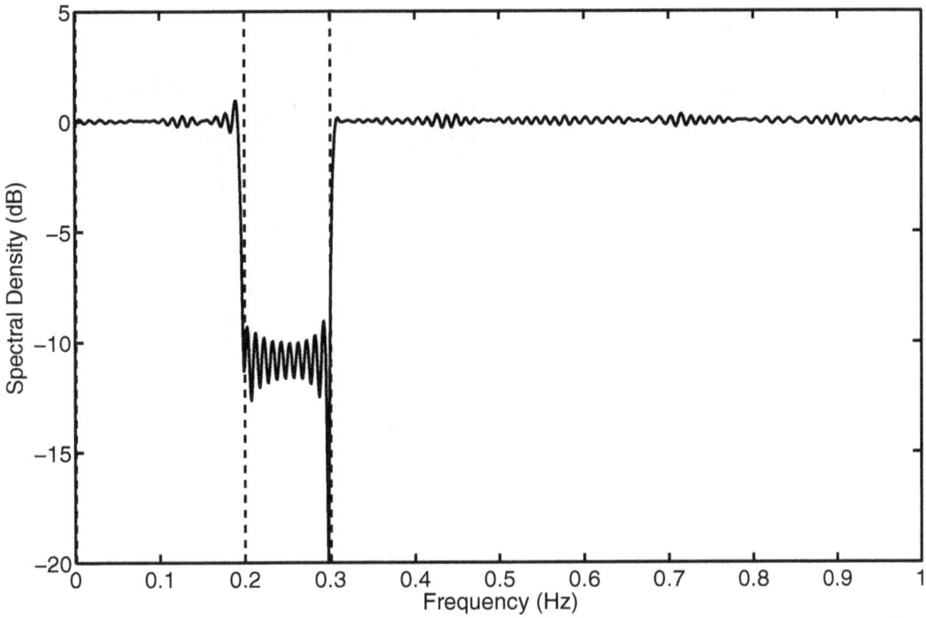

(a)

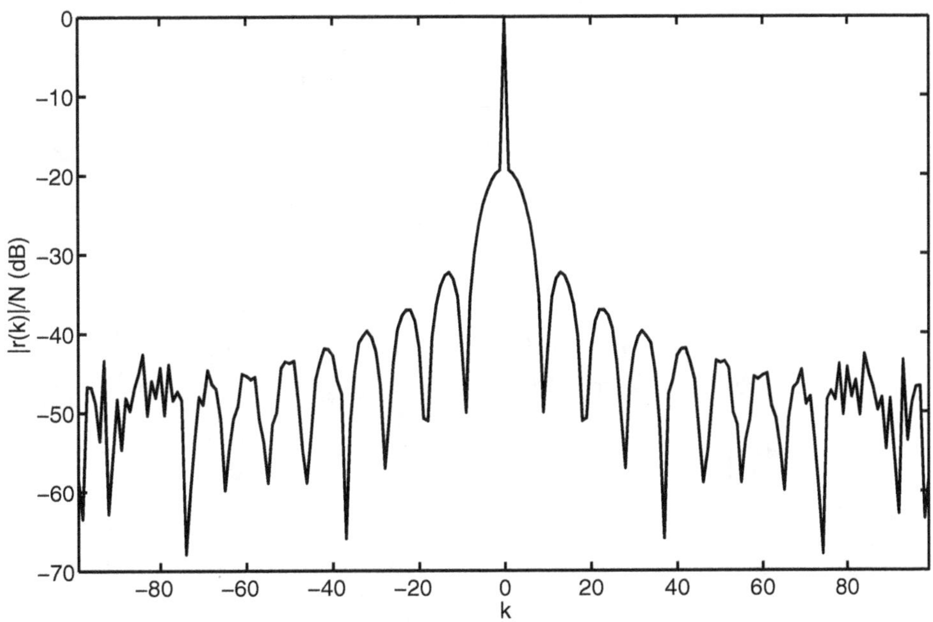

(b)

Figure 5.5 The spectral power and correlation level of a SCAN sequence generated with the same parameters as in Figure 5.1, except that the constraint PAR ≤ 2 is imposed instead of unimodularity. (a) The spectral power and (b) the correlation level.

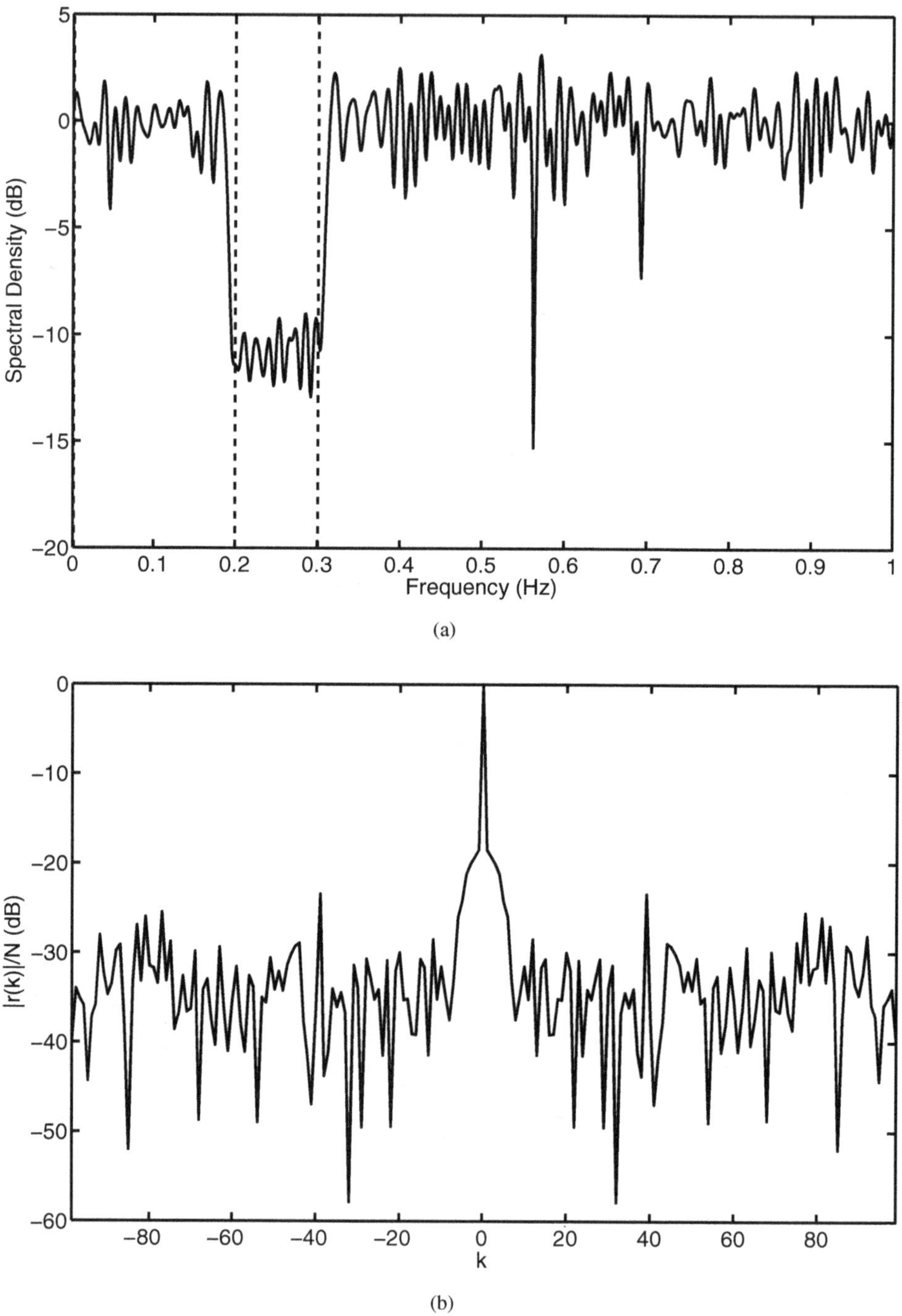

Figure 5.6 The spectral power and correlation level of a SCAN sequence generated with the same parameters as in Figure 5.1, except that (5.26) is used in lieu of (5.4) in the SCAN algorithm. (a) The spectral power and (b) the correlation level.

formance. However, in comparison with Figure 5.1 the spectral power in Figure 5.5(a) jitters less and the correlation level in Figure 5.5(b) is smaller for large time lags. Thus such a design is worth considering if PAR > 1 is allowed in the real system.

5.3.4 Using a different frequency formulation

Finally we show an example in which the discrete-frequency formulation (5.4) is replaced by the continuous-frequency one in (5.26). We still use the parameter settings of Figure 5.1. In this case the rank of **R** equals 22. The spectral power and correlation level of the resulting SCAN sequence are shown in Figure 5.6, where $P_{\text{stop}} = -9.0$ dB and $P_{\text{corr}} = -18.5$ dB; we observe that there is no performance gain from using a relatively more involved continuous-frequency stopband formulation.

5.4 Conclusions

In this chapter We have presented two algorithms, namely SCAN and WeSCAN, for the design of unimodular sequences with low spectral power in predetermined stopbands and low time correlation sidelobes as well. The SCAN algorithm is based on FFT operations and can efficiently handle very long sequence designs (up to 10^6). The WeSCAN algorithm involves more computations but is more flexible in terms of the stopband–correlation trade-off. Such designed sequences are desired in cognitive radar as well as in other general radar applications. Both the SCAN and WeSCAN algorithms can be extended to design a set of unimodular or PAR-constrained sequences with desired cross-correlations.

6 Ambiguity function (AF)

The ambiguity function (AF) measures the response of a matched filter to a signal with various time delays and Doppler frequency shifts. Let $u(t)$ denote a probing signal with time support $[0, T]$ (i.e., $u(t)$ is assumed to be zero outside $[0, T]$). The continuous-time AF of $u(t)$ is defined as

$$\chi(\tau, f) = \int_{-\infty}^{\infty} u(t)u^*(t-\tau)e^{-j2\pi f(t-\tau)}\,dt \tag{6.1}$$

where τ is the time delay and f is the Doppler frequency shift. Note that (6.1) is in fact the ambiguity function for narrowband signals, which is the focus of this chapter as well as the next. See Appendix 6A for a discussion on ambiguity functions for wideband signals.

There exists an extensive literature on radar ambiguity functions, for example [Woodward 1957][Sussman 1962][Wolf et al. 1968][Costas 1984][Guey & Bell 1998][Levanon & Mozeson 2004][Gladkova & Chebanov 2004][Bonami et al. 2007][Chen & Vaidyanathan 2008][Abramovich & Frazer 2008] and [Benedetto et al. 2009].

6.1 AF properties

Figure 6.1 shows the ambiguity function of a chirp signal (see (1.19) in Chapter 1) with parameters $T = 10$ s and $B = 5$ Hz. Note that in the figure the absolute value of $\chi(\tau, f)$ is normalized in such a way that the peak value at the origin is 1, the delay τ is normalized by T and the Doppler shift f is normalized by $1/T$. Such normalizations will also be used in most other AF plots in order to provide a consistent scaling.

Two AF features can be easily observed from Figure 6.1. The first is that the maximum value of $|\chi(\tau, f)|$ is achieved by $|\chi(0, 0)|$, which in fact equals the energy of $u(t)$. The second is the symmetry with respect to the origin, i.e., $|\chi(\tau, f)| = |\chi(-\tau, -f)|$, so that it suffices to show the AF only for half the (τ, f) plane (as in Figure 6.1(a)).

Another prominent feature of the AF, which is less obvious than the two mentioned above, is the constant volume property:

$$\int_{-\infty}^{\infty}\int_{-\infty}^{\infty} |\chi(\tau, f)|^2 d\tau df = E^2, \tag{6.2}$$

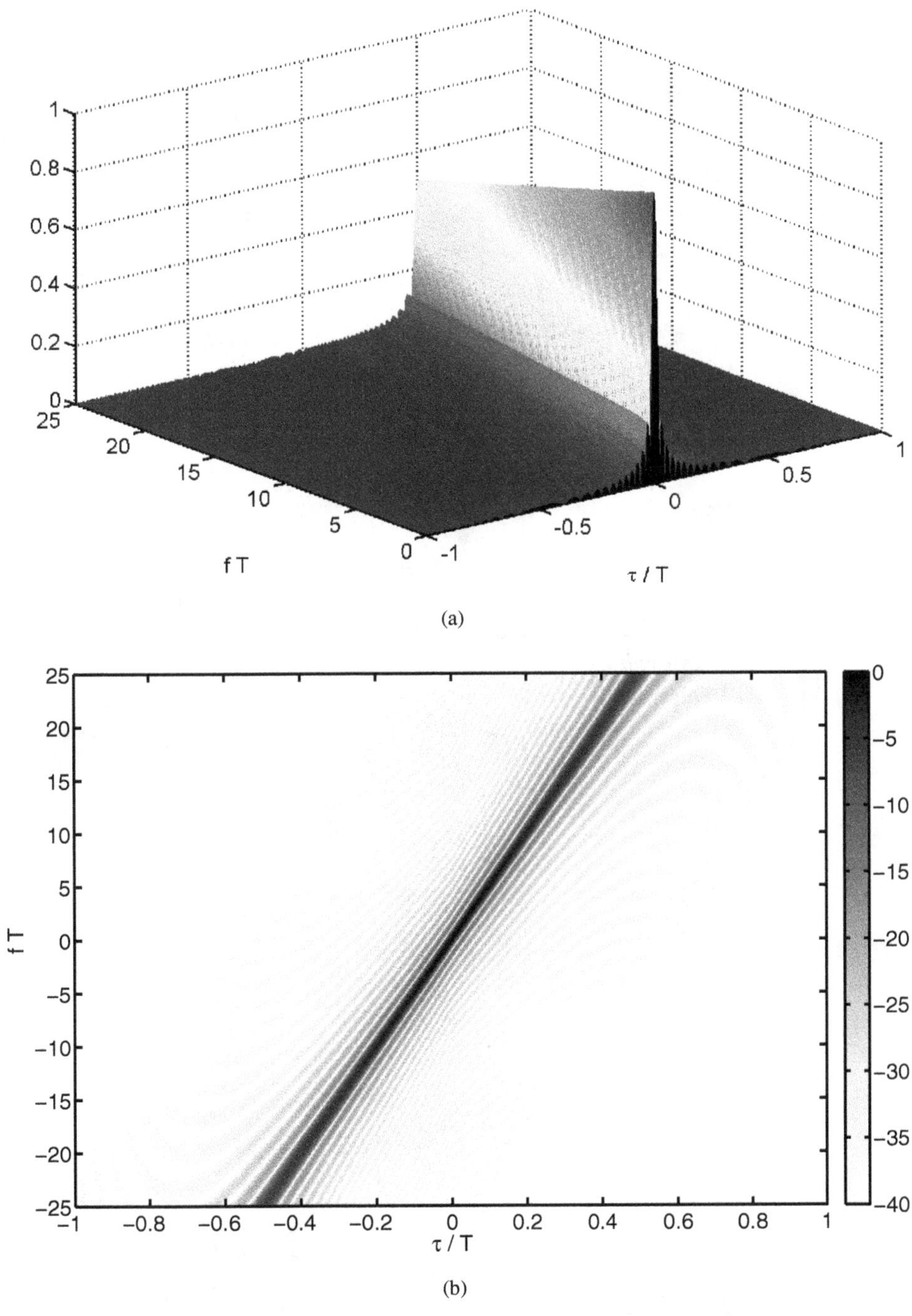

Figure 6.1 The AF of a chirp signal with $T = 10$ s and $B = 5$ Hz. (a) 3D plot of the positive Doppler plane and (b) 2D plot of the entire plane.

where

$$E = \int_{-\infty}^{\infty} |u(t)|^2 dt \tag{6.3}$$

is the energy of $u(t)$. These aforementioned three properties are listed below together with proofs for completeness.

- *Maximum value at the origin*
  By using the Cauchy–Schwartz inequality we obtain

$$|\chi(\tau, f)|^2 \leq \int_{-\infty}^{\infty} |u(t)|^2 dt \int_{-\infty}^{\infty} |u^*(t-\tau)e^{-j2\pi f(t-\tau)}|^2 dt = E^2, \tag{6.4}$$

  where E denotes the energy of $u(t)$ (see (6.3)). Since $|\chi(0,0)| = E$, it follows that the maximum value of $|\chi(\tau, f)|$ is achieved at the origin.
- *Symmetry*
  A simple variable change ($t \leftarrow t + \tau$) shows that

$$\begin{aligned}\chi(-\tau, -f) &= \int_{-\infty}^{\infty} u(t)u^*(t+\tau)e^{j2\pi f(t+\tau)} dt \\ &= \int_{-\infty}^{\infty} u(t-\tau)u^*(t)e^{j2\pi f t} dt,\end{aligned} \tag{6.5}$$

  which implies the symmetry property: $|\chi(\tau, f)| = |\chi(-\tau, -f)|$.
- *Constant volume*
  The volume of $|\chi(\tau, f)|^2$ is given by

$$\begin{aligned}V &= \int_{-\infty}^{\infty}\int_{-\infty}^{\infty} |\chi(\tau, f)|^2 d\tau df \\ &= \int_{-\infty}^{\infty}\int_{-\infty}^{\infty} \left|\int_{-\infty}^{\infty} u(t)u^*(t-\tau)e^{-j2\pi f t} dt\right|^2 d\tau df.\end{aligned} \tag{6.6}$$

  Let $W_\tau(f)$ denote the Fourier transform of $u(t)u^*(t-\tau)$. Using the Parseval equality we get

$$\begin{aligned}\int_{-\infty}^{\infty} |W_\tau(f)|^2 df &= \int_{-\infty}^{\infty} \left|\int_{-\infty}^{\infty} u(t)u^*(t-\tau)e^{-j2\pi f t} dt\right|^2 df \\ &= \int_{-\infty}^{\infty} |u(t)u^*(t-\tau)|^2 dt.\end{aligned} \tag{6.7}$$

  Therefore

$$\begin{aligned}V &= \int_{-\infty}^{\infty}\int_{-\infty}^{\infty} |u(t)u^*(t-\tau)|^2 dt d\tau \\ &= \int_{-\infty}^{\infty}\int_{-\infty}^{\infty} |u(x)u^*(y)|^2 dx dy \\ &= \int_{-\infty}^{\infty} |u(x)|^2 dx \int_{-\infty}^{\infty} |u(y)|^2 dy \\ &= E^2,\end{aligned} \tag{6.8}$$

  where the change of variables $\{x = t,\ y = t - \tau\}$ are used.

The AF of the chirp signal illustrated in Figure 6.1 has the Doppler-tolerance property, in the sense that a mismatch in the Doppler frequency can still lead to a peak in the matched-filtering process although at the cost of a time-delay estimation error. Therefore, even if a bank of filters matched to different Doppler frequencies is not available at the receiver end, targets with unknown Doppler shifts may still be detected owing to this delay-Doppler coupling property.

As discussed in Section 1.3, many sequences such as the Golomb and Frank sequences can be derived from the chirp signal. Not surprisingly, their AFs inherit the Doppler-tolerance property. Figure 6.2 shows the AF of the waveform (1.1) when $\{x(n)\}$ is a length-50 Golomb sequence ($N = 50$ and $T = Nt_p$) and $p_n(t)$ is the rectangular shaping pulse. (Later, when we refer to the AF of a sequence, we implicitly mean the AF of the underlying coded waveform.) Comparing Figures 6.1 and 6.2, it is interesting to observe how part of the AF volume in the central ridge is moved to the edge in Figure 6.2 (the total volume being constant, see (6.2)).

As discussed in Chapter 2, the CAN algorithm can be used to design sequences with low correlation sidelobes. The auto-correlation function is nothing but the zero-Doppler cut of AF. Figure 6.3 shows the AF of a length-50 CAN sequence when the Golomb sequence of the same length is used to initialize the algorithm. When compared with the AF in Figure 6.2, the AF in Figure 6.3 exhibits a similar Doppler tolerance property due to the initialization by the Golomb sequence. We also observe a horizontal white stripe at zero-Doppler frequency, which indicates low correlation sidelobes.

Figure 6.4 shows the AF of a length-50 random-phase sequence, i.e., $x(n) = e^{j\phi(n)}$ where $\{\phi(n)\}$ are independent random variables uniformly distributed in $[0, 2\pi]$. We then use this random-phase sequence to initialize the CAN algorithm; the AF of the resulting CAN sequence is shown in Figure 6.5. We observe that the AF in Figure 6.5 is thumbtack-shaped, which leads to high resolution for both delay and Doppler estimation. This type of waveform is usually referred to as a Doppler-sensitive waveform. Another example of a Doppler-sensitive waveform is the m-sequence (see Section 1.3), whose AF is shown in Figure 6.6 when the sequence length is 63. As an aside note, the CAN algorithm will converge after only one iteration if an m-sequence is used for initialization; in fact, the so-obtained CAN sequence is the same as the m-sequence. Such a phenomenon indicates that an m-sequence is probably already at a local minimum of the ISL criterion (see (2.1)).

Another fact worth noting is that the zero-delay cut of AF is the Fourier transform of $u(t)u^*(t)$. For the waveforms discussed above that have unit modulus, the zero-delay cut of AF can be easily calculated as

$$\begin{aligned} \chi(0, f) &= \int_{-\infty}^{\infty} u(t)u^*(t)e^{-j2\pi ft}dt = \int_{0}^{T} e^{-j2\pi ft}dt \\ &= \frac{1 - e^{-j2\pi fT}}{j2\pi f} = e^{-j\pi fT}T\,\mathrm{sinc}(\pi fT), \end{aligned} \tag{6.9}$$

where $\mathrm{sinc}(x) = \sin(x)/x$. Therefore $|\chi(0, f)| = |T\,\mathrm{sinc}(\pi fT)|$ regardless of $u(t)$ (provided that $u(t)$ is unimodular). Since $\mathrm{sinc}(\pi fT)$ dies out quickly as f increases, the

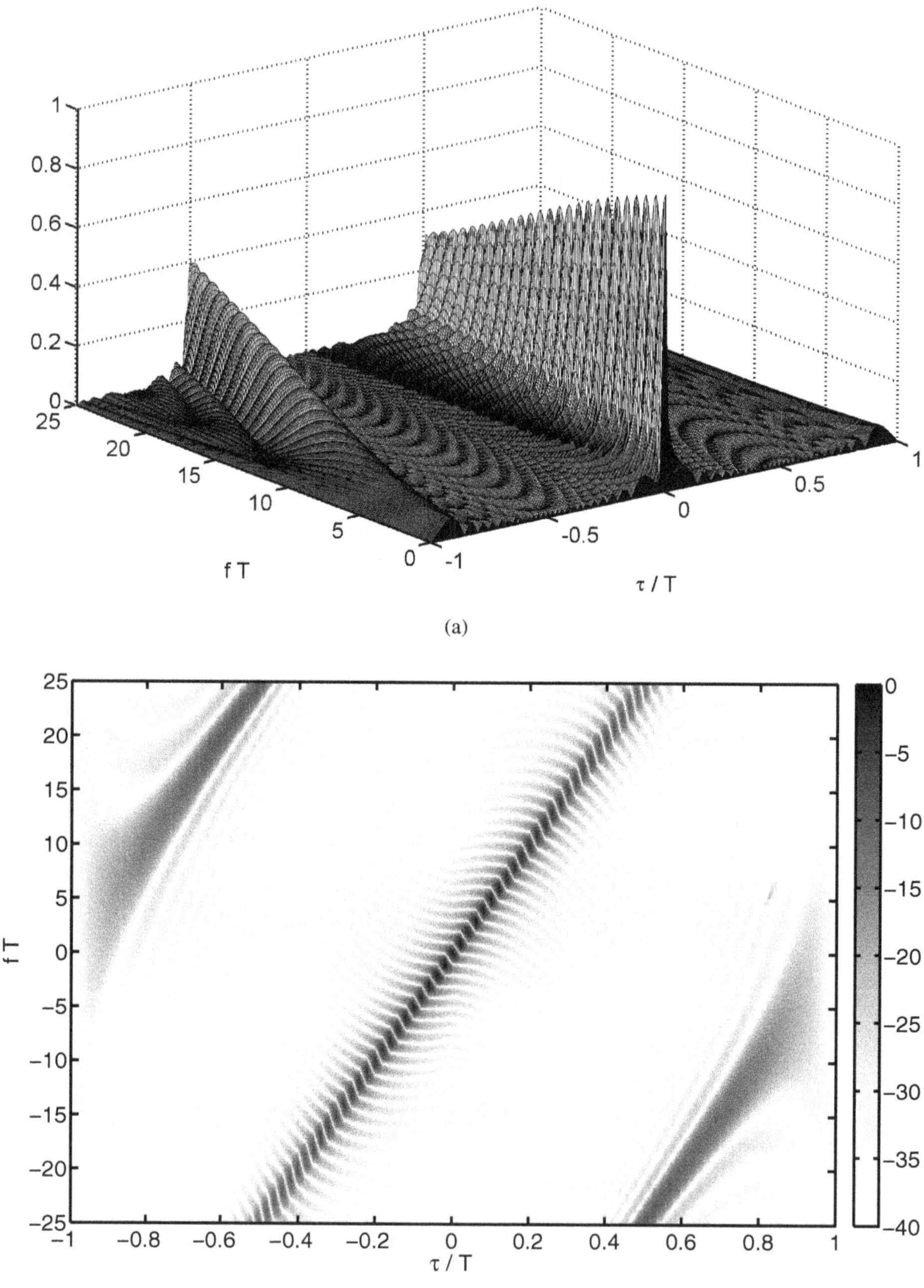

(a)

(b)

Figure 6.2 The AF of a length-50 Golomb sequence. (a) 3D plot of the positive Doppler plane and (b) 2D plot of the entire plane.

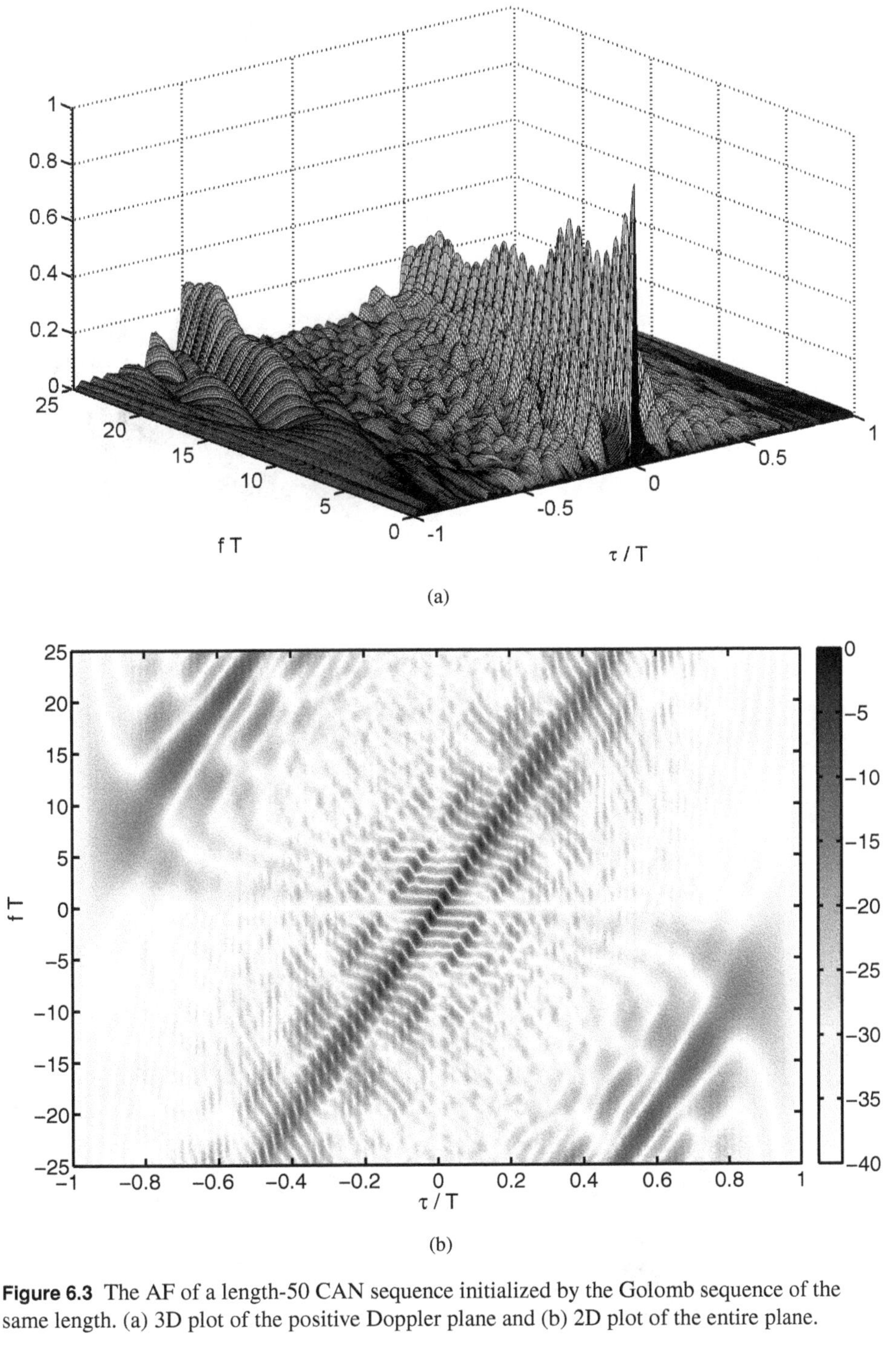

Figure 6.3 The AF of a length-50 CAN sequence initialized by the Golomb sequence of the same length. (a) 3D plot of the positive Doppler plane and (b) 2D plot of the entire plane.

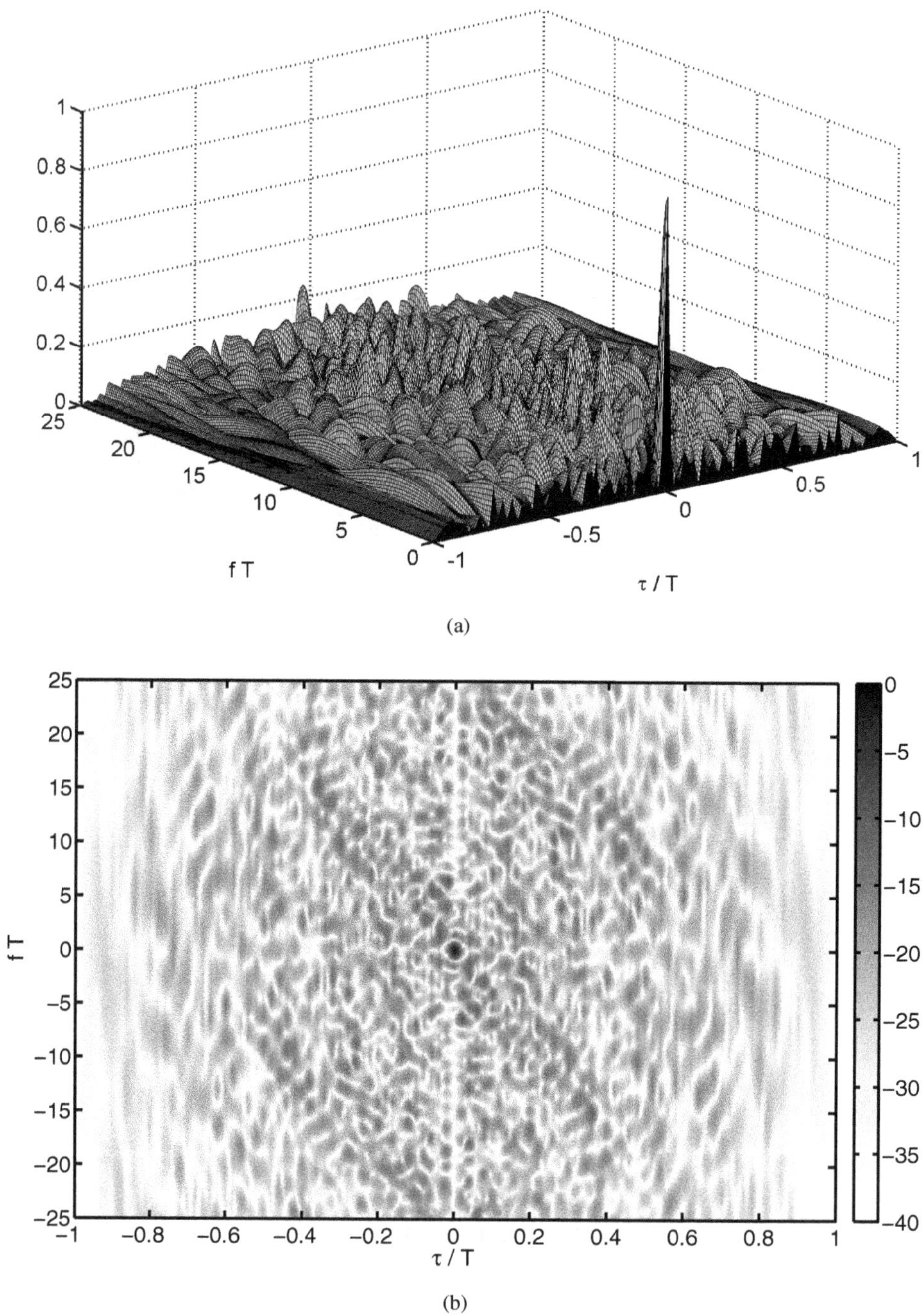

Figure 6.4 The AF of a length-50 random-phase sequence. (a) 3D plot of the positive Doppler plane and (b) 2D plot of the entire plane.

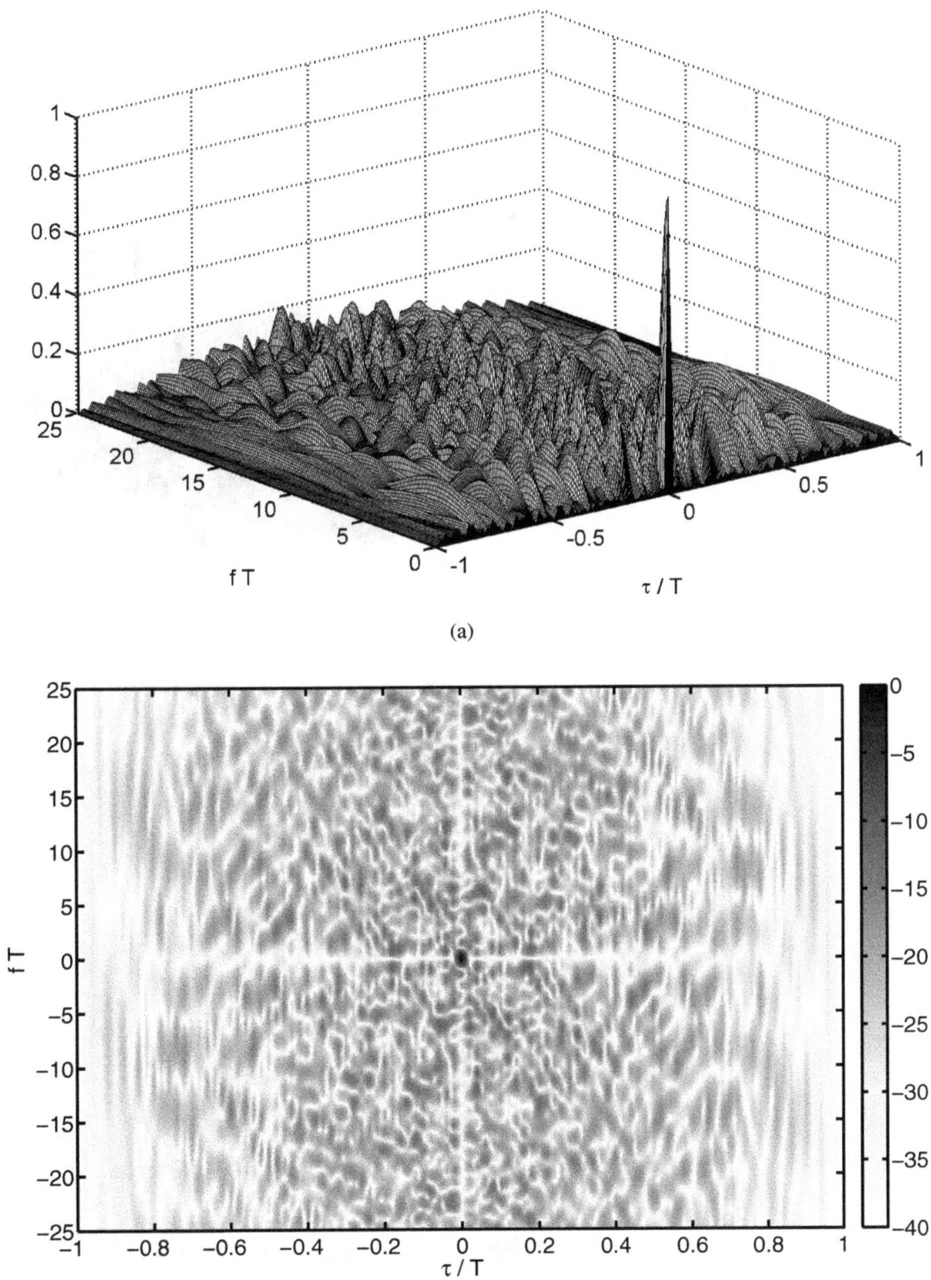

(a)

(b)

Figure 6.5 The AF of a length-50 CAN sequence initialized by the random-phase sequence associated with Figure 6.4. (a) 3D plot of the positive Doppler plane and (b) 2D plot of the entire plane.

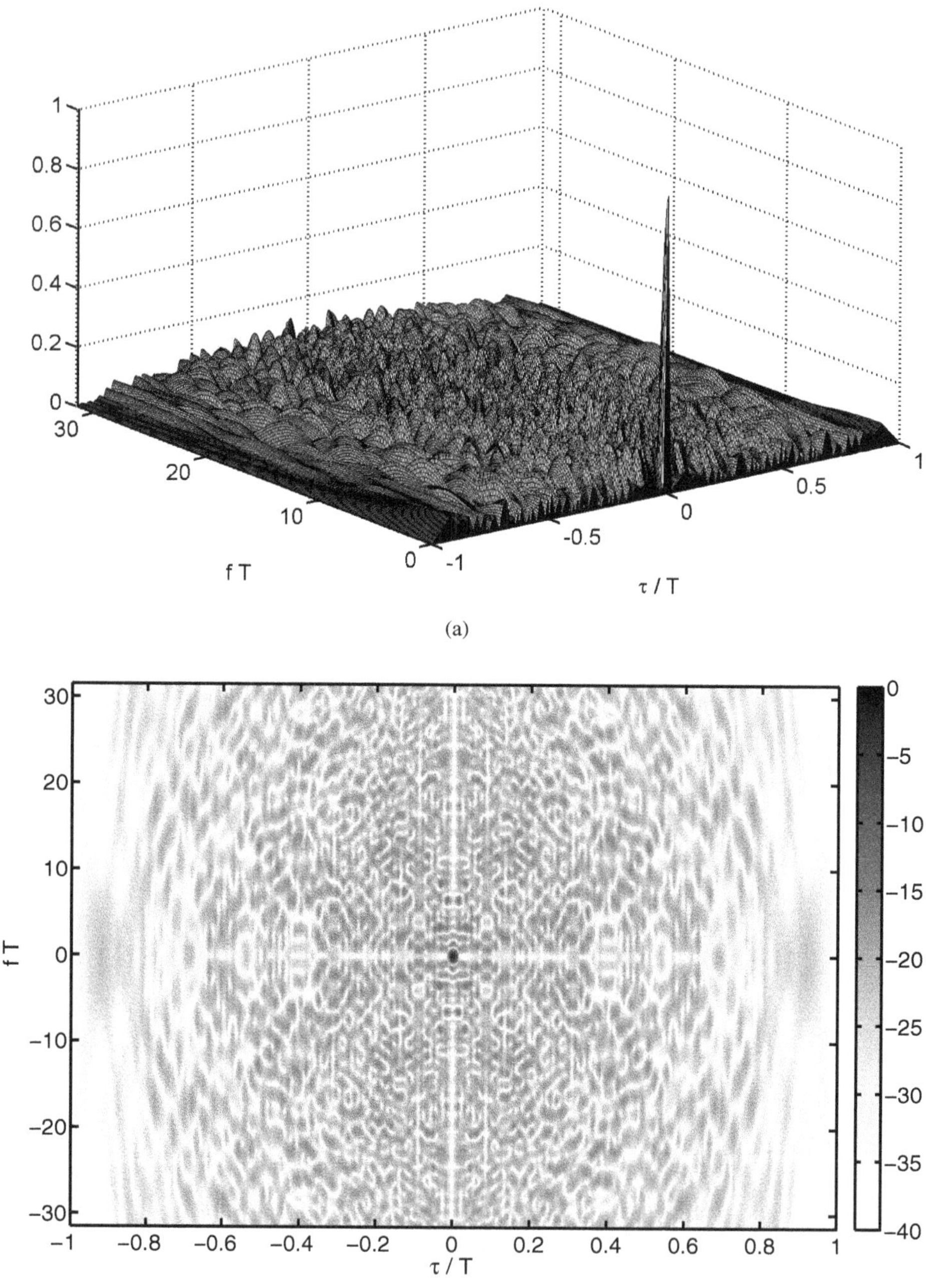

(a)

(b)

Figure 6.6 The AF of a length-63 m-sequence. (a) 3D plot of the positive Doppler plane and (b) 2D plot of the entire plane.

zero-delay cut appears as a vertical stripe of small values in Figures 6.3 and 6.5, as well as in Figures 6.1 and 6.2 (though it is somewhat less obvious).

We have illustrated the ambiguity functions of several waveforms. How to realize a desired ambiguity function using practical signals has been a classical problem in the waveform design area and there exists a considerable literature on this topic [Sussman 1962][Wolf et al. 1968][Stein 1981][Costas 1984][Levanon & Mozeson 2004][Gladkova & Chebanov 2004][Bonami et al. 2007][Sharma 2010]. Despite this extensive literature, apparently there is no universal method that can synthesize an arbitrary ambiguity function. In fact, matching only the zero-Doppler cut of an ambiguity function or minimizing the sidelobes of the auto-correlation function is a difficult problem in itself, as discussed in Chapters 2 and 3. In what follows, we introduce the discrete-AF concept and show that the sidelobes of a discrete-AF can be suppressed in a region close to the origin. In Chapter 7 we will consider the *cross* AF (CAF) synthesis problem which has more degrees of freedom than the AF synthesis problem.

6.2 Discrete-AF

We restrain the attention to baseband waveforms modulated in the following way (see also (1.1) and (1.2)):

$$u(t) = \sum_{n=1}^{N} x(n)p_n(t), \quad 0 \leq t \leq T \tag{6.10}$$

where $\{x(n)\}_{n=1}^{N}$ is the modulating code sequence that is to be designed (which is assumed to be zero when $n \notin [1, N]$) and $p_n(t)$ is an ideal rectangular shaping pulse of time length t_p (thus $T = Nt_p$). The energy of $\{x(n)\}_{n=1}^{N}$ is constrained to be N, i.e.,

$$\sum_{n=1}^{N} |x(n)|^2 = N. \tag{6.11}$$

As we have pointed out several times before, it is usually desirable to have a unimodular sequence:

$$x(n) = e^{j\phi_n}, \quad n = 1, \ldots, N, \tag{6.12}$$

where $\{\phi_n\}$ are the phases.

For the waveform defined in (6.10), the ambiguity function (6.1) can be simplified as follows. Substitute (6.10) into (6.1) to obtain

$$\begin{aligned}\chi(\tau, f) &= \int_0^T \left(\sum_{n=1}^{N} x(n)p_n(t)\right)\left(\sum_{m=1}^{N} x^*(m)p_m(t-\tau)\right) e^{-j2\pi f(t-\tau)}\, dt \\ &= \sum_{m=1}^{N}\sum_{n=1}^{N} x^*(m)\left(\int_0^T p_n(t)p_m(t-\tau)e^{-j2\pi f(t-\tau)}\, dt\right) x(n).\end{aligned} \tag{6.13}$$

Consider the time grid $\{\tau = kt_p\}$ $(k = -N+1, \ldots, 0, \ldots, N-1)$ whose points are integer multiples of the subpulse length t_p. It is not difficult to calculate $\chi(\tau, f)$ at $\tau = kt_p$:

$$\chi(kt_p, f) = \sum_{m=1}^{N}\sum_{n=1}^{N} x^*(m)\left(\int_{(n-1)t_p}^{nt_p} |p_n(t)|^2 e^{-j2\pi f(t-kt_p)}\, dt\right)\delta_{m+k,n} x(n) \tag{6.14}$$

$$= \frac{e^{j\pi f t_p}\sin(\pi f t_p)}{\pi f t_p}\sum_{n=1}^{N} x(n)x^*(n-k)e^{-j2\pi f t_p(n-k)}. \tag{6.15}$$

Note that $\chi(kt_p, f)$ equals zero if k lies outside $[-N+1, N-1]$. In a similar way to the time grid, we consider $\{f = p/Nt_p\}$ for an integer p. Then we obtain

$$\chi\left(kt_p, \frac{p}{Nt_p}\right) = e^{j\pi\frac{p}{N}}\,\mathrm{sinc}\left(\pi\frac{p}{N}\right)\bar{r}(k, p) \tag{6.16}$$

where $\mathrm{sinc}(x) = \sin(x)/x$ and $\bar{r}(k, p)$ is what we call the discrete-AF:

$$\bar{r}(k, p) = \sum_{n=1}^{N} x(n)x^*(n-k)e^{-j2\pi\frac{(n-k)p}{N}}, \tag{6.17}$$

$$k = -N+1, \ldots, N-1, \quad p = -\frac{N}{2}, \ldots, \frac{N}{2} - 1.$$

The range of p is chosen as $-N/2, \ldots, N/2 - 1$ because it corresponds to the largest Doppler frequency range that can be unambiguously identified (note that the bandwidth of $u(t)$ is approximately equal to $1/t_p$). Also note that without loss of generality, we have implicitly assumed and will assume hereafter that N is even in this chapter. An odd N would lead to the range $p = -(N-1)/2, \ldots, (N-1)/2$, which would make little difference to the discussion below.

In practice the Doppler frequency f can be much smaller than the bandwidth of the probing waveform. For example, assuming an X-band radar operating at wavelength $\lambda = 3$ cm, a fighter jet moving at a speed Mach 3 ($v = 1020$ m/s) induces a Doppler frequency of only

$$f = \frac{2v}{\lambda} = \frac{2 \times 1020 \text{ m/s}}{0.03 \text{ m}} = 68 \text{ kHz}; \tag{6.18}$$

this is much smaller than the bandwidth that is on the order of many MHz. As another example, for a sonar operating at a frequency $f = 20$ kHz (corresponding to a wavelength of $1500/20 = 75$ mm), a fast-moving submarine at the speed of 25 knots ($v = 13$ m/s) induces a Doppler frequency of

$$f = \frac{2v}{\lambda} = \frac{2 \times 13 \text{ m/s}}{0.075 \text{ m}} = 346 \text{ Hz}, \tag{6.19}$$

which can also be considered to be very small compared with the several-kHz bandwidth normally used. In such scenarios, we can confine our attention to values of $|p| \ll N$, in which case $\mathrm{sinc}(\pi p/N) \approx 1$ and thus

$$\left|\chi\left(kt_p, \frac{p}{Nt_p}\right)\right| \approx |\bar{r}(k, p)|, \quad k = -N+1, \ldots, N-1, \quad |p| \ll N. \tag{6.20}$$

Since, almost always, it is only the magnitude of AF that matters in target detection applications, (6.20) shows that it suffices to consider the discrete-AF defined in (6.17).

The well-known AF properties of symmetry and constant volume also hold for the discretized version in (6.17). It is easy to show the symmetry property:

$$\bar{r}(-k,-p) = \bar{r}(k,p)^* e^{j2\pi \frac{kp}{N}} \Rightarrow |\bar{r}(k,p)| = |\bar{r}(-k,-p)|. \tag{6.21}$$

The constant volume property, i.e., $\sum_{k=-N+1}^{N+1} \sum_{p=-N/2}^{N/2-1} |\bar{r}(k,p)|^2 = N^3$ will be proved for the more general cross-AF case in Chapter 7. In addition, $\bar{r}(0,0)$ is always equal to N because of the energy constraint in (6.11); we refer to $\bar{r}(k,p)$ for $k \neq 0$ or $p \neq 0$ as the sidelobes. Note that the above properties follow directly from the discrete-AF definition in (6.17) in spite of its approximate relationship with the original AF as shown in (6.20).

In the next section we will be concerned mainly with designing the sequence $\{x(n)\}_{n=1}^{N}$ so as to minimize the sidelobes of the discrete-AF in a certain region:

$$\min_{\{x(n)\}} C_1 = \sum_{k \in \mathcal{K}} \sum_{p \in \mathcal{P}} |\bar{r}(k,p)|^2 \tag{6.22}$$

where $\mathcal{K}$ and $\mathcal{P}$ are the index sets specifying the region of interest. Because the total volume of $\bar{r}(k,p)$ is fixed, (6.22) is meaningful only when $\mathcal{K}$ and $\mathcal{P}$ are strictly subsets of $\{-N+1,\ldots,N-1\}$ and $\{-N/2,\ldots,N/2-1\}$, respectively.

6.3 Minimizing the discrete-AF sidelobes

Following the notation specified in (6.22), assume that the time delay set of interest is given by $\mathcal{K} = \{0, \pm 1, \ldots, \pm(K-1)\}$ and that the Doppler frequency set of interest is given by $\mathcal{P} = \{0, \pm 1, \ldots, \pm(P-1)\}$. Define a set of P sequences $\{x_m(n)\}_{m=1}^{P}$ as follows:

$$\begin{aligned} &\{x_1(n) = x(n)\}_{n=1}^{N}, \\ &\{x_2(n) = x(n) e^{j2\pi \frac{n}{N}}\}_{n=1}^{N}, \\ &\quad \vdots \\ &\{x_P(n) = x(n) e^{j2\pi \frac{n(P-1)}{N}}\}_{n=1}^{N}. \end{aligned} \tag{6.23}$$

Note that $\{x_m(n)\}_{m=1}^{P}$ are zero when $n \notin [1,N]$. Let $\{r_{ml}(k)\}$ denote the correlation between $\{x_m(n)\}$ and $\{x_l(n)\}$:

$$\begin{aligned} r_{ml}(k) &= \sum_{n=1}^{N} x_m(n) x_l^*(n-k) \\ &= e^{j2\pi \frac{(m-1)k}{N}} \sum_{n=1}^{N} x(n) x^*(n-k) e^{-j2\pi \frac{(n-k)(l-m)}{N}}, \quad k \in \mathcal{K} \\ &\quad m, l = 1, \ldots, P. \end{aligned} \tag{6.24}$$

It is straightforward to verify that all values of $\{|\bar{r}(k,p)|\}$ ($k \in \mathcal{K}$ and $p \in \mathcal{P}$) are contained in the set $\{|r_{ml}(k)|\}$ ($k \in \mathcal{K}$ and $m, l = 1, \ldots, P$). Interestingly, m and l do not need to increase stepwise from 1 to P. For example, $\{|r_{ml}(k)|\}$ $(m, l = 1, 2, 5, 7)$ already covers all values of $\{|\bar{r}(k, p)|\}$ $(p = 0, \ldots, 6)$ (a fact which bears a resemblance to the minimum redundancy linear array [Van Trees 2002]). This observation saves computation but does not improve the algorithms performance, so (6.24) will be used for $m, l = 1, \ldots, P$ to keep the notation simple.

The foregoing discussion implies that by minimizing the correlations of the sequence set in (6.23), we equivalently minimize the discrete-AF sidelobes, i.e., the criterion C_1 in (6.22). The Multi-CAO algorithm discussed in Chapter 3 which was used to design sets of waveforms with good correlations, can be adapted to the problem of minimizing C_1, as explained in what follows.

Define

$$\mathbf{X} = \begin{bmatrix} \mathbf{X}_1 & \cdots & \mathbf{X}_P \end{bmatrix}_{(N+K-1)\times KP}, \tag{6.25}$$

where

$$\mathbf{X}_m = \begin{bmatrix} x_m(1) & & \mathbf{0} \\ \vdots & \ddots & \\ \vdots & & x_m(1) \\ x_m(N) & & \vdots \\ & \ddots & \vdots \\ \mathbf{0} & & x_m(N) \end{bmatrix}_{(N+K-1)\times K}, \quad m = 1, \ldots, P, \tag{6.26}$$

and $\{x_m(n)\}$ are defined in (6.23). It is not difficult to see that all $\{r_{ml}(k)\}$ ($k \in \mathcal{K}$ and $m, l = 1, \ldots, P$) appear in the matrix $\mathbf{X}^H\mathbf{X}$. Also note that the diagonal elements of $\mathbf{X}^H\mathbf{X}$ are equal to N because of the energy constraint in (6.11). Therefore, the correlations of the sequence set (6.23) can be made small through minimizing the following criterion:

$$\hat{C}_1 = \left\| \mathbf{X}^H\mathbf{X} - N\mathbf{I}_{KP} \right\|^2. \tag{6.27}$$

Note that the criterion $\hat{C}_1$ equals zero if the matrix $\mathbf{X}$ is a semi-unitary matrix scaled by $\sqrt{N}$, an observation that leads to the following minimization problem, which has a simpler form than (6.27):

$$\begin{aligned} \min_{\mathbf{X},\mathbf{U}} \; & \|\mathbf{X} - \sqrt{N}\mathbf{U}\|^2, \\ \text{s.t. } & |x(n)| = 1, \quad n = 1, \ldots, N \\ & x_m(n) = x(n)e^{j2\pi\frac{n(m-1)}{N}}, \quad m = 1, \ldots, P, \quad n = 1, \ldots, N, \\ & \mathbf{U}^H\mathbf{U} = \mathbf{I} \quad (\mathbf{U} \text{ is an } (N + K - 1) \times KP \text{ matrix}). \end{aligned} \tag{6.28}$$

The minimization problem (6.28) can be solved by the cyclic algorithm described in Table 6.1.

As an example, consider a scenario with $N = 100, K = 10$ and $P = 3$. We use the algorithm in Table 6.1 to design a unimodular sequence $\{x(n)\}_{n=1}^N$. The so-obtained discrete-AF, $|\bar{r}(k, p)|$, is shown in Figure 6.7. The “white” area in the center indicates that the sidelobes near the origin were successfully suppressed. Note that due to the

Table 6.1. The cyclic algorithm for minimizing discrete-AF sidelobes

Step 0: Randomly initialize the sequence $\{x(n)\}$.
Step 1: For fixed $\mathbf{X}$, the minimizer $\mathbf{U}$ is given by (see Section 3.3)

$$\mathbf{U} = \mathbf{U}_2\mathbf{U}_1^H, \tag{6.29}$$

where the matrices $\mathbf{U}_1$ $(KP \times KP)$ and $\mathbf{U}_2$ $((N + K - 1) \times KP)$ come from the SVD (see also (3.33)) of $\mathbf{X}^H$, i.e., $\mathbf{X}^H = \mathbf{U}_1\mathbf{\Sigma}\mathbf{U}_2^H$.
Step 2: For fixed $\mathbf{U}$, the criterion (6.28) can be written as follows (note the unit-modulus constraint (6.12)):

$$\left\|\mathbf{X} - \sqrt{N}\mathbf{U}\right\|^2 = \sum_{n=1}^{N}\sum_{l=1}^{KP} |\mu_{nl}x(n) - f_{nl}|^2 = \text{const} - 2\sum_{n=1}^{N} \text{Re}\left[\left(\sum_{l=1}^{KP} \mu_{nl}^* f_{nl}\right) x^*(n)\right] \tag{6.30}$$

where the constant does not depend on $\{x(n)\}$, $\{\mu_{nl}\}$ are the elements of $\mathbf{X}$ that contain $x(n)$,

$$[\mu_{n1} \ \cdots \ \mu_{n,KP}] = \Big[\underbrace{1 \cdots 1}_{K} \ \underbrace{e^{j2\pi\frac{n}{N}} \cdots e^{j2\pi\frac{n}{N}}}_{K} \ \cdots \ \underbrace{e^{j2\pi\frac{n(P-1)}{N}} \cdots e^{j2\pi\frac{n(P-1)}{N}}}_{K}\Big]_{1\times KP} \tag{6.31}$$

and $\{f_{nl}\}$ are the elements of $\sqrt{N}\mathbf{U}$ whose positions are the same as those of $\{\mu_{nl}\}$ in $\mathbf{X}$. The minimizer $x(n)$ (more exactly its phase) is obtained immediately:

$$\phi_n = \arg\left(\sum_{l=1}^{KP} \mu_{nl}^* f_{nl}\right), \quad n = 1, \ldots, N. \tag{6.32}$$

Iteration: Repeat steps 1 and 2 until convergence.

limited number of degrees of freedom (N elements in $\{x_n\}$) and a possibly large number of constraints ($\sim 2KP$), the low-sidelobe region is confined to a small area (i.e., with small K and P) close to the origin.

6.4 Conclusions

The concept of an ambiguity function (AF) can be considered as an extension of the correlation concept discussed in detail in previous chapters. The ambiguity function is an important tool that depicts the response of a matched filter to signals that are time delayed and Doppler shifted. Several prominent properties of AF have been shown, one of which is the constant volume property, which cumbers free manipulation of an AF's shape. For phase-coded waveforms, the definition of a discrete-AF can be naturally derived. The Multi-CAO algorithm introduced previously was adapted as an algorithm for designing sequences with a thumbtack-shaped discrete-AF near the origin of the delay–Doppler plane.

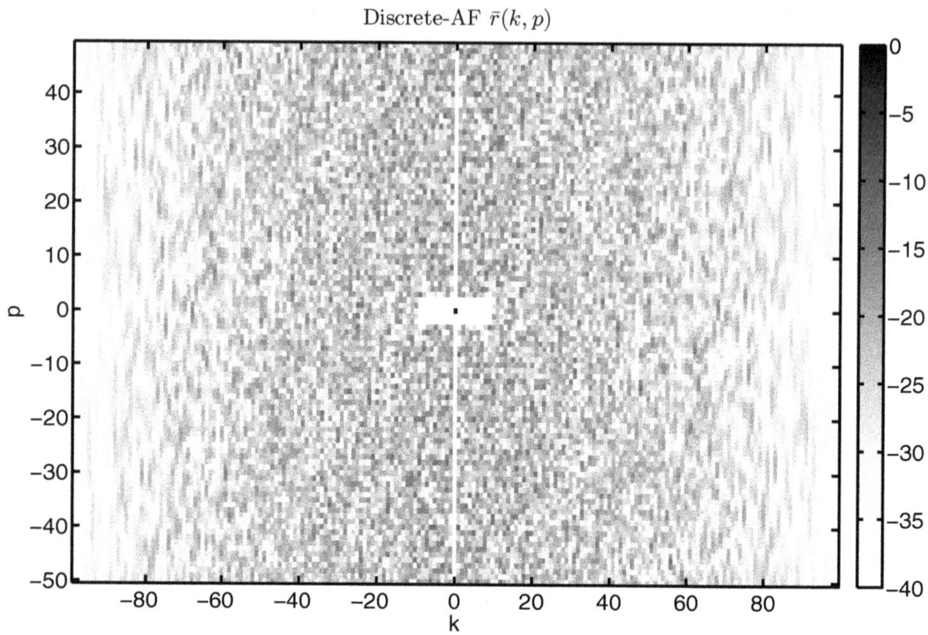

Figure 6.7 A synthesized discrete-AF: $|\bar{r}(k,p)|$.

Appendix 6A Wideband ambiguity function

General AF

Suppose that the speed of the (electromagnetic or sound) wave is c, the transmitted signal is $u(t)$ and the target that reflects $u(t)$ has speed v (a positive v indicates a target moving towards the radar or sonar). The returned signal is a delayed and time-scaled version of $u(t)$ [Lin 1988][Levanon & Mozeson 2004]:

$$v(t) = \sqrt{\eta}u(\eta(t-\tau)) \tag{6.33}$$

where τ is the time delay and the scaling factor η is given by

$$\eta = \frac{1+v/c}{1-v/c} = \frac{c+v}{c-v}. \tag{6.34}$$

The factor $\sqrt{\eta}$ in (6.33) comes from the energy conservation in the time scaling:

$$E \triangleq \int |u(t)|^2\, dt = \int |\sqrt{\eta}u(\eta t')|^2\, dt'. \tag{6.35}$$

The ambiguity function equals the matched filter output, which is given by the cross-correlation of $u(t)$ and $v(t)$:

$$\bar{\chi}(\tau,\eta) = \sqrt{\eta}\int_{-\infty}^{\infty} u^*(t)u(\eta(t-\tau))dt. \tag{6.36}$$

Equation (6.36) is valid for general signals.

Narrowband AF

Two assumptions are needed to specialize (6.36) to the familiar narrowband AF, which can be written as

$$\chi(\tau, f) = \int_{-\infty}^{\infty} u^*(t)u(t-\tau)e^{j2\pi f(t-\tau)}dt. \tag{6.37}$$

These two assumptions are:

- $v \ll c$;
- $u(t)$ is narrowband so that a time scaling can be considered as merely a frequency shift of the carrier.

We will show that, under the above two assumptions, (6.36) and (6.37) are equivalent because

$$\sqrt{\eta}u(\eta(t-\tau)) \approx u(t-\tau)e^{j2\pi f(t-\tau)} \tag{6.38}$$

for a certain f (see below).

To prove (6.38) note that $v \ll c$ leads to

$$\begin{aligned}\eta =& \frac{1+v/c}{1-v/c} = \left(1+\frac{v}{c}\right)\left(1+\frac{v}{c}+\frac{v^2}{c^2}+\cdots\right)\\ \approx& 1+\frac{2v}{c}.\end{aligned} \tag{6.39}$$

Let $u(t) = A(t)e^{j2\pi f_c t}$ where f_c is the carrier frequency and $A(t)$ is the envelop (slowly varying compared with f_c). Then we obtain

$$\begin{aligned}\sqrt{\eta}u(\eta(t-\tau)) =& \sqrt{1+\frac{2v}{c}}A\left(\left(1+\frac{2v}{c}\right)(t-\tau)\right)e^{j2\pi f_c(1+\frac{2v}{c})(t-\tau)}\\ \approx& A(t-\tau)e^{j2\pi f_c(1+\frac{2v}{c})(t-\tau)}\\ =& A(t-\tau)e^{j2\pi(f_c+\frac{2v}{\lambda_c})(t-\tau)} \quad (\lambda_c \text{ is the carrier wavelength})\\ =& \left[A(t-\tau)e^{j2\pi f_c(t-\tau)}\right]e^{j2\pi f(t-\tau)}\\ =& u(t-\tau)e^{j2\pi f(t-\tau)},\end{aligned} \tag{6.40}$$

where f is given by the (narrowband) Doppler shift formula:

$$f = \frac{2v}{\lambda_c}. \tag{6.41}$$

This completes the proof of (6.38), as well as of the statement that (6.36) becomes (6.37) when the signal is narrowband and $v \ll c$.

If we substitute $u(t) = A(t)e^{j2\pi f_c t}$ into the narrowband AF in (6.37), we obtain

$$\chi(\tau, f) = e^{-j2\pi f_c \tau}\int_{-\infty}^{\infty} A^*(t)A(t-\tau)e^{j2\pi f(t-\tau)}dt \tag{6.42}$$

$$\triangleq e^{-j2\pi f_c \tau}\chi_B(\tau, f). \tag{6.43}$$

Therefore when we plot or study an ambiguity function, we can actually plot or analyze the modulus of $\chi_B(\tau, f)$ – the baseband AF. Using $\chi_B(\tau, f)$ instead of $\chi(\tau, f)$ is justified because $|\chi_B(\tau, f)| = |\chi(\tau, f)|$. In fact, (6.1) is the same as the AF $\chi_B(\tau, f)$ here, i.e., the $u(t)$ in (6.1) corresponds to the baseband signal $A(t)$ here.

Wideband AF

While considering only the envelop (the baseband component) is sufficient for computing the AF of a narrowband signal (hereafter we assume that $v \ll c$ holds true), this is not the case for a wideband AF. If we substitute $u(t) = A(t)e^{j2\pi f_c t}$ into the general AF in (6.36), we obtain

$$\bar{\chi}(\tau, \eta) = e^{-j2\pi f_c \eta\tau} \sqrt{\eta} \int_{-\infty}^{\infty} A^*(t) A(\eta(t - \tau)) e^{j2\pi f_c(\eta-1)t} dt \tag{6.44}$$

$$\triangleq e^{-j2\pi f_c \eta\tau} \bar{\chi}_B(\tau, \eta). \tag{6.45}$$

Therefore the correct formula for the AF of a wideband signal (in its baseband form) is given by

$$\bar{\chi}_B(\tau, \eta) = \sqrt{\eta} \int_{-\infty}^{\infty} A^*(t) A(\eta(t - \tau)) e^{j2\pi f_c(\eta-1)t} dt \tag{6.46}$$

Note that for a narrowband AF the following well-known symmetry holds:

$$\chi(\tau, f) = \chi^*(-\tau, -f). \tag{6.47}$$

However, for the general AF in (6.36) or (6.46), there is no such perfect symmetry. Instead, it is not difficult to show that

$$\bar{\chi}(\tau, v) = \bar{\chi}^*(-\eta\tau, -v), \tag{6.48}$$

$$|\bar{\chi}_B(\tau, v)| = |\bar{\chi}_B(-\eta\tau, -v)|, \tag{6.49}$$

where η is replaced by v according to (6.34) for clarity. Thus if we want to plot the magnitude of the general AF in (6.36) or (6.46), we need to show all four quadrants, although the figure looks almost symmetric because the scaling factor η is close to 1.

Figure 6.8 shows the AF of the chirp signal

$$u(t) = e^{j2\pi(f_c t + kt^2)}, \quad t \in [0, 10] \text{ s} \tag{6.50}$$

where $f_c = 10$ Hz and $k = 0.25$ s^{-2}. It is easy to calculate the bandwidth $B = 2kT = 5$ Hz, where $T = 10$ s is the waveform duration. Note the differences between the wideband AF shown here and the narrowband AF shown in Figure 6.1.

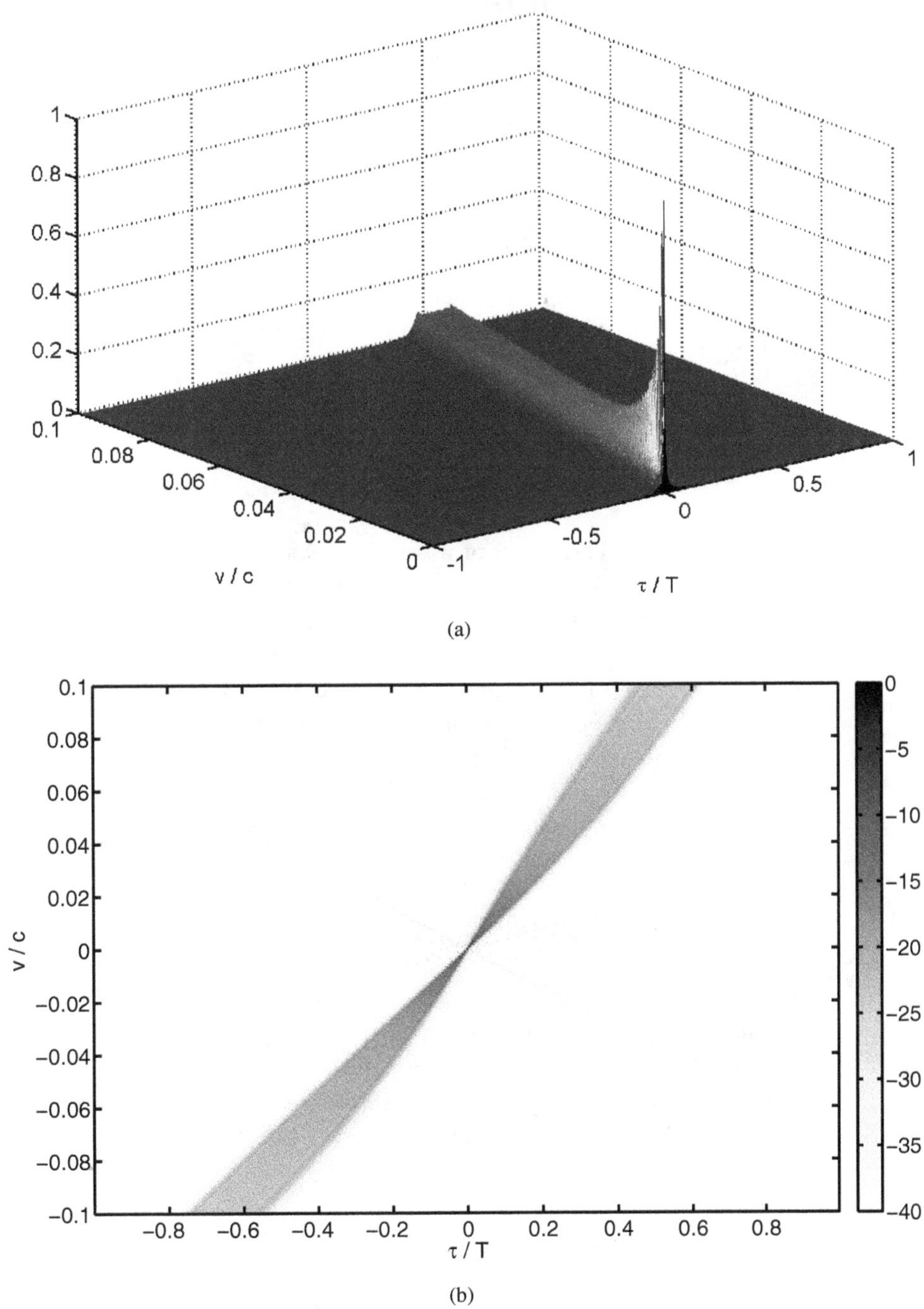

Figure 6.8 The wideband AF of a chirp signal with $f_c = 10$ Hz, $T = 10$ s and $B = 5$ Hz. (a) The 3D plot and (b) 2D plot.

7 Cross ambiguity function (CAF)

In this chapter we focus on the synthesis of the *cross* ambiguity function (CAF), which is defined as follows:

$$\chi(\tau, f) = \int_{-\infty}^{\infty} v(t)u^*(t-\tau)e^{-j2\pi f(t-\tau)}dt \tag{7.1}$$

$$= \left(\int_{-\infty}^{\infty} u(t)v^*(t+\tau)e^{j2\pi ft}dt \right)^*, \tag{7.2}$$

where $u(t)$ and $v(t)$ are a pair of waveforms. Note that (7.1) and the complex conjugate of (7.2) are two of the most often used formulae for CAF; either will be used in this chapter depending on the ease of expression in the context. The CAF appears naturally when $u(t)$ is the transmit signal and $v(t)$ is the receive filter [Stein 1981], or when both $u(t)$ and $v(t)$ are transmit signals in the case of a MIMO radar [Sharma 2010]. Note that the definition of CAF has more degrees of freedom than that of the conventional ambiguity function, a case where $v(t)$ equals $u(t)$.

We start from the discrete-CAF synthesis, which is easier to deal with. Then we move on to the problem of continuous-time CAF synthesis.

7.1 Discrete-CAF synthesis

As seen from (6.17), the ambiguity function can be interpreted as the matched filter output to the time-delayed and Doppler-shifted signal $x(n-k)e^{j2\pi p(n-k)/N}$. In the more general case, we can use any sequence, denoted as $\{y(n)\}_{n=1}^N$, as a mismatched filter at the receiver end; $\{y(n)\}$ is also called an instrumental variable (IV) filter (see [Stoica, Li & Xue 2008] and also Chapter 8). Such a filtering process leads to the following discrete-CAF:

$$\bar{r}_{xy}(k,p) = \sum_{n=1}^{N} y(n)x^*(n-k)e^{-j2\pi\frac{(n-k)p}{N}}, \tag{7.3}$$

$$k = -N+1, \ldots, N-1; \quad p = -\frac{N}{2}, \ldots, \frac{N}{2} - 1.$$

As usual, we assume that $y(n) = 0$ for $n \notin [1, N]$ and also that $\{y(n)\}_{n=1}^{N}$ has a fixed energy:

$$\sum_{n=1}^{N} |y(n)|^2 = N. \tag{7.4}$$

Interestingly, the energy of $\bar{r}_{xy}(k, p)$ is also a constant no matter how we choose $\{x(n)\}$ or $\{y(n)\}$, as long as the energies of $\{x(n)\}$ and $\{y(n)\}$ are fixed; see the proof in Appendix 7A. However, unlike for (6.17), the value of the discrete-CAF at the origin (i.e., $\bar{r}_{xy}(0, 0)$) is no longer equal to N^2. In fact, it always holds that $|\bar{r}_{xy}(0, 0)| \le N^2$ where the equality is satisfied if and only if $\{y(n) = x(n)\}$. Should we want the value N^2 at the origin in the discrete-CAF case, the energy of $\{y(n)\}$ must be allowed to be larger than N, which increases the volume of the discrete-CAF.

The problem of interest is to design a pair of sequences $\{x(n)\}_{n=1}^{N}$ and $\{y(n)\}_{n=1}^{N}$ to match a given discrete-CAF:

$$\min_{\{x(n)\},\{y(n)\},\{\phi_{kp}\}} C_2 = \sum_{k=-N+1}^{N-1} \sum_{p=-N/2}^{N/2-1} \left|g_{kp} e^{j\phi_{kp}} - \bar{r}_{xy}(k, p)\right|^2, \tag{7.5}$$

where g_{kp} is the given magnitude of the desired discrete-CAF. Note that, because the phases of the discrete-CAF do not matter in practice, the $\{\phi_{kp}\}$ are allowed to vary freely and so that become part of the minimization variables.

7.1.1 The proposed algorithm

Equation (7.3) can be written in a more compact form as follows:

$$\bar{r}_{xy}(k, p) = \mathbf{x}^H \mathbf{J}_{kp} \mathbf{y}, \tag{7.6}$$

where

$$\mathbf{x} = \left[x(1)\ x(2)\ \cdots\ x(N)\right]^T, \quad \mathbf{y} = \left[y(1)\ y(2)\ \cdots\ y(N)\right]^T \tag{7.7}$$

and

$$\mathbf{J}_{kp} = \begin{bmatrix} \overbrace{0 \cdots 0}^{k}\ e^{-j2\pi\frac{p}{N}} & & \mathbf{0} \\ & \ddots & \\ & & e^{-j2\pi\frac{(N-k)p}{N}} \\ \mathbf{0} & & \mathbf{0} \end{bmatrix}_{N\times N} \quad (k \ge 0), \tag{7.8}$$

$$\mathbf{J}_{kp} = \begin{bmatrix} 0 & & \mathbf{0} \\ \vdots & & \\ 0 & & \\ e^{-j2\pi\frac{(1-k)p}{N}} & & \\ & \ddots & \\ \mathbf{0} & e^{-j2\pi\frac{Np}{N}} & \mathbf{0} \end{bmatrix}_{N\times N} \quad (k < 0). \tag{7.9}$$

Table 7.1. The cyclic algorithm for discrete-CAF synthesis

Step 0: Randomly initialize $\mathbf{x}$ and $\mathbf{y}$.

Step 1: For fixed $\mathbf{x}$ and $\mathbf{y}$, the minimizer ϕ_{kp} can be easily obtained as

$$\phi_{kp} = \arg\left(\mathbf{x}^H \mathbf{J}_{kp}\mathbf{y}\right). \tag{7.14}$$

Step 2: For fixed $\{\phi_{kp}\}$ and $\mathbf{y}$, similarly to step 2 of Table 6.1, the minimizer $\{x(n)\}$ is determined as follows, depending on which constraint is imposed. Under the unit-modulus constraint, we have

$$\mathbf{x} = \exp(j \arg\{\mathbf{B}\mathbf{y}\}). \tag{7.15}$$

Under the more general PAR constraint, $\{x(n)\}$ is obtained by solving the following "nearest-vector" problem:

$$\min_{\mathbf{x}} \|\mathbf{x} - \mathbf{B}\mathbf{y}\|^2 \quad \text{s.t. } \mathrm{PAR}(\mathbf{x}) \le \rho. \tag{7.16}$$

Step 3: For fixed $\{\phi_{kp}\}$ and $\mathbf{x}$, the minimizer $\mathbf{y}$ (on which the only constraint is $\|\mathbf{y}\|^2 = N$) can be obtained from (7.12) by using the Cauchy–Schwartz inequality:

$$\mathbf{y} = \sqrt{N}\frac{\mathbf{B}^H\mathbf{x}}{\|\mathbf{B}^H\mathbf{x}\|}. \tag{7.17}$$

Iteration: Repeat steps 1, 2 and 3 until convergence.

According to the criterion (7.5), we aim to design $\mathbf{x}$ and $\mathbf{y}$ so that the magnitude of the discrete-CAF $\mathbf{x}^H\mathbf{J}_{kp}\mathbf{y}$ approximates a desired one, g_{kp}, for $k = -N+1, \ldots, N-1$ and $p = -N/2, \ldots, N/2-1$.

It follows from (7.6) that the criterion C_2 in (7.5) can be written as

$$C_2 = \sum_{k=-N+1}^{N-1} \sum_{p=-N/2}^{N/2-1} \left|g_{kp}e^{j\phi_{kp}} - \mathbf{x}^H\mathbf{J}_{kp}\mathbf{y}\right|^2 \tag{7.10}$$

$$= \sum_k \sum_p |g_{kp}|^2 - \mathbf{x}^H\mathbf{B}\mathbf{y} - \mathbf{y}^H\mathbf{B}^H\mathbf{x} + \sum_k \sum_p \left|\mathbf{x}^H\mathbf{J}_{kp}\mathbf{y}\right|^2 \tag{7.11}$$

$$= \sum_k \sum_p |g_{kp}|^2 + N^3 - 2\,\mathrm{Re}(\mathbf{x}^H\mathbf{B}\mathbf{y}), \tag{7.12}$$

where

$$\mathbf{B} = \sum_k \sum_p g_{kp}e^{-j\phi_{kp}}\mathbf{J}_{kp} \tag{7.13}$$

(note that the last term in (7.11) equals N^3 because of the constant volume property; see (7.53)). The criterion C_2 can be minimized by the cyclic algorithm outlined in Table 7.1.

Remark: Since the discrete-CAF has a constant volume (see (7.53)), it might seem reasonable to require that the desired discrete-CAF $\{g_{kp}\}$ also satisfies $\sum_k \sum_p |g_{kp}|^2 = N^3$. However, we do not require this because a scaling of g_{kp} (i.e., replacing g_{kp} by αg_{kp} for any positive number α) does not change the updating formulae in Table 7.1. □

For the algorithm in Table 7.1, all the values of the discrete-CAF $\bar{r}_{xy}(k, p)$ are considered equally important, which is often not the case in practice. To take into account the fact that matching some areas of $\bar{r}_{xy}(k, p)$ may be more important than matching other areas, we introduce weights $\{w_{kp}\}$ (real-valued and non-negative) into (7.5) to obtain the following criterion:

$$C_3(\mathbf{x}, \mathbf{y}, \{\phi_{kp}\}) = \sum_{k=-N+1}^{N-1} \sum_{p=-N/2}^{N/2-1} w_{kp} \left| g_{kp} e^{j\phi_{kp}} - \mathbf{x}^H \mathbf{J}_{kp} \mathbf{y} \right|^2 . \tag{7.18}$$

The above function can be rewritten as

$$C_3 = \sum_k \sum_p w_{kp} |g_{kp}|^2 - \mathbf{x}^H \mathbf{B} \mathbf{y} - \mathbf{y}^H \mathbf{B}^H \mathbf{x} + \sum_k \sum_p w_{kp} \left| \mathbf{x}^H \mathbf{J}_{kp} \mathbf{y} \right|^2 \tag{7.19}$$

where now

$$\mathbf{B} = \sum_k \sum_p w_{kp} g_{kp} e^{-j\phi_{kp}} \mathbf{J}_{kp}. \tag{7.20}$$

Note that, in contrast with (7.11), the last term in (7.19) is no longer equal to N^3 because of the weights $\{w_{kp}\}$, which makes the minimization of C_3 much harder than that of C_2. Indeed, under the energy constraint on $\mathbf{x}$ and $\mathbf{y}$ and the PAR constraint on $\mathbf{x}$, it does not seem possible to find a simple solution for $\mathbf{x}$ similar to that in step 2 of Table 7.1. To get around this problem we omit the PAR constraint on $\mathbf{x}$ and thus let $\mathbf{x}$, like $\mathbf{y}$, vary freely. Under this relaxed condition, we can use the cyclic algorithm outlined in Table 7.2 to minimize C_3.

7.1.2 Numerical examples

We aim to synthesize the desired discrete-CAF shown in Figure 7.1(a), where the value of the central peak is $N = 50$, the two horizontal stripes at Doppler shifts $p = 0$ and $p = 9$ have zero sidelobes, and the volume of the discrete-CAF is evenly distributed elsewhere. (The figure is normalized by the maximum value and the gray scale is shown in dB.) The synthesized discrete-CAF $|\bar{r}_{xy}(k, p)|$ obtained from using the algorithm in Table 7.1 with $|x(n)| = 1$ is shown in Figure 7.1(b), which roughly approximates the desired one in Figure 7.1(a). In this case the algorithm in Table 7.1 with a relaxed PAR constraint does not lead to a visually improved performance and the corresponding result is therefore omitted.

In the next two examples we use the algorithm in Table 7.2. Figure 7.2(a) shows the desired discrete-CAF and Figure 7.2(b) shows the employed weights. Note that the central "weight stripe" in Figure 7.2(b) is chosen to be wider than the "sidelobe stripe" in Figure 7.2(a). The synthesized discrete-CAF is shown in Figure 7.3(a). The central peak in Figure 7.3(a) has a relatively small value (around 7) while the highest sidelobe is almost 12. For the so-obtained $\mathbf{x}$ and $\mathbf{y}$, the energy or PAR constraint is not satisfied:

$$\|\mathbf{x}\|^2 = 24.08, \ \mathrm{PAR}(\mathbf{x}) = 2.05, \tag{7.29}$$

$$\|\mathbf{y}\|^2 = 95.63 \tag{7.30}$$

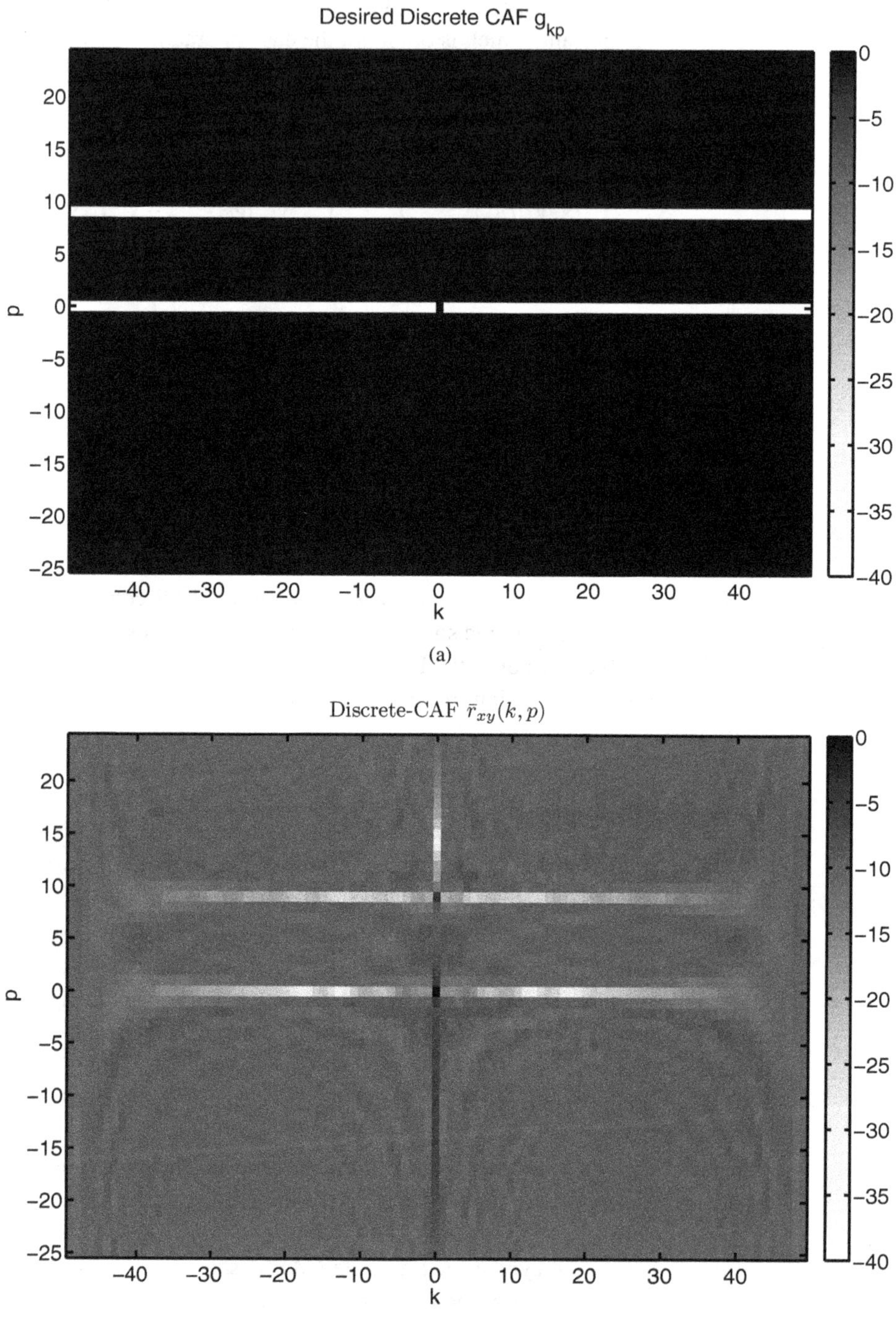

Figure 7.1 (a) The desired discrete-CAF g_{kp}. (b) The discrete-CAF, $|\bar{r}_{xy}(k, p)|$, synthesized using the algorithm in Table 7.1 with $|x(n)| = 1$.

Table 7.2. The cyclic algorithm for weighted discrete-CAF synthesis with weights

Step 0: Randomly initialize $\mathbf{x}$ and $\mathbf{y}$.
Step 1: For fixed $\mathbf{x}$ and $\mathbf{y}$, the minimizer ϕ_{kp} is given by (7.14).
Step 2: For fixed $\{\phi_{kp}\}$ and $\mathbf{y}$, we can write the criterion C_3 as follows:

$$C_3 = \mathbf{x}^H \mathbf{D}_1 \mathbf{x} - \mathbf{x}^H \mathbf{d}_1 - \mathbf{d}_1^H \mathbf{x} + \text{const} \tag{7.21}$$

where

$$\mathbf{d}_1 = \mathbf{B}\mathbf{y} \tag{7.22}$$

and

$$\mathbf{D}_1 = \sum_k \sum_p w_{kp} \mathbf{J}_{kp} \mathbf{y}\mathbf{y}^H \mathbf{J}_{kp}^H. \tag{7.23}$$

It follows from (7.21) that the minimizer $\mathbf{x}$ is given by

$$\mathbf{x} = \mathbf{D}_1^{-1} \mathbf{d}_1. \tag{7.24}$$

Step 3: For fixed $\{\phi_{kp}\}$ and $\mathbf{x}$, we have

$$C_3 = \mathbf{y}^H \mathbf{D}_2 \mathbf{y} - \mathbf{y}^H \mathbf{d}_2 - \mathbf{d}_2^H \mathbf{y} + \text{const} \tag{7.25}$$

where

$$\mathbf{d}_2 = \mathbf{B}^H \mathbf{x} \tag{7.26}$$

and

$$\mathbf{D}_2 = \sum_k \sum_p w_{kp} \mathbf{J}_{kp}^H \mathbf{x}\mathbf{x}^H \mathbf{J}_{kp}. \tag{7.27}$$

Similarly to Step 2, the minimizer $\mathbf{y}$ is given by

$$\mathbf{y} = \mathbf{D}_2^{-1} \mathbf{d}_2. \tag{7.28}$$

Iteration: Repeat Steps 1, 2 and 3 until convergence.

(see Figure 7.3(b) for $\{|x_n|\}$). However, the PAR value 2.05 of the transmit waveform may be acceptable for some modern systems. In addition, $\mathbf{x}$ can always be scaled so that its energy equals a desired number (such as N) as long as $\mathbf{y}$ is scaled accordingly. Note that the energy and the PAR values of $\mathbf{y}$ do not have to satisfy the hard constraints since $\mathbf{y}$ is used only at the receiver.

Figure 7.4(a) shows another desired discrete-CAF (with a wide mainlobe) and Figure 7.4(b) shows the employed weights. The synthesized discrete-CAF $|\bar{r}_{xy}(k, p)|$ is shown in Figure 7.5.

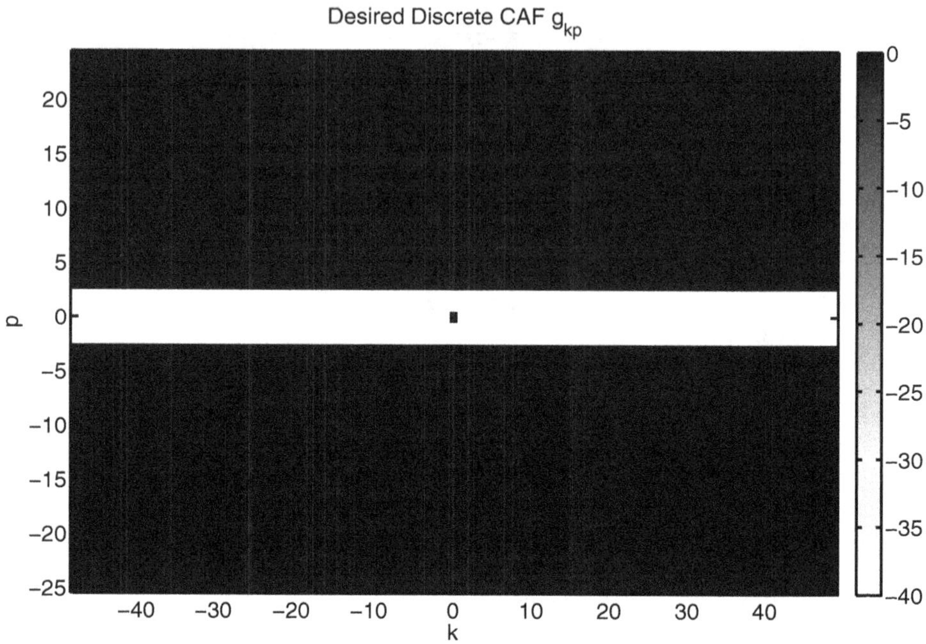

(a)

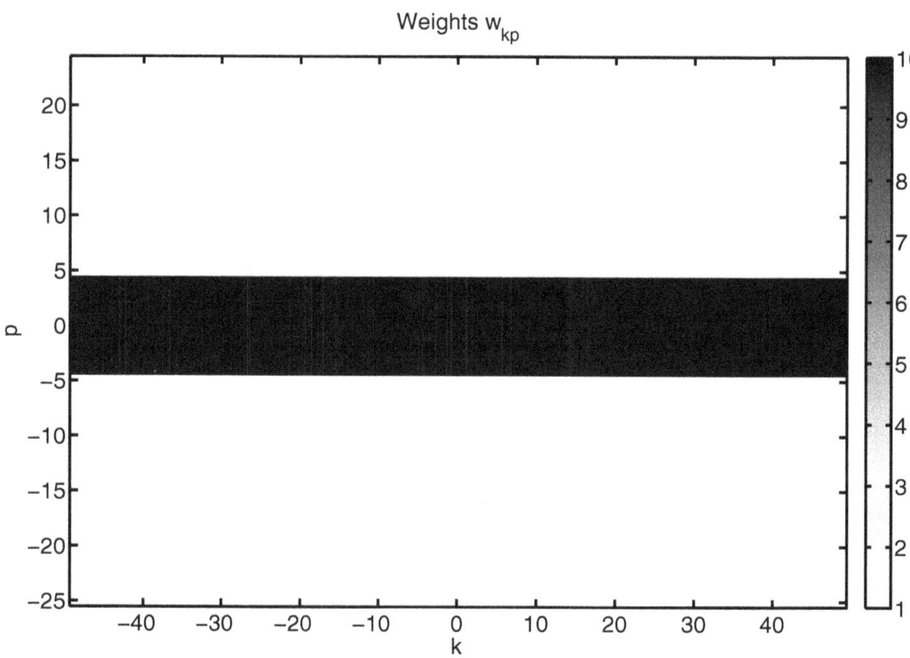

(b)

Figure 7.2 (a) The desired discrete-CAF g_{kp} and (b) the weights w_{kp}.

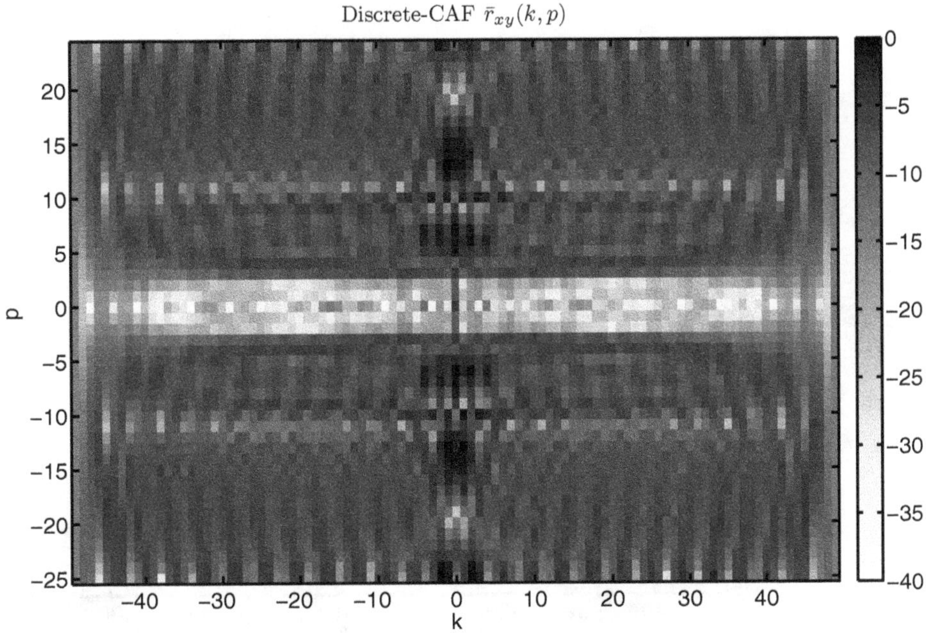

(a)

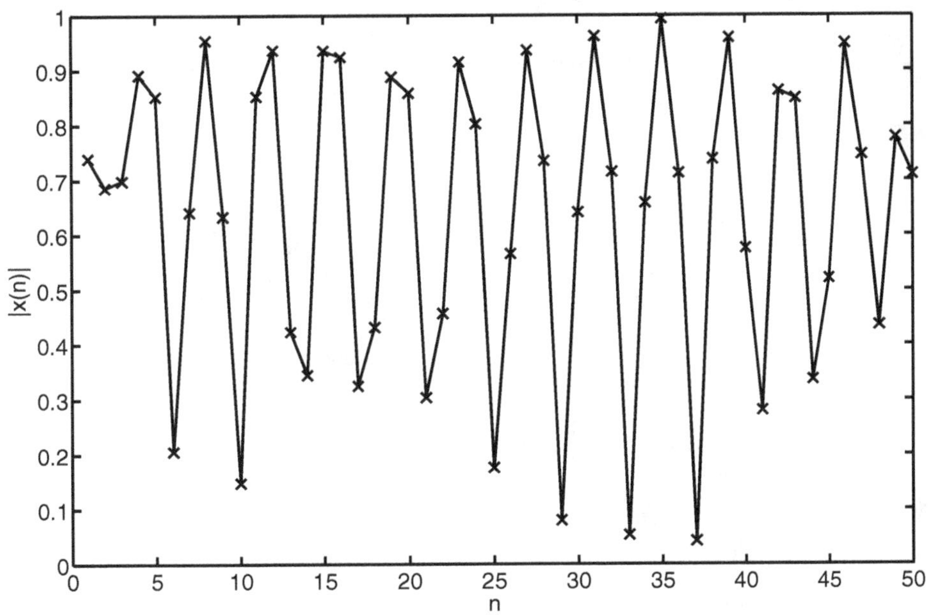

(b)

Figure 7.3 (a) The discrete-CAF $|\bar{r}_{xy}(k, p)|$ synthesized using the algorithm in Table 7.2 (see Figure 7.2 for the desired discrete-CAF and the weights used in the algorithm). (b) The magnitude of the waveform $\{x(n)\}$.

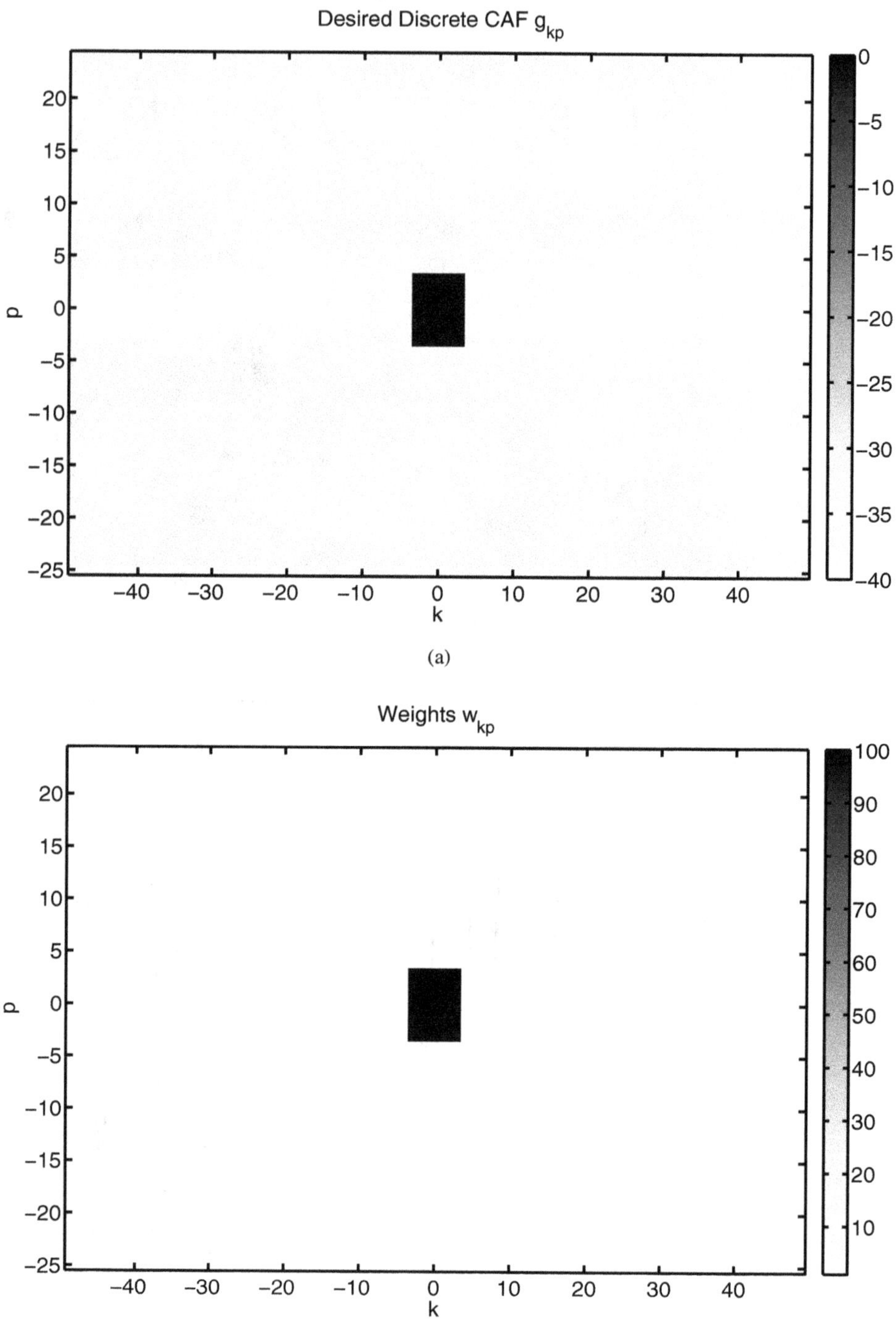

Figure 7.4 (a) The desired discrete-CAF g_{kp} and (b) The weights w_{kp}.

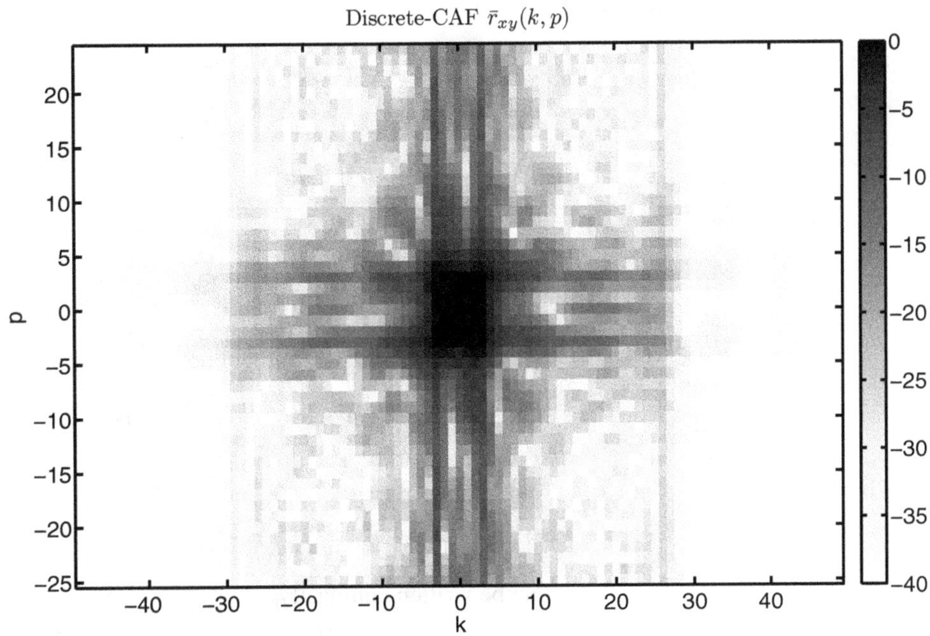

Figure 7.5 The synthesized discrete-CAF (see Figure 7.4 for the desired discrete-CAF and the weights used).

7.2 CAF synthesis

In this section we propose an algorithm that designs $u(t)$ and $v(t)$ in (7.1) jointly so that their (continuous-time) CAF $\chi(\tau, f)$ approximates a desired one. The desired CAF need not be a valid function associated with a pair of underlying signals; it can be any 2D function that represents a general desirable shape, such as a peak at the origin and low sidelobes in a certain area. Note that the volume of a CAF, which is defined as $V = \int_{-\infty}^{\infty}\int_{-\infty}^{\infty} |\chi(\tau, f)|^2 d\tau df$, is always equal to the product of the energy of $u(t)$ and that of $v(t)$. Such a property essentially prevents the synthesis of an ideal CAF with a narrow peak at the origin and zero sidelobes everywhere else. Instead we will demonstrate the synthesis of a CAF with a clear area around the origin as well as of a CAF with a high diagonal ridge but low sidelobes in the rest of the (τ, f) plane.

Suppose that both $u(t)$ and $v(t)$ consist of N subpulses

$$u(t) = \sum_{k=1}^{N} x(k)p_k(t), \quad v(t) = \sum_{l=1}^{N} y(l)p_l(t) \tag{7.31}$$

where $\{p_n(t)\}_{n=1}^{N}$ are pulse-shaping functions. For example, $p_n(t)$ can be the ideal rectangular shaping pulse described in (1.2).

For the above waveform setting, the CAF in (7.2) becomes

$$\begin{aligned}\chi^*(\tau, f) &= \int_{-\infty}^{\infty}\left(\sum_{k=1}^{N} x(k)p_k(t)\right)\left(\sum_{l=1}^{N} y^*(l)p_l^*(t+\tau)\right)e^{j2\pi ft}dt \\ &= \sum_{k=1}^{N}\sum_{l=1}^{N} x(k)y^*(l)\int_{-\infty}^{\infty} p_k(t)p_l^*(t+\tau)e^{j2\pi ft}dt. \end{aligned} \tag{7.32}$$

Let

$$\bar{\chi}_{kl}(\tau, f) = \int_{-\infty}^{\infty} p_k(t)p_l^*(t+\tau)e^{j2\pi ft}dt \tag{7.33}$$

denote the CAF of the pulse-shaping functions and define

$$\mathbf{x} = \left[x(1) \cdots x(N)\right]^T, \ \mathbf{y} = \left[y(1) \cdots y(N)\right]^T, \tag{7.34}$$

$$\mathbf{K}(\tau, f) = \begin{bmatrix} \bar{\chi}_{1,1}(\tau, f) & \cdots & \bar{\chi}_{N,1}(\tau, f) \\ \vdots & & \vdots \\ \bar{\chi}_{1,N}(\tau, f) & \cdots & \bar{\chi}_{N,N}(\tau, f) \end{bmatrix}. \tag{7.35}$$

Then the CAF in (7.32) can be written compactly as

$$\chi^*(\tau, f) = \mathbf{y}^H\mathbf{K}(\tau, f)\mathbf{x}. \tag{7.36}$$

Assume that the desired CAF modulus is given by $d(\tau, f)$ (real-valued and non-negative). The CAF phases are usually not of interest because of carrier phase incoherence. We aim to design a pair of signals $u(t)$ and $v(t)$ (more precisely $\mathbf{x}$ and $\mathbf{y}$, because the pulse shaping function is fixed) such that the modulus of their CAF approximates $d(\tau, f)$ as closely as possible. Such an optimization problem can be formulated as

$$\min_{\mathbf{x},\mathbf{y}} \ g(\mathbf{x}, \mathbf{y}) = \int_{-\infty}^{\infty}\int_{-\infty}^{\infty} w(\tau, f)\cdot\left[d(\tau, f) - |\mathbf{y}^H\mathbf{K}(\tau, f)\mathbf{x}|\right]^2 d\tau df \tag{7.37}$$

where $w(\tau, f)$ is the weighting function that specifies which area of CAF needs to be emphasized. For example, the values of $\chi(\tau, f)$ for large f usually do not matter (see the discussions around (6.18) and (6.19)) and can be safely ignored by choosing a zero weight for them.

7.2.1 The proposed algorithm

To solve (7.37), we first introduce auxiliary phases $\phi(\tau, f)$ and write the criterion as

$$\bar{g}(\mathbf{x}, \mathbf{y}, \phi) = \int_{-\infty}^{\infty}\int_{-\infty}^{\infty} w(\tau, f)\left|d(\tau, f)e^{j\phi(\tau,f)} - \mathbf{y}^H\mathbf{K}(\tau, f)\mathbf{x}\right|^2 d\tau df. \tag{7.38}$$

It is not difficult to see that

$$\min_{\phi(\tau,f)} \ \bar{g}(\mathbf{x}, \mathbf{y}, \phi) = g(\mathbf{x}, \mathbf{y}) \tag{7.39}$$

and that the minimizer $\phi(\tau, f)$ is given by

$$\phi(\tau, f) = \arg\left[\mathbf{y}^H\mathbf{K}(\tau, f)\mathbf{x}\right]. \tag{7.40}$$

The minimization of (7.38) can be done in a cyclic way: two variables out of $\{\mathbf{x}, \mathbf{y}, \phi(\tau, f)\}$ are fixed and $\bar{g}$ is minimized with respect to the third variable. For fixed $\mathbf{x}$ and $\mathbf{y}$, the minimizer $\phi(\tau, f)$ is as given in (7.40). For fixed $\phi(\tau, f)$ and $\mathbf{y}$, the criterion $\bar{g}$ can be written as

$$\begin{aligned}\bar{g}(\mathbf{x}) =& \mathbf{x}^H\mathbf{D}_1\mathbf{x} - \mathbf{x}^H\mathbf{B}\mathbf{y} - \mathbf{y}^H\mathbf{B}^H\mathbf{x} + \int_{-\infty}^{\infty}\int_{-\infty}^{\infty} w(\tau, f)|d(\tau, f)|^2 d\tau df \\ =& (\mathbf{x} - \mathbf{D}_1^{-1}\mathbf{B}\mathbf{y})^H\mathbf{D}_1(\mathbf{x} - \mathbf{D}_1^{-1}\mathbf{B}\mathbf{y}) + \text{const}_1,\end{aligned} \tag{7.41}$$

where

$$\mathbf{D}_1 = \int_{-\infty}^{\infty}\int_{-\infty}^{\infty} w(\tau, f)\mathbf{K}^H(\tau, f)\mathbf{y}\mathbf{y}^H\mathbf{K}(\tau, f)d\tau df, \tag{7.42}$$

$$\mathbf{B} = \int_{-\infty}^{\infty}\int_{-\infty}^{\infty} w(\tau, f)d(\tau, f)e^{j\phi(\tau, f)}\mathbf{K}^H(\tau, f)d\tau df \tag{7.43}$$

and const_1 is a term that does not depend on $\mathbf{x}$. It follows from (7.41) that the minimizer $\mathbf{x}$ is given by

$$\mathbf{x} = \mathbf{D}_1^{-1}\mathbf{B}\mathbf{y}. \tag{7.44}$$

Similarly, for fixed $\phi(\tau, f)$ and $\mathbf{x}$, the criterion $\bar{g}$ can be written as

$$\bar{g}(\mathbf{y}) = (\mathbf{y} - \mathbf{D}_2^{-1}\mathbf{B}^H\mathbf{x})^H\mathbf{D}_2(\mathbf{y} - \mathbf{D}_2^{-1}\mathbf{B}^H\mathbf{x}) + \text{const}_2 \tag{7.45}$$

where

$$\mathbf{D}_2 = \int_{-\infty}^{\infty}\int_{-\infty}^{\infty} w(\tau, f)\mathbf{K}(\tau, f)\mathbf{x}\mathbf{x}^H\mathbf{K}^H(\tau, f)\, d\tau df \tag{7.46}$$

and const_2 is a term that does not depend on $\mathbf{y}$. The minimizer $\mathbf{y}$ is thus given by

$$\mathbf{y} = \mathbf{D}_2^{-1}\mathbf{B}^H\mathbf{x}. \tag{7.47}$$

The above steps are summarized in Table 7.3. Each step of the algorithm decreases the minimization criterion and thus the algorithm is guaranteed to converge to a local minimum. Note that $\phi(\tau, f)$ and $\mathbf{K}(\tau, f)$ are functions of the time delay τ and Doppler frequency shift f, both of which range from $-\infty$ to ∞, theoretically. In practice, however, the maximum time delay can be chosen as Nt_p (which is the time duration of the signal) and the maximum Doppler frequency shift can be chosen as the signal bandwidth (which is approximately equal to $1/t_p$).

When the rectangular pulse (1.2) is used, each element of $\mathbf{K}(\tau, f)$ has a closed-form expression. Furthermore, (1.2) is an orthogonal shaping pulse (i.e., $\int_{-\infty}^{\infty} p_k(t)p_l(t)dt$ equals 1 when $k = l$ and equals 0 otherwise), which can be shown to yield $\mathbf{D}_1 = \|\mathbf{y}\|^2\mathbf{I}_N$ and $\mathbf{D}_2 = \|\mathbf{x}\|^2\mathbf{I}_N$ if $w(\tau, f)$ is equal to 1 for all (τ, f). Such properties can be leveraged to greatly speed up the computation of $\mathbf{K}(\tau, f)$, $\mathbf{D}_1$ and $\mathbf{D}_2$, which generally requires numerical approximations using fine sampling.

If $\{x(n)\}$ is required to be a unimodular sequence, the following operation

$$x(n) \leftarrow e^{j\arg[x(n)]}, \quad n = 1, \ldots, N, \tag{7.48}$$

Table 7.3. The cyclic algorithm for CAF synthesis

Step 0: Initialize $\mathbf{x}$ and $\mathbf{y}$ using randomly generated sequences.
Step 1: $\phi(\tau, f) = \arg\{\mathbf{y}^H \mathbf{K}(\tau, f)\mathbf{x}\}$.
Step 2: $\mathbf{x} = \mathbf{D}_1^{-1}\mathbf{B}\mathbf{y}$.
Step 3: $\mathbf{y} = \mathbf{D}_2^{-1}\mathbf{B}^H\mathbf{x}$.
The definitions of $\mathbf{D}_1$, $\mathbf{B}$ and $\mathbf{D}_2$ are given in (7.42), (7.43) and (7.46), respectively.
Iteration: Repeat steps 1, 2 and 3 until a certain stop criterion is met (e.g., until $\|\mathbf{x}^{(i)} - \mathbf{x}^{(i+1)}\|^2 + \|\mathbf{y}^{(i)} - \mathbf{y}^{(i+1)}\|^2 < 10^{-3}$, where $(\cdot)^{(i)}$ indicates the ith iteration).

can be added after $\mathbf{x} = \mathbf{D}_1^{-1}\mathbf{B}\mathbf{y}$ in step 2 of Table 7.3. Note that step 2 combined with (7.48) provides only an approximate minimizer and hence it does not necessarily decrease the criterion $\bar{g}$ unless $\mathbf{D}_1$ is proportional to $\mathbf{I}_N$ (a case mentioned in the previous paragraph). For the numerical examples in Section 7.2.2, (7.48) will be used only if so stated.

7.2.2 Numerical examples

Assume that there are $N = 50$ subpulses based on the rectangular pulse shaping function (1.2). The time duration of each subpulse is t_p and that of the total waveform is $T = Nt_p$. In the simulations, the time delay τ is normalized by T and the Doppler frequency f is normalized by $1/T$, so that the particular value of t_p does not affect the final sequences obtained from the algorithm. Randomly generated sequences are used for initialization. The algorithm usually converges within a few hundred iterations and takes tens of minutes on an ordinary PC.

Suppose that a thumbtack CAF is desired, i.e.,

$$d(\tau, f) = \begin{cases} N, & (\tau, f) = (0, 0), \\ 0, & \text{elsewhere.} \end{cases} \tag{7.49}$$

The weighting function is selected as

$$w(\tau, f) = \begin{cases} 1, & (\tau, f) \in \Omega, \ \& \ (\tau, f) \notin \Omega_m \\ 0, & \text{elsewhere,} \end{cases} \tag{7.50}$$

where $\Omega = \{(\tau, f),\ |\tau| \le 10t_p\ \&\ |f| \le 2/T\}$ is the region of interest and $\Omega_m = \{(\tau, f),\ |\tau| \le t_p\ \&\ |f| \le 1/T\ \&\ \tau f \ne 0\}$ is the mainlobe area excluding the origin; Ω_m is excluded from the region of interest in order to compensate for the sharp change in $d(\tau, f)$ near the origin.

Figure 7.6(a) shows the CAF of the waveform resulting from the sequences $\mathbf{x}$ and $\mathbf{y}$. The white rectangular area close to the origin is the desired low sidelobe region. The PAR of $\mathbf{x}$ is 3.3 and that of $\mathbf{y}$ is 3.5. This relatively high (albeit acceptable in certain applications) PAR is due to the fact that the modulus of $\mathbf{x}$ is not constrained in the algorithm. To ensure a constant-modulus $\mathbf{x}$, we can add (7.48) to step 2 of Table 7.3. The resulting CAF is shown in Figure 7.6(b), which has somewhat higher sidelobes in the region of interest than Figure 7.6(a).

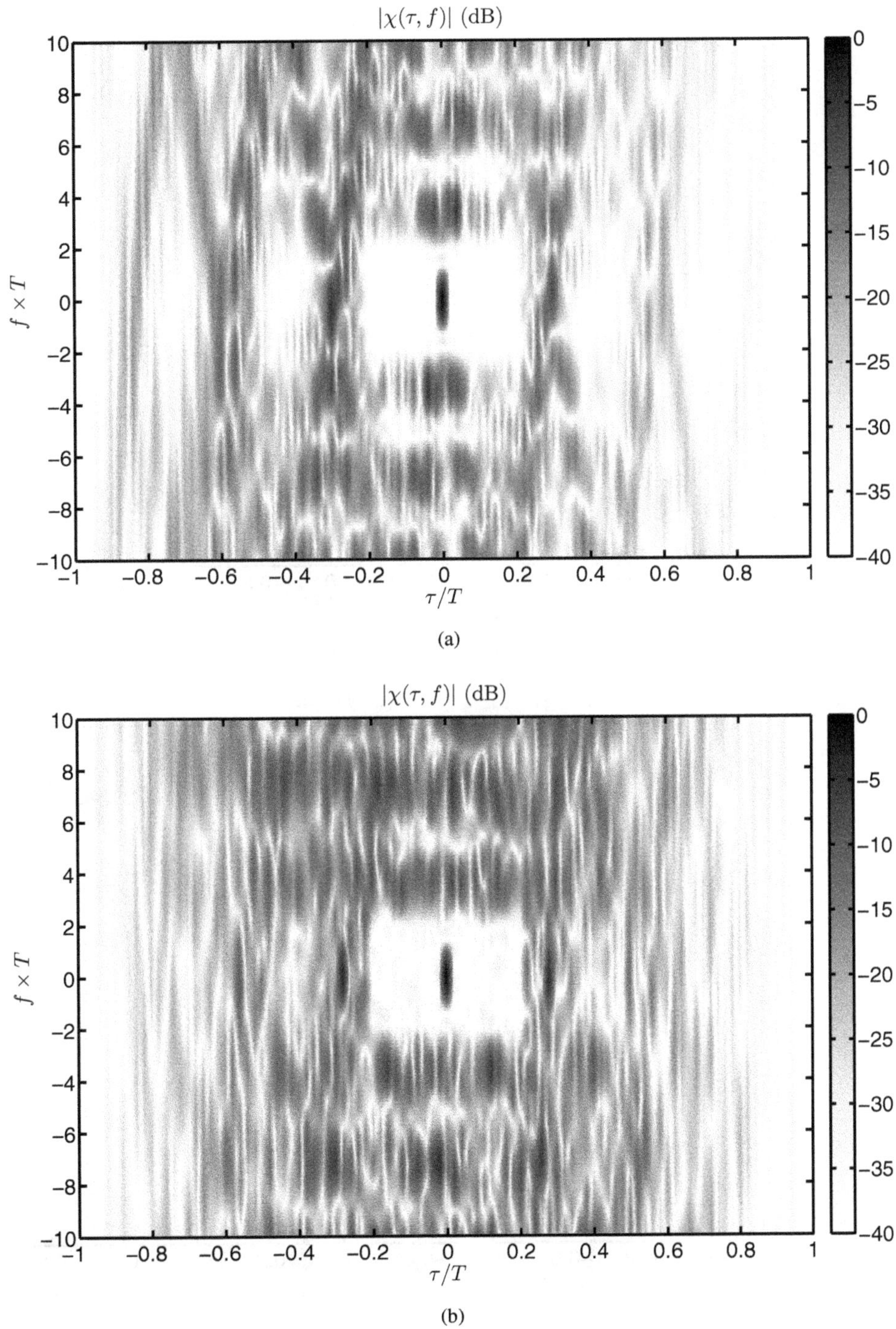

Figure 7.6 (a) The synthesized CAF when $d(\tau, f)$ and $w(\tau, f)$ are given in (7.49) and (7.50), respectively. (b) The same as (a) except that (7.48) is imposed in step 2 of Table 7.3.

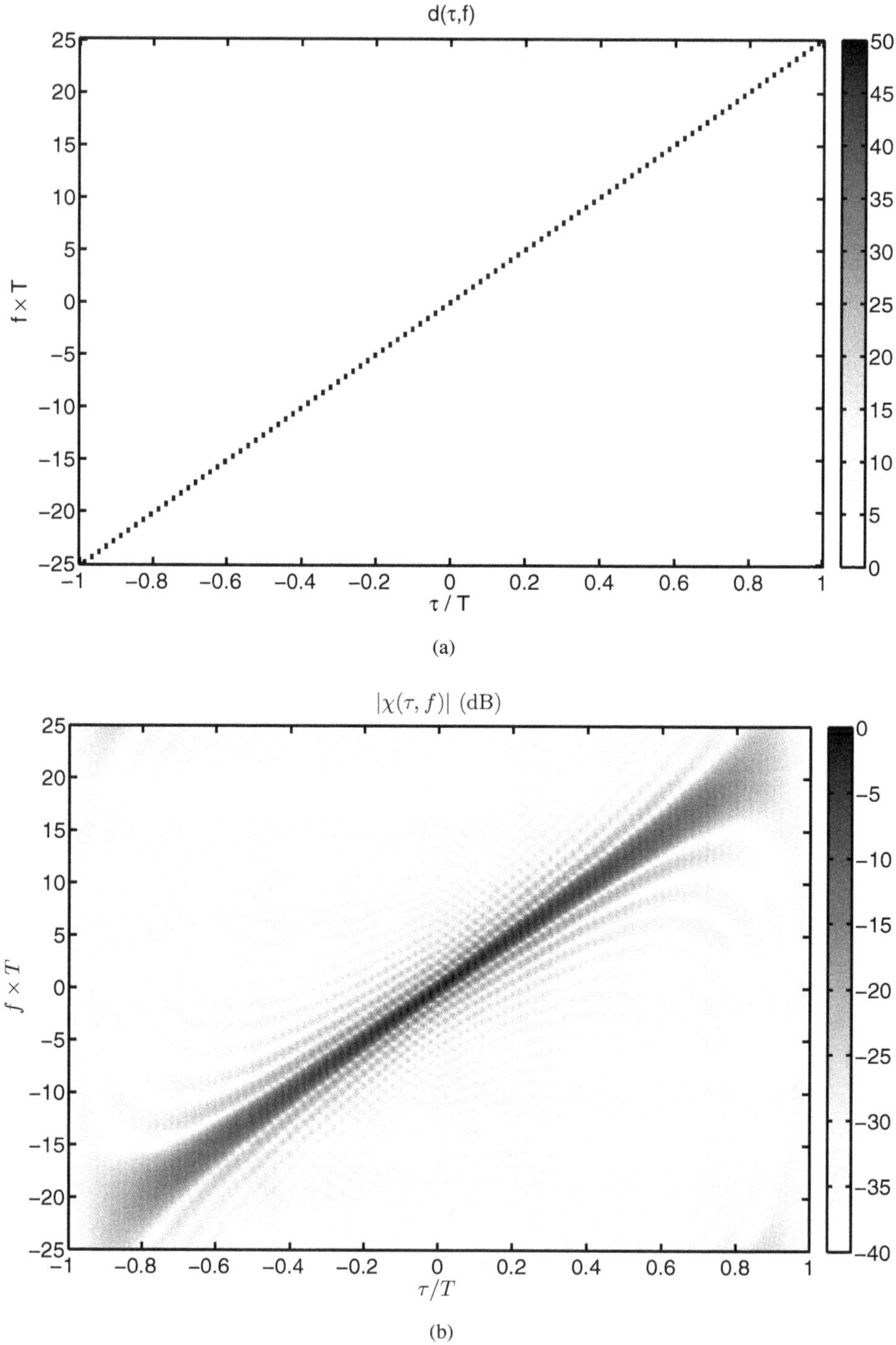

Figure 7.7 (a) The desired CAF and (b) the synthesized CAF.

The thumbtack-shaped CAF in Figure 7.6 leads to good range and Doppler resolution within the clear area and is often sought for in practice. In the next example, however, we try to synthesize the CAF in Figure 7.7(a) that has a diagonal ridge of height N and equals zero elsewhere. Such a CAF is tolerant to Doppler frequency shifts and is desired when a bank of filters for different frequencies is too expensive [Skolnik 2008]; see also the discussions on the AF of chirp waveforms in Section 6.1.

We set the weighting function $w(\tau, f) = 1$ for all (τ, f) and add (7.48) to step 2 of Table 7.3 so that $\mathbf{x}$ has constant modulus. The resulting CAF is shown in Figure 7.7(b), which approximates Figure 7.7(a) well.

7.3 Conclusions

Two algorithms have been presented to synthesize a given cross-ambiguity function (CAF) by minimizing the integrated square error between a desired CAF and a realizable CAF. The first algorithm is for discrete-CAF synthesis while the second is for continuous-time CAF synthesis. A desired CAF can be any two-dimensional function and is not required to be a valid ambiguity function. Weighting was introduced in the design problem formulation to account for more interest in one region of the (τ, f) plane than in other regions. The waveforms thus-generated have relatively low peak-to-average power ratios and they can be made unimodular by a modification of the algorithms if desired.

Appendix 7A Constant volume property of discrete-CAF

It follows from (7.3) that

$$\begin{aligned}
&\sum_{k=-N+1}^{N-1} \sum_{p=-N/2}^{N/2-1} |\bar{r}_{xy}(k, p)|^2 \qquad (7.51)\\
&= \sum_k \sum_p \left| \sum_{n=1}^{N} y(n) x^*(n-k) e^{-j2\pi \frac{(n-k)p}{N}} \right|^2 \\
&= \sum_k \sum_p \sum_{n_1=1}^{N} \sum_{n_2=1}^{N} y(n_1) y^*(n_2) x^*(n_1 - k) x(n_2 - k) e^{-j2\pi \frac{(n_1-n_2)p}{N}} \\
&= N \sum_k \sum_{n=1}^{N} |y(n)|^2 |x(n-k)|^2,
\end{aligned}$$

where we have used the fact that $\sum_{p=-N/2}^{N/2-1} e^{-j2\pi\frac{(n_1-n_2)p}{N}} = N\delta_{n_1 n_2}$. Since $x(n)$ or $y(n)$ equals zero when $n \notin [1, N]$, (7.51) can be rearranged as follows:

$$\begin{aligned}\sum_{k=-N+1}^{N-1}\sum_{p=-N/2}^{N/2-1} |\bar{r}_{xy}(k,p)|^2 &= N\sum_{n=1}^{N}\left(|y(n)|^2 \sum_{k=-N+1}^{N-1} |x(n-k)|^2\right) \\ &= N\sum_{n=1}^{N}\left(|y(n)|^2 \sum_{m=n-N+1}^{n-1+N} |x(m)|^2\right) \\ &= N\sum_{n=1}^{N}|y(n)|^2 \sum_{m=1}^{N}|x(m)|^2,\end{aligned} \tag{7.52}$$

which, when combined with the energy constraints in (6.11) and (7.4), leads to the constant-volume property of the discrete-CAF:

$$\sum_{k=-N+1}^{N-1}\sum_{p=-N/2}^{N/2-1} |\bar{r}_{xy}(k,p)|^2 = N^3. \tag{7.53}$$

8 Joint design of transmit sequence and receive filter

There has been a long-standing interest in designing radar transmit signals and receive filters *jointly* for clutter/interference rejection; see, e.g., [Rummler 1967][DeLong Jr. & Hofstetter 1967][Spafford 1968][Stutt & Spafford 1968][Blunt & Gerlach 2006][Kay 2007] and [Stoica, Li & Xue 2008]. Clutter (or reverberation in sonar terminology) refers to unwanted echoes that are usually correlated with the transmitted signal, while interference is a term used for noise as well as (adverse) jamming signals. Since the negative impact from clutter and interference should be minimized at the receiver side, a natural criterion for designing transmit signals and receive filters would be to maximize the signal-to-clutter-plus-interference ratio (SCIR) of the receiver output at the time of target detection.

It is well known that a matched filter maximizes the signal-to-noise ratio (SNR) in the presence of additive white noise (see Chapter 1). The matched filter can be implemented as a correlator that multiplies the received signal with a delayed replica of the transmitted signal. The peak of the receiver output indicates the time delay of the target signal. If the received signal also has a Doppler frequency shift due to the relative movement between the target and the platform, a bank of filters is needed, each of which is matched to a specific Doppler frequency.

A matched filter is not able to take care of clutter or jamming suppression, a feature that is left to the transmit signal design. Jamming signals can come from an adversary or interfering radio applications and usually operate in particular frequency bands. Their negative effects can be largely avoided if we put little energy of the transmit signal into those frequency bands. As for clutter which appears as signal-like returns but with different delays or frequency shifts from the signal of interest, its effects can be analyzed using the ambiguity function (AF). In order to minimize clutter effects, the AF sidelobes need to be minimized (see Chapter 6). When the intra-pulse Doppler shifts are negligible, clutter suppression can be related to a more thoroughly studied problem, namely the minimization of the auto-correlation sidelobes of the transmit signal (see Chapter 2). Regarding the dynamic range of the transmit signals, owing to hardware constraints such as the maximum clipping of power amplifiers and analog-to-digital converters [Skolnik 2008], it is desirable that these signals have constant modulus or low peak-to-average power ratios (PAR). Such constraints are difficult to take into account in the design and are not always addressed properly in the literature.

Alternatively, we can use a "mismatched" filter, also called an instrumental variable (IV) filter [Stoica, Li & Xue 2008], at the receiver side. Using such a filter instead of a matched filter means trading off SNR for improved SCIR. This trade-off is legitimate if the radar detection performance is actually clutter or jamming limited. One benefit of using an IV filter is that it is not subject to the constant-modulus or low PAR constraint, which allows for more degrees of freedom (DOF) in the design. However, a joint design of the IV filter and the transmit signal leads to a more complex optimization problem that involves cross ambiguity functions (see Chapter 7) or cross correlations in the negligible Doppler case [Spafford 1968][DeLong Jr. & Hofstetter 1969].

8.1 Data model and problem formulation

Let

$$\mathbf{x} = \begin{bmatrix} x_1 \; x_2 \; \cdots \; x_N \end{bmatrix}^T \tag{8.1}$$

denote the probing sequence that modulates the transmitted train of subpulses; see (1.1). Note that in previous chapters we have always used $\{x(n)\}$ to denote a single sequence and $\{x_m(n)\}$ to denote a set of sequences. In this chapter where we are only concerned with a single sequence, however, we choose to drop the index n in parentheses and instead use a subscript $_n$ for notational simplicity, as shown in (8.1).

We assume a digital system so that only the discrete-time signal is of concern. The energy of $\{x_k\}_{k=1}^N$ is constrained to be N:

$$\|\mathbf{x}\|^2 = N \tag{8.2}$$

without any loss of generality. Such an energy constraint is always imposed in the design because not limiting the signal energy could lead to meaningless results; e.g., the jamming interference could be ignored if the transmit power were infinite.

We constrain $\{x_k\}_{k=1}^N$ to satisfy either

1. Unit-modulus constraint,

$$\begin{aligned} |x_k| = 1 \iff x_k = e^{j\phi_k}, \\ \phi_k \in [0, 2\pi), \quad k = 1, \ldots, N, \end{aligned} \tag{8.3}$$

   (see also (1.4)) or
2. PAR constraint,

$$\text{PAR}(\mathbf{x}) \le \mu, \quad \mu \in [1, N], \tag{8.4}$$

where PAR($\mathbf{x}$) is as defined in (4.14), and μ is a predefined parameter that specifies the maximum allowed PAR. Note that when $\mu = N$, (8.4) imposes no constraint at all; and that when $\mu = 1$, (8.4) coincides with the unit-modulus constraint in (8.3).

Under the assumptions that the intra-pulse Doppler is negligible and that the sampling is synchronized to the pulse rate, the received discrete-time baseband data vector, after alignment to the range cell of interest, can be written as ([Blunt & Gerlach 2006][Kay

2007][Stoica, Li & Xue 2008]):

$$\mathbf{y} = \alpha_0 \begin{bmatrix} x_1 \\ \vdots \\ x_{N-1} \\ x_N \end{bmatrix} + \alpha_1 \begin{bmatrix} 0 \\ x_1 \\ \vdots \\ x_{N-1} \end{bmatrix} + \cdots + \alpha_{N-1} \begin{bmatrix} 0 \\ \vdots \\ 0 \\ x_1 \end{bmatrix} + \alpha_{-1} \begin{bmatrix} x_2 \\ \vdots \\ x_N \\ 0 \end{bmatrix} + \cdots + \alpha_{-N+1} \begin{bmatrix} x_N \\ 0 \\ \vdots \\ 0 \end{bmatrix} + \boldsymbol{\epsilon} \quad (8.5)$$

where α_0 is the scattering coefficient of the current range cell, $\{\alpha_k\}_{k\neq 0}$ are the scattering coefficients of the adjacent range cells that contribute clutter or reverberation components to $\mathbf{y}$, and $\boldsymbol{\epsilon}$ is an interference term that comprises measurement noise as well as other disturbances such as jamming. Note that if the Doppler effects are not negligible (owing to a fast relative motion between the platform and the target), the data model in (8.5) needs to be modified to accommodate frequency shift [Stoica, Li & Xue 2008] and the analysis should involve two-dimensional ambiguity functions; see [Spafford 1968][DeLong Jr. & Hofstetter 1969] and also Chapter 6. However, such an analysis is not undertaken in this chapter, where we focus on the simpler data model in (8.5).

We assume that the covariance matrix of $\boldsymbol{\epsilon}$, viz.,

$$\mathrm{E}\{\boldsymbol{\epsilon}\boldsymbol{\epsilon}^H\} = \boldsymbol{\Gamma} \quad (8.6)$$

is Toeplitz (which is a weak assumption given the uniform sampling that led to (8.5)). We also assume that the clutter coefficients $\{\alpha_k\}_{k\neq 0}$ in (8.5) are independent of one another and of $\boldsymbol{\epsilon}$, and that

$$\mathrm{E}\{|\alpha_k|^2\} = \beta, \quad k \neq 0. \quad (8.7)$$

Both $\boldsymbol{\Gamma}$ and β are assumed to be known. In active sensing applications, information on $\boldsymbol{\Gamma}$ and β can be obtained by some form of preprocessing and is usually assumed available in the so-called cognitive systems; see, e.g., [Haykin 2006].

One of the principal goals of data processing for (8.5) is the estimation of α_0. The instrumental variable (IV) estimate of α_0, also more commonly called the mismatched filter estimate (we prefer to call it IV rather than mismatched filter owing to the negative connotation implied by the adjective ‘mismatched’ in the latter name; see [Stoica, Li & Xue 2008]) is the following linear function of $\mathbf{y}$:

$$\hat{\alpha}_0 = \frac{\mathbf{w}^H\mathbf{y}}{\mathbf{w}^H\mathbf{x}} \quad (8.8)$$

where $\mathbf{w}$ is the $N \times 1$ IV vector. An important special case of (8.8) is that of the matched-filter estimate

$$\hat{\alpha}_0 = \frac{\mathbf{x}^H\mathbf{y}}{||\mathbf{x}||^2}, \quad (8.9)$$

which corresponds to using the sequence itself as the IV vector:

$$\mathbf{w} = \mathbf{x}. \quad (8.10)$$

Under the stated assumptions (see (8.6) and (8.7)), the mean-squared error (MSE) of (8.8) can be readily derived:

$$\mathrm{MSE}(\hat{\alpha}_0) = \mathrm{E}\left\{\left|\frac{\mathbf{w}^H\mathbf{y}}{\mathbf{w}^H\mathbf{x}} - \alpha_0\right|^2\right\} = \frac{\mathbf{w}^H\mathbf{R}\mathbf{w}}{|\mathbf{w}^H\mathbf{x}|^2} \tag{8.11}$$

where

$$\mathbf{R} = \beta \sum_{\substack{k=-N+1 \\ k\neq 0}}^{N-1} \mathbf{J}_k\mathbf{x}\mathbf{x}^H\mathbf{J}_k^H + \boldsymbol{\Gamma} \tag{8.12}$$

and $\mathbf{J}_k$ denotes the following shift matrix:

$$\mathbf{J}_k = \mathbf{J}_{-k}^H = \begin{bmatrix} \overbrace{\qquad\qquad 1}^{k+1} & \mathbf{0} \\ & \ddots \\ & 1 \\ \mathbf{0} & \end{bmatrix}_{N\times N}^H, \quad k = 0, \ldots, N-1. \tag{8.13}$$

The main goal of this chapter is to jointly design the receive filter $\mathbf{w}$ and the transmit sequence $\mathbf{x}$ so as to minimize the MSE in (8.11). Observe that the denominator in (8.11) is the power of the signal component in the receiver output, and the numerator is the power of the clutter and interference. Hence minimizing the MSE (8.11) is equivalent to maximizing the SCIR. Note also that the IV estimate has $2N$ additional (real-valued) degrees of freedom (DOFs) as compared with the matched filter; these extra DOFs of (8.8) should allow a more accurate estimation of α_0 than what is possible using (8.9).

In what follows we present three algorithms for the design of $\mathbf{w}$ and $\mathbf{x}$. These algorithms are referred to as the CREW (cognitive receiver and waveform) designs, where "cognitive" implies the availability of prior knowledge (see (8.6) and (8.7)).

8.2 A gradient approach

The minimization of (8.11) with respect to (w.r.t.) $\mathbf{w}$, for fixed $\mathbf{x}$, yields the following well-known closed-form expression for the minimizing vector:

$$\mathbf{w} = \mathbf{R}^{-1}\mathbf{x} \tag{8.14}$$

to within a multiplicative constant. The proof of (8.14) is immediate. Let $\mathbf{R}^{1/2}$ denote a Hermitian square root of $\mathbf{R}$. By the Cauchy–Schwartz inequality, we have that

$$\left|\mathbf{w}^H\mathbf{x}\right|^2 = \left|\mathbf{w}^H\mathbf{R}^{1/2}\mathbf{R}^{-1/2}\mathbf{x}\right|^2 \leq \left(\mathbf{w}^H\mathbf{R}\mathbf{w}\right)\left(\mathbf{x}^H\mathbf{R}^{-1}\mathbf{x}\right). \tag{8.15}$$

It follows from (8.11) and (8.15) that

$$\mathrm{MSE}(\hat{\alpha}_0) = \frac{\mathbf{w}^H\mathbf{R}\mathbf{w}}{|\mathbf{w}^H\mathbf{x}|^2} \geq \frac{1}{\mathbf{x}^H\mathbf{R}^{-1}\mathbf{x}}, \tag{8.16}$$

Table 8.1. The CREW(gra) algorithm with unit-modulus constraint

Step 0: Use an available sequence or a randomly generated one to initialize $\mathbf{x}$. **Step 1**: Solve (8.17) for $\mathbf{x}$ using any gradient-based solver (the expression of the first-order derivative is given in (8.19)), such as the "fminunc" function in MATLAB, which utilizes the BFGS algorithm. **Step 2**: Compute $\mathbf{w}$ using (8.14) with $\mathbf{x}$ provided by step 1.

where the lower bound is obtained at (8.14).

Next we aim to minimize $1/(\mathbf{x}^H\mathbf{R}^{-1}\mathbf{x})$ w.r.t. $\mathbf{x}$ under the unit-modulus constraint (8.3) or, equivalently,

$$\min_{\boldsymbol{\phi}} \left[-\mathbf{x}^H(\boldsymbol{\phi})\mathbf{R}^{-1}(\boldsymbol{\phi})\mathbf{x}(\boldsymbol{\phi})\right] \triangleq f(\boldsymbol{\phi}), \tag{8.17}$$

where

$$\boldsymbol{\phi} = \left[\phi_1 \cdots \phi_N\right]^T \tag{8.18}$$

comprises the phases of $\mathbf{x}$. Gradient-based methods can be used to solve (8.17), such as the Broyden–Fletcher–Goldfarb–Shanno (BFGS) algorithm [Fletcher 1970]. In contrast with the classical Newton method, in BFGS the Hessian matrix of the second-order derivatives does not need to be evaluated directly; only the first-order derivative vector $\nabla f(\boldsymbol{\phi})$ is needed. The elements of $\nabla f(\boldsymbol{\phi})$ can be computed as:

$$\frac{\partial f(\boldsymbol{\phi})}{\partial \phi_k} = -\left(\frac{\partial \mathbf{x}^H}{\partial \phi_k}\mathbf{R}^{-1}\mathbf{x} + \mathbf{x}^H\frac{\partial \mathbf{R}^{-1}}{\partial \phi_k}\mathbf{x} + \mathbf{x}^H\mathbf{R}^{-1}\frac{\partial \mathbf{x}}{\partial \phi_k}\right), \quad k = 1, \ldots, N, \tag{8.19}$$

where

$$\begin{aligned} \frac{\partial \mathbf{x}}{\partial \phi_k} &= \Big[\underbrace{0 \cdots 0}_{k-1}\ je^{j\phi_k}\ \underbrace{0 \cdots 0}_{N-k}\Big]^T, \\ \frac{\partial \mathbf{R}^{-1}}{\partial \phi_k} &= -\,\mathbf{R}^{-1}\frac{\partial \mathbf{R}}{\partial \phi_k}\mathbf{R}^{-1} \end{aligned} \tag{8.20}$$

and

$$\frac{\partial \mathbf{R}}{\partial \phi_k} = \beta \sum_{\substack{l=-(N-1)\\ l\neq 0}}^{N-1} \mathbf{J}_l\left(\frac{\partial \mathbf{x}}{\partial \phi_k}\mathbf{x}^H + \mathbf{x}\frac{\partial \mathbf{x}^H}{\partial \phi_k}\right)\mathbf{J}_l^H. \tag{8.21}$$

The above expression for $\nabla f(\boldsymbol{\phi})$ is all that is needed by a number of available numerical solvers that implement BFGS, such as the "fminunc" function in MATLAB. The so-obtained gradient-based algorithm is called CREW(gra) and is summarized in Table 8.1. Note that CREW(gra) can only deal with the unit-modulus constraint. Also note that Step 1 of Table 8.1 may require many iterations, during each of which all N elements of $\nabla f(\boldsymbol{\phi})$ need to be recomputed. This makes CREW(gra) computationally expensive in the case of large N (e.g., larger than 400 on an ordinary PC).

8.3 A frequency-domain approach

We begin this section by including the zero-delay sequence into the matrix $\mathbf{R}$ as defined in (8.12):

$$\begin{aligned}\mathbf{R} + \beta \mathbf{x}\mathbf{x}^H &= \beta \sum_{k=-N+1}^{N-1} \mathbf{J}_k \mathbf{x}\mathbf{x}^H \mathbf{J}_k^H + \boldsymbol{\Gamma} \\ &= \beta \mathbf{A}^H \mathbf{A} + \boldsymbol{\Gamma}\end{aligned} \tag{8.22}$$

where

$$\mathbf{A}^H = \begin{bmatrix} x_1 & 0 & \cdots & 0 & x_N & x_{N-1} & \cdots & x_2 \\ x_2 & x_1 & & \vdots & 0 & x_N & & \vdots \\ \vdots & \vdots & \ddots & 0 & \vdots & & & x_N \\ x_N & x_{N-1} & \cdots & x_1 & 0 & 0 & \cdots & 0 \end{bmatrix} \tag{8.23}$$

is an $N \times (2N-1)$ matrix. Using (8.22) we rewrite the MSE in (8.11) as

$$\text{MSE}(\hat{\alpha}_0) = \frac{\mathbf{w}^H(\beta \mathbf{A}^H \mathbf{A} + \boldsymbol{\Gamma})\mathbf{w}}{|\mathbf{w}^H \mathbf{x}|^2} - \beta. \tag{8.24}$$

Note that

$$\mathbf{A}^H \mathbf{A} = \begin{bmatrix} r_0 & r_1^* & \cdots & r_{N-1}^* \\ r_1 & r_0 & & r_{N-2}^* \\ \vdots & & \ddots & \vdots \\ r_{N-1} & r_{N-2} & & r_0 \end{bmatrix} \tag{8.25}$$

(see Appendix 8A for a proof), where $\{r_k\}_{k=0}^{N-1}$ are the correlations of $\{x_n\}_{n=1}^{N}$ (see (1.15)).

Next we aim to express the objective function (8.24) in the frequency domain. To this end we will make use of some standard properties of Toeplitz and circulant matrices. Let γ_{i-j} denote the (i, j)th element of $\boldsymbol{\Gamma}$, which is an $N \times N$ Toeplitz matrix. The matrix $\boldsymbol{\Gamma}$ can be readily embedded in a $(2N-1) \times (2N-1)$ circulant matrix:

$$\mathbf{C}^i = \begin{bmatrix} \gamma_0 & \gamma_1^* & \cdots & \gamma_{N-1}^* & \gamma_{N-1} & \gamma_{N-2} & \cdots & \gamma_1 \\ \gamma_1 & \gamma_0 & & \vdots & \gamma_{N-1}^* & \gamma_{N-1} & & \vdots \\ \vdots & & \ddots & \gamma_1^* & \vdots & & & \gamma_{N-1} \\ \gamma_{N-1} & \cdots & \gamma_1 & \gamma_0 & \gamma_1^* & \cdots & & \gamma_{N-1}^* \\ \gamma_{N-1}^* & \gamma_{N-1} & \cdots & \gamma_1 & \gamma_0 & \gamma_1^* & \cdots & \gamma_{N-2}^* \\ \vdots & & & \vdots & & & & \vdots \\ \gamma_1^* & \cdots & \gamma_{N-1}^* & \gamma_{N-1} & \gamma_{N-2} & \cdots & & \gamma_0 \end{bmatrix} \tag{8.26}$$

where the superscript i indicates that the matrix accounts for interference. Somewhat similarly, we can also embed $\mathbf{A}^H\mathbf{A}$ in a $(2N-1)\times(2N-1)$ square circulant matrix:

$$\mathbf{C}^s = \begin{bmatrix} r_0 & r_1^* & \cdots & r_{N-1}^* & r_{N-1} & r_{N-2} & \cdots & r_1 \\ r_1 & r_0 & & \vdots & r_{N-1}^* & r_{N-1} & & \vdots \\ \vdots & & \ddots & r_1^* & \vdots & & & r_{N-1} \\ r_{N-1} & \cdots & r_1 & r_0 & r_1^* & \cdots & & r_{N-1}^* \\ r_{N-1}^* & r_{N-1} & \cdots & r_1 & r_0 & r_1^* & \cdots & r_{N-2}^* \\ \vdots & & & \vdots & & & & \vdots \\ r_1^* & \cdots & r_{N-1}^* & r_{N-1} & r_{N-2} & \cdots & & r_0 \end{bmatrix} \tag{8.27}$$

where the superscript s indicates that the matrix accounts for signal and signal-dependent clutter.

In addition, we let

$$\tilde{\mathbf{w}} = \begin{bmatrix} \mathbf{w} \\ \mathbf{0} \end{bmatrix}_{(2N-1)\times 1} \tag{8.28}$$

and

$$\tilde{\mathbf{x}} = \begin{bmatrix} \mathbf{x} \\ \mathbf{0} \end{bmatrix}_{(2N-1)\times 1} \tag{8.29}$$

denote, respectively, the receive filter and probing sequence vectors padded with $N-1$ zeros. Using this notation we can write the objective function (8.24) as

$$\text{MSE}(\hat{\alpha}_0) + \beta = \frac{\tilde{\mathbf{w}}^H(\beta\mathbf{C}^s + \mathbf{C}^i)\tilde{\mathbf{w}}}{|\tilde{\mathbf{w}}^H\tilde{\mathbf{x}}|^2}. \tag{8.30}$$

Let $\mathbf{F}$ denote the unitary DFT (discrete Fourier transform) matrix with elements given by

$$F_{kp} = \frac{1}{\sqrt{2N-1}} e^{j\frac{2\pi}{2N-1}(k-1)(p-1)}, \quad k, p = 1, \ldots, 2N-1. \tag{8.31}$$

Finally, define the (normalized) DFT of the zero-padded filter as

$$\mathbf{h} = \begin{bmatrix} h_1 & \cdots & h_{2N-1} \end{bmatrix}^T = \mathbf{F}^H\tilde{\mathbf{w}}, \tag{8.32}$$

that of the zero-padded probing sequence as

$$\boldsymbol{\xi} = \begin{bmatrix} \xi_1 & \cdots & \xi_{2N-1} \end{bmatrix}^T = \mathbf{F}^H\tilde{\mathbf{x}} \tag{8.33}$$

and the (normalized) power spectrum of the interference term $\boldsymbol{\epsilon}$ in (8.5) at the frequencies in $\mathbf{F}$ (see, e.g., [Stoica & Moses 2005] for the "correlogram" formula) as

$$\Phi_p = \frac{1}{2N-1} \sum_{k=-N+1}^{N-1} \gamma_k e^{-j\frac{2\pi}{2N-1}k(p-1)}, \quad p = 1, \ldots, 2N-1. \tag{8.34}$$

Note that $\{\Phi_p\}_{p=1}^{2N-1}$ is the (real-valued and nonnegative) DFT of the first column of $\mathbf{C}^i$. Indeed:

$$\begin{aligned}
&\sum_{k=0}^{N-1} \gamma_k e^{-j\frac{2\pi}{2N-1}kp} + \sum_{k=N}^{2N-2} \gamma^*_{(2N-1)-k} e^{-j\frac{2\pi}{2N-1}kp} \\
&= \sum_{k=0}^{N-1} \gamma_k e^{-j\frac{2\pi}{2N-1}kp} + \sum_{k=N}^{2N-2} \gamma_{k-(2N-1)} e^{-j\frac{2\pi}{2N-1}(k-2N+1)p} \\
&= \sum_{k=0}^{N-1} \gamma_k e^{-j\frac{2\pi}{2N-1}kp} + \sum_{k=-N+1}^{-1} \gamma_k e^{-j\frac{2\pi}{2N-1}kp}
\end{aligned} \tag{8.35}$$

which equals (8.34) up to the multiplicative constant $1/(2N-1)$. Similarly, the (normalized) power spectrum of the probing sequence $\mathbf{x}$, given by $\{|\xi_p|^2\}_{p=1}^{2N-1}$ (see, e.g., [Stoica & Moses 2005] for the "periodogram" formula), is the DFT of the first column of $\mathbf{C}^s$ up to the multiplicative constant $1/(2N-1)$.

Making use of the above notation along with a standard property of circulant matrices (see [Golub & Van Loan 1984] and also [Stoica et al. 2009*b*]) allows us to rewrite (8.30) in a frequency-domain form:

$$\begin{aligned}
&\frac{\tilde{\mathbf{w}}^H(\beta\mathbf{C}^s + \mathbf{C}^i)\tilde{\mathbf{w}}}{|\tilde{\mathbf{w}}^H\tilde{\mathbf{x}}|^2} \\
&= \frac{2N-1}{|\tilde{\mathbf{w}}^H\mathbf{F}^H\mathbf{F}\tilde{\mathbf{x}}|^2}\tilde{\mathbf{w}}^H\left(\beta\mathbf{F}\begin{bmatrix} |\xi_1|^2 & & \mathbf{0} \\ & \ddots & \\ \mathbf{0} & & |\xi_{2N-1}|^2 \end{bmatrix}\mathbf{F}^H + \mathbf{F}\begin{bmatrix} \Phi_1 & & \mathbf{0} \\ & \ddots & \\ \mathbf{0} & & \Phi_{2N-1} \end{bmatrix}\mathbf{F}^H\right)\tilde{\mathbf{w}} \\
&= \frac{2N-1}{\left|\sum_{p=1}^{2N-1} h_p^*\xi_p\right|^2}\sum_{p=1}^{2N-1}|h_p|^2(\beta|\xi_p|^2 + \Phi_p).
\end{aligned} \tag{8.36}$$

The energy constraint can also be written in the frequency domain (cf. the Parseval equality):

$$\|\mathbf{x}\|^2 = \|\mathbf{F}\tilde{\mathbf{x}}\|^2 = \|\boldsymbol{\xi}\|^2 = N. \tag{8.37}$$

We can therefore think of formulating the design problem as follows:

$$\begin{aligned}
&\min_{\{h_p\},\{\xi_p\}} \frac{1}{\left|\sum_{p=1}^{2N-1} h_p^*\xi_p\right|^2}\sum_{p=1}^{2N-1}|h_p|^2(\beta|\xi_p|^2 + \Phi_p) \\
&\text{s.t.} \sum_{p=1}^{2N-1}|\xi_p|^2 = N.
\end{aligned} \tag{8.38}$$

Note that this is a relaxed version of the original design problem. Indeed, while for given $\mathbf{w}$ and $\mathbf{x}$ we can uniquely determine $\{h_p\}$ and $\{\xi_p\}$ from (8.32) and (8.33), the converse is not necessarily true. In other words, there might exist no $\mathbf{w}$ and $\mathbf{x}$ that can synthesize exactly a set of given $\{h_p\}$ and $\{\xi_p\}$. Some form of approximation will therefore be necessary when converting the result of the optimization problem in (8.38) to the original design variables $\mathbf{w}$ and $\mathbf{x}$, as detailed later in this section.

Continuing with (8.38), we note that the minimization of the objective function of this problem w.r.t. $\{h_p\}$ for fixed $\{\xi_p\}$ can be simply done as follows. It is a consequence of the Cauchy–Schwartz inequality that

$$\frac{\sum_{p=1}^{2N-1}|h_p|^2(\beta|\xi_p|^2+\Phi_p)}{\left|\sum_{p=1}^{2N-1}h_p^*\xi_p\right|^2}=\frac{\sum_{p=1}^{2N-1}|h_p|^2(\beta|\xi_p|^2+\Phi_p)}{\left|\sum_{p=1}^{2N-1}h_p^*(\beta|\xi_p|^2+\Phi_p)^{1/2}\frac{\xi_p}{(\beta|\xi_p|^2+\Phi_p)^{1/2}}\right|^2}$$
$$\geq\frac{1}{\sum_{p=1}^{2N-1}\frac{|\xi_p|^2}{\beta|\xi_p|^2+\Phi_p}} \tag{8.39}$$

where the lower bound is attained for

$$h_p=\frac{\xi_p}{\beta|\xi_p|^2+\Phi_p}. \tag{8.40}$$

Remark: It follows from (8.40) that at any frequency where the clutter dominates the interference, i.e., at the values of p where Φ_p in (8.40) is negligible w.r.t. $\beta|\xi_p|^2$, we have

$$|h_p|=\frac{1}{\beta|\xi_p|}, \tag{8.41}$$

which provides a theoretical explanation for an empirical observation that appears in the literature. □

Making use of (8.39) and (8.40) reduces (8.38) to the following maximization problem w.r.t. $\{z_p=|\xi_p|^2\}$:

$$\max_{\{z_p\}}\sum_{p=1}^{2N-1}\frac{z_p}{\beta z_p+\Phi_p} \tag{8.42}$$

$$\text{s.t.}\ \sum_{p=1}^{2N-1}z_p=N, \tag{8.43}$$

$$z_p\geq 0.\quad (z_p=|\xi_p|^2) \tag{8.44}$$

A straightforward calculation, based once again on the Cauchy–Schwartz inequality and the constraint on $\{z_p\}$ in (8.43), shows that

$$\sum_{p=1}^{2N-1}\frac{z_p}{\beta z_p+\Phi_p}=\frac{2N-1}{\beta}-\frac{1}{\beta}\sum_{p=1}^{2N-1}\frac{\Phi_p}{\beta z_p+\Phi_p} \tag{8.45}$$
$$=\frac{2N-1}{\beta}-\frac{1}{\beta}\left(\sum_{p=1}^{2N-1}\frac{\Phi_p}{\beta z_p+\Phi_p}\right)\frac{\sum_{p=1}^{2N-1}(\beta z_p+\Phi_p)}{\beta N+\sum_{p=1}^{2N-1}\Phi_p}$$
$$\leq\frac{2N-1}{\beta}-\frac{1}{\beta}\frac{\left(\sum_{p=1}^{2N-1}\frac{\Phi_p^{1/2}}{(\beta z_p+\Phi_p)^{1/2}}(\beta z_p+\Phi_p)^{1/2}\right)^2}{\beta N+\sum_{p=1}^{2N-1}\Phi_p}=\frac{2N-1}{\beta}-\frac{1}{\beta}\frac{\left(\sum_{p=1}^{2N-1}\Phi_p^{1/2}\right)^2}{\beta N+\sum_{p=1}^{2N-1}\Phi_p},$$

where the upper bound on the right-hand side is achieved for

$$z_p=\frac{\rho\Phi_p^{1/2}-\Phi_p}{\beta},\quad \rho=\frac{\beta N+\sum_{p=1}^{2N-1}\Phi_p}{\sum_{p=1}^{2N-1}\Phi_p^{1/2}}. \tag{8.46}$$

Table 8.2. The bisection method for (8.49)

Step 0: Let $f(\lambda) = \frac{1}{\beta}\sum_{p=1}^{2N-1} \max\{\lambda\rho\Phi_p^{1/2} - \Phi_p, 0\}$. Let $f_{\text{obj}} = N$, $\lambda_{\text{left}} = 0$ and $\lambda_{\text{right}} = 1$.
Step 1: Set $\lambda = (\lambda_{\text{left}} + \lambda_{\text{right}})/2$ and compute $f(\lambda)$.
Step 2: If $f(\lambda) < f_{\text{obj}}$, then $\lambda_{\text{left}} \leftarrow \lambda$; else $\lambda_{\text{right}} \leftarrow \lambda$.
Iteration: Repeat steps 1 and 2 until $|f(\lambda) - f_{\text{obj}}| \leq \epsilon$ where ϵ is a predefined threshold (e.g., 10^{-2}).

Note that the above $\{z_p\}$ satisfy, as they should, the energy constraint:

$$\sum_{p=1}^{2N-1} z_p = \frac{1}{\beta}\left(\rho \sum_{p=1}^{2N-1} \Phi_p^{1/2} - \sum_{p=1}^{2N-1} \Phi_p\right) = N. \tag{8.47}$$

If the above $\{z_p\}$ also satisfy the constraint $\{z_p \geq 0\}$ then (8.46) is the sought solution. Otherwise the solution can be shown, by using the Lagrange approach (see Appendix 8B, and also [Kay 2007] for a continuous version of the problem (8.42), with the sum replaced by an integral), to have the form

$$|\xi_p|^2 = z_p = \frac{1}{\beta}\max\{\lambda\rho\Phi_p^{1/2} - \Phi_p, 0\} \quad p = 1, \ldots, 2N-1, \tag{8.48}$$

where λ is determined by the condition that $\{z_p\}$ satisfy the energy constraint

$$\frac{1}{\beta}\sum_{p=1}^{2N-1} \max\{\lambda\rho\Phi_p^{1/2} - \Phi_p, 0\} = N. \tag{8.49}$$

Because some of the $\{z_p\}$ in (8.46), which satisfy (8.47), can be negative and as the left-hand side of (8.49) is an increasing function of λ, it follows that the solution of (8.49) cannot be larger than one, that is,

$$\lambda \in [0, 1]. \tag{8.50}$$

Using this fact one can easily solve (8.49), for example by means of the bisection method as outlined in Table 8.2.

Remark: The expression in (8.48) for the optimum $\{|\xi_p|\}$ has the flavor of the water-filling results commonly encountered in the communication literature (as also noted in [Kay 2007] for a related problem). More specifically, one can see from (8.48) that the optimum probing sequence does not contain any power at those frequencies where the interference is strong, i.e., it satisfies $\Phi_p^{1/2} \geq \lambda\rho$. □

It follows from (8.30), (8.36) and (8.39) that the following *lower bound* on $\text{MSE}(\hat{\alpha}_0)$ holds:

$$B_{\text{MSE}} = \frac{2N-1}{\sum_{p=1}^{2N-1} \frac{\hat{z}_p}{\beta\hat{z}_p + \Phi_p}} - \beta, \tag{8.51}$$

where $\{\hat{z}_p\}_{p=1}^{2N-1}$ denotes the optimum power spectrum obtained from (8.48) and (8.49). This lower bound can be attained only if $\{\hat{z}_p\}_{p=1}^{2N-1}$ can be exactly synthesized by selecting $\{x_k\}_{k=1}^{N}$; see below for details on this synthesis problem.

Once λ is found, $\{|\xi_p|\}$ is given by the closed-form expression in (8.48). Note that the phases of $\{\xi_p\}$, say $\{\psi_p\}$, are not variables of the design problem in (8.42) and therefore they can be chosen freely. The original design variable $\mathbf{x}$ can be obtained by solving (8.33) in a least-squares sense:

$$\min_{\mathbf{x},\{\psi_p\}} \|\boldsymbol{\xi} - \mathbf{F}^H\tilde{\mathbf{x}}\|^2. \tag{8.52}$$

Solving (8.52) w.r.t. $\mathbf{x}$ and $\{\psi_p\}$ is a task that in general requires an iterative solver. We use a cyclic algorithm for this task (see below for details) which is reminiscent of the Sussman–Gerchberg–Saxton procedure [Sussman 1962][Gerchberg & Saxton 1972] and of the CAN algorithm (see Chapter 2).

Consider first the minimization of the objective function in (8.52) w.r.t. $\mathbf{x}$, for fixed $\{\psi_p\}$ (i.e., $\boldsymbol{\xi}$ is fixed). Under the energy constraint, this minimization can be done in closed form

$$\mathbf{x} = \sqrt{N}\frac{\boldsymbol{\nu}}{\|\boldsymbol{\nu}\|}, \quad \boldsymbol{\nu} = \tilde{\mathbf{F}}\boldsymbol{\xi}, \tag{8.53}$$

where $\tilde{\mathbf{F}}$ is the $N \times (2N-1)$ matrix consisting of the first N rows of $\mathbf{F}$. To prove (8.53) note that (recall that $\|\boldsymbol{\xi}\|^2 = \|\mathbf{x}\|^2 = \|\tilde{\mathbf{x}}\|^2 = N$)

$$\|\boldsymbol{\xi} - \mathbf{F}^H\tilde{\mathbf{x}}\|^2 = 2N - 2\,\mathrm{Re}\left(\boldsymbol{\xi}^H\mathbf{F}^H\tilde{\mathbf{x}}\right) \tag{8.54}$$

where

$$\begin{aligned}\mathrm{Re}\{\boldsymbol{\xi}^H\mathbf{F}^H\tilde{\mathbf{x}}\} &\le |\boldsymbol{\xi}^H\mathbf{F}^H\tilde{\mathbf{x}}| = |\boldsymbol{\xi}^H\tilde{\mathbf{F}}^H\mathbf{x}| \\ &\le \|\mathbf{x}\|\,\|\tilde{\mathbf{F}}\boldsymbol{\xi}\| = \sqrt{N}\|\tilde{\mathbf{F}}\boldsymbol{\xi}\|.\end{aligned} \tag{8.55}$$

The upper bound in (8.55), and hence the minimum value of (8.54), is achieved for the $\mathbf{x}$ in (8.53), which also satisfies the energy constraint and is therefore the sought solution.

If, instead of the energy constraint, the unit-modulus constraint is imposed on $\mathbf{x}$, the minimizing $\mathbf{x}$ will be given by

$$x_k = e^{j\arg(\nu_k)}, \quad k = 1, \ldots, N. \tag{8.56}$$

This can be seen from the following derivation:

$$\mathrm{Re}\left(\boldsymbol{\xi}^H\mathbf{F}^H\tilde{\mathbf{x}}\right) = \sum_{k=1}^{N} |\nu_k| \cos\left[\arg(x_k) - \arg(\nu_k)\right] \le \sum_{k=1}^{N} |\nu_k|, \tag{8.57}$$

where equality is achieved for (8.56).

If the PAR constraint is imposed, note that $\|\boldsymbol{\xi} - \mathbf{F}^H\tilde{\mathbf{x}}\|^2 = \|\mathbf{F}\boldsymbol{\xi} - \tilde{\mathbf{x}}\|^2 = \|\boldsymbol{\nu} - \mathbf{x}\|^2$, and thus (8.52) can be cast as

$$\begin{aligned}&\min_{\mathbf{x}} \|\mathbf{x} - \boldsymbol{\nu}\|^2 \\ &\text{s.t. } \mathrm{PAR}(\mathbf{x}) \le \mu,\end{aligned} \tag{8.58}$$

which can be efficiently solved via a global finite-step "nearest-vector" algorithm; see the discussions following (4.18).

Table 8.3. The CREW(fre) algorithm with unit-modulus or PAR constraint

Step 0: Solve (8.49) for λ (see Table 8.2). Determine $\{|\xi_p|\}$ from (8.48). Use an available sequence or a randomly generated one to initialize $\mathbf{x}$.
Step 1: Compute

$$\psi_p = \arg(\tilde{\mathbf{f}}_p^H \mathbf{x}), \quad p = 1, \ldots, 2N-1. \tag{8.61}$$

Step 2: Compute

$$\boldsymbol{\nu} = \tilde{\mathbf{F}}\boldsymbol{\xi}, \tag{8.62}$$

where $\{\xi_p = |\xi_p|e^{j\psi_p}\}$ with the $\{|\xi_p|\}$ provided by step 0. If only the energy constraint is imposed, use (8.53) to update $\mathbf{x}$; if the unit-modulus constraint is imposed, use (8.56); and if the PAR constraint is enforced, use (8.58).
Iteration: Repeat steps 1 and 2 until convergence.
Step 3: Compute $\mathbf{w}$ using (8.14) with the vector $\mathbf{x}$ provided by the above iteration.

For a given $\mathbf{x}$, the minimization of (8.52) w.r.t. to $\{\psi_p\}$ also has a simple closed-form solution. Let $\mathbf{f}_p^H$ denote the pth row of $\mathbf{F}^H$ and rewrite (8.52) as

$$\min_{\mathbf{x}, \{\psi_p\}} \sum_{p=1}^{2N-1} \left| |\xi_p| e^{j\psi_p} - \mathbf{f}_p^H \tilde{\mathbf{x}} \right|^2. \tag{8.59}$$

The minimization of the above function w.r.t. $\{\psi_p\}$, for any fixed $\mathbf{x}$, can be easily achieved in closed form (the proof is similar to (8.57)):

$$\psi_p = \arg\left(\mathbf{f}_p^H \tilde{\mathbf{x}}\right), \quad p = 1, \ldots, 2N-1. \tag{8.60}$$

Note that the calculation of $\boldsymbol{\nu}$ in (8.53) and of $\{\mathbf{f}_p^H \tilde{\mathbf{x}}\}$ in (8.60) can be done efficiently via an FFT (fast Fourier transform). After obtaining $\mathbf{x}$, the receive filter $\mathbf{w}$ is given by (8.14). The so-obtained frequency domain-based algorithm, which is called CREW(fre) for short, is summarized in Table 8.3. Note that CREW(fre) can handle both unit-modulus and PAR constraints and is computationally efficient due to the leveraged FFT operations.

8.4 Specialization for matched filtering

In this section we specialize the CREW(fre) algorithm of Section 8.3 to the matched-filter case in which $\mathbf{w} = \mathbf{x}$ and thus $\mathbf{h} = \boldsymbol{\xi}$. It follows from (8.30) and (8.36) that

$$\text{MSE}(\hat{\alpha}_0) = \frac{2N-1}{N^2} \sum_{p=1}^{2N-1} |\xi_p|^2 (\beta |\xi_p|^2 + \Phi_p) - \beta. \tag{8.63}$$

Therefore the problem of minimizing the MSE metric becomes (using the variables $z_p = |\xi_p|^2$, as in (8.42)):

$$\min_{\{z_p\}} \sum_{p=1}^{2N-1} (\beta z_p^2 + \Phi_p z_p) \tag{8.64}$$
$$\text{s.t.} \sum_{p=1}^{2N-1} z_p = N \quad (z_p \geq 0).$$

The above problem is convex. Its Lagrangian has the following decoupled form:

$$\sum_{p=1}^{2N-1} \left(\beta z_p^2 + \Phi_p z_p - \lambda z_p + \frac{N}{2N-1}\lambda \right) \tag{8.65}$$

and is clearly minimized over $\{z_p \geq 0\}$ at

$$z_p = \frac{1}{2\beta} \max\{\lambda - \Phi_p, 0\}. \tag{8.66}$$

The multiplier λ in (8.66) is the solution of the constraint equation

$$\frac{1}{2\beta} \sum_{p=1}^{2N-1} \max\{\lambda - \Phi_p, 0\} = N. \tag{8.67}$$

Let λ_0 satisfy the equation

$$\frac{1}{2\beta} \sum_{p=1}^{2N-1} (\lambda_0 - \Phi_p) = N \tag{8.68}$$

which yields

$$\lambda_0 = \frac{2N\beta + \sum_{p=1}^{2N-1} \Phi_p}{2N-1}. \tag{8.69}$$

If $\lambda_0 \geq \Phi_p$ (for $p = 1, \ldots, 2N-1$) then the solution of (8.67) is $\lambda = \lambda_0$. In all other cases λ must satisfy $\lambda < \lambda_0$ (because the left-hand side of (8.67) is an increasing function of λ that is equal to 0 at $\lambda = 0$ and is larger than or equal to N at $\lambda = \lambda_0$). It follows that the solution to (8.67) must lie in the interval

$$\lambda \in (0, \lambda_0] \tag{8.70}$$

and it can therefore be conveniently found by the bisection method (see (8.49)).

Remark: Note again the water-filling character of (8.66): the optimum probing sequence does not have any power at the frequencies where the interference is strong enough for $\lambda - \Phi_p$ to be negative (see the remark following (8.50) for a similar comment on the CREW(fre) solution). □

Once λ is found, the $\{|\xi_p| = z_p^{1/2}\}$ are determined from (8.66). The problem that remains is to find $\mathbf{x}$ from the solution of (8.52), which can be done as explained in the previous section. The so-obtained matched filter-based algorithm, named CREW(mat), is summarized in Table 8.4. Note that CREW(mat) can be interpreted as an extension

Table 8.4. The CREW(mat) algorithm with unit-modulus or PAR constraint

Step 0: Calculate λ from (8.67) using a bisection method. Then determine $\{|\xi_p|\}$ from (8.66). Use an available sequence or a randomly generated one to initialize $\mathbf{x}$.
Step 1, Step 2, Iteration: Same as those in Table 8.3.

of the CAN algorithm (see Chapter 2) to the case in which some $\{\Phi_p\}$ are different from zero (indeed, CREW(mat) reduces to CAN in the no-interference case of $\{\Phi_p \equiv 0\}$).

8.5 Numerical examples

The covariance matrix of the interference term, viz. $\mathbf{\Gamma}$ (see (8.6)), is constructed as follows:

$$\mathbf{\Gamma} = \sigma_{\text{J}}^2 \mathbf{\Gamma}_{\text{J}} + \sigma^2 \mathbf{I} \tag{8.71}$$

where σ_{J}^2 and σ^2 are the jamming and noise powers, respectively, and

$$\mathbf{\Gamma}_{\text{J}} = \begin{bmatrix} q_0 & q_1^* & \cdots & q_{N-1}^* \\ q_1 & q_0 & & q_{N-2}^* \\ \vdots & & \ddots & \vdots \\ q_{N-1} & q_{N-2} & \cdots & q_0 \end{bmatrix} \quad (q_0 = 1) \tag{8.72}$$

is the normalized jamming covariance matrix. The matrix $\mathbf{\Gamma}_{\text{J}}$ can be determined directly by the IDFT (inverse DFT) of a desired jamming power spectrum. More concretely, let $\{\eta_p\}_{p=1}^{2N-1}$ denote a given jamming power spectrum at frequencies $\{(p-1)/(2N-1)\}_{p=1}^{2N-1}$. Similarly to (8.35), the IDFT of $\{\eta_p\}_{p=1}^{2N-1}$ leads to

$$\begin{bmatrix} q_0 \; q_1 \; \cdots \; q_{N-1} \; q_{N-1}^* \; \cdots \; q_1^* \end{bmatrix}^T, \tag{8.73}$$

which needs to be normalized so that $q_0 = 1$.

Throughout the following examples we choose $\sigma_{\text{J}}^2 = 100$, $\sigma^2 = 0.1$ and $\beta = 1$ (see (8.7)). The proposed algorithms will be compared with the CAN algorithm (see Chapters 2 and 4). The CAN sequence is designed to suppress the correlation sidelobes. In the case of a matched filter (i.e., $\mathbf{w} = \mathbf{x}$), low correlation sidelobes lead to small MSE($\hat{\alpha}_0$) (see (8.11)) if the clutter dominates the interference (i.e., $\mathbf{\Gamma}$ is not considered in the design). The CAN algorithm can handle both the unit-modulus and PAR constraints. As already mentioned, the proposed CREW(mat) algorithm extends CAN by taking $\mathbf{\Gamma}$ into account (a fact that should lead to a smaller MSE than that for CAN).

Note that, while the CAN sequence is designed under the implicit assumption of using a matched filter, it can also be used with the IV filter as in (8.14). The combination of a CAN sequence with the corresponding optimum IV filter is referred to as CAN-IV.

In what follows, the jamming type and the sequence length will be varied to examine the effect on algorithm performance in terms of MSE($\hat{\alpha}_0$). The considered algorithms are listed in Table 8.5 for easy reference. The Golomb sequence (see (1.21)) will be used

Table 8.5. Algorithms to be compared

	transmit sequence $\mathbf{x}$	receive filter $\mathbf{w}$	power constraints
CAN	the CAN sequence	the matched filter	unit-modulus/PAR
CAN-IV	the CAN sequence	the IV filter	unit-modulus/PAR
CREW(gra)	the $\mathbf{x}$ from Table 8.1	the IV filter	unit-modulus
CREW(fre)	the $\mathbf{x}$ from Table 8.3	the IV filter	unit-modulus/PAR
CREW(mat)	the $\mathbf{x}$ from Table 8.4	the matched filter	unit-modulus/PAR

for initialization of all algorithms. This sequence appears to be a good choice for initialization, although we did observe occasionally that a randomly generated sequence can lead to a lower MSE than the Golomb sequence. The MSE lower bound (see (8.51)) will be used as the benchmark. Note that CAN, CAN-IV and the lower bound will be shown in every considered case; however, for CREW(gra), CREW(fre) and CREW(mat) we only show the one that yields the lowest MSE to avoid over-crowding the figures. Also recall that CREW(gra) can only deal with the unit-modulus constraint and also that it cannot handle large values of N due to its relatively expensive computation (consequently when we show the MSE of CREW(gra) we limit N to 300).

8.5.1 Spot jamming

Assume a spot jamming located at frequency f_0 whose power spectrum is given by

$$\eta_p = \begin{cases} 1, & p = \lfloor (2N-1) f_0 \rfloor, \\ 0, & \text{elsewhere}, \end{cases} \qquad p = 1, \ldots, 2N-1. \tag{8.74}$$

We choose $f_0 = 0.2$ Hz.

Figure 8.1(a) shows the MSEs of CAN, CAN-IV and CREW(gra) for $N = 25, 50, 100, 200, 300$ under the unit-modulus constraint. It can be observed that CAN-IV provides a significantly smaller MSE than CAN and that CREW(gra) further improves over CAN-IV. Figure 8.1(b) shows the MSEs of CAN, CAN-IV and CREW(mat) for $N = 25, 50, 100, 200, 300, 500, 1000$ under the constraint PAR ≤ 2. It can be observed that CREW(mat) gives the smallest MSE for all N values.

Figure 8.2 shows the power spectrum of the length-100 CREW(gra) sequence associated with Figure 8.1(a) and the frequency response of the corresponding receive filter. Observe the spectral notch at the jamming frequency $f_0 = 0.2$ Hz for both the probing sequence and the receive filter. Except at frequency f_0, the spectrum of the probing sequence and the frequency response of the receive filter are approximately the reciprocal of each other, as explained in the Remark following (8.40). Figure 8.3 shows the cross-correlation (see (3.1) for the definition) of this CREW(gra) sequence and the corresponding receive filter. We observe low correlation sidelobes (similar to the auto-correlation sidelobes shown in Figure 2.4), which indicate good clutter suppression.

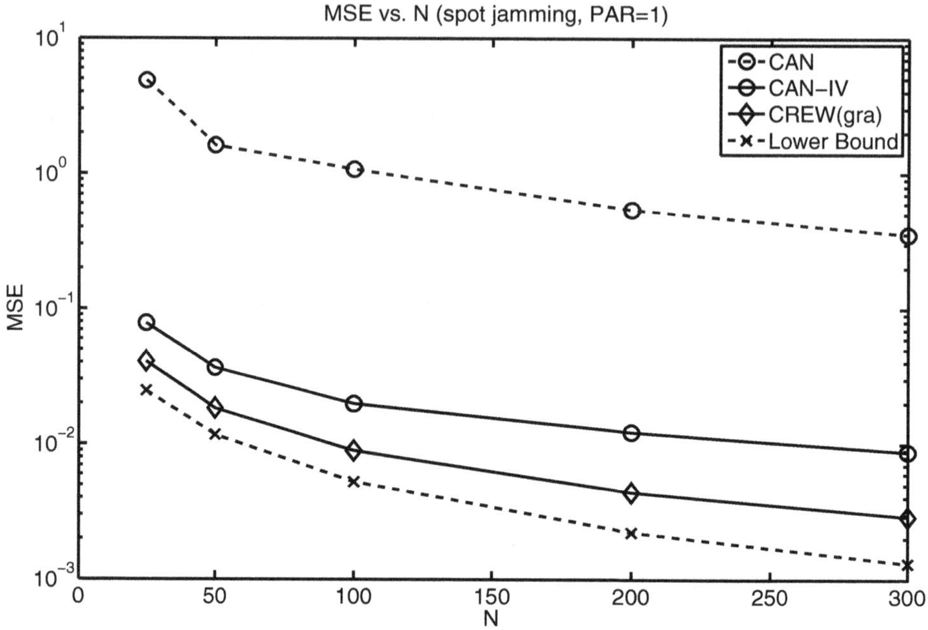

(a)

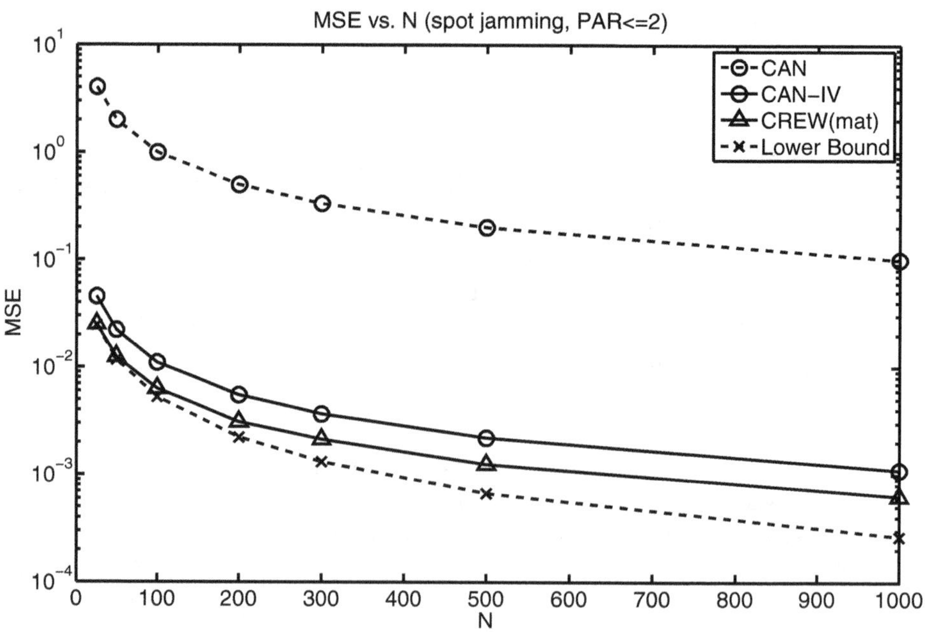

(b)

Figure 8.1 MSE comparison in the case of spot jamming. (a) MSE of CAN, CAN-IV and CREW(gra) under the unit-modulus constraint. (b) MSE of CAN, CAN-IV and CREW(mat) under the constraint PAR ≤ 2.

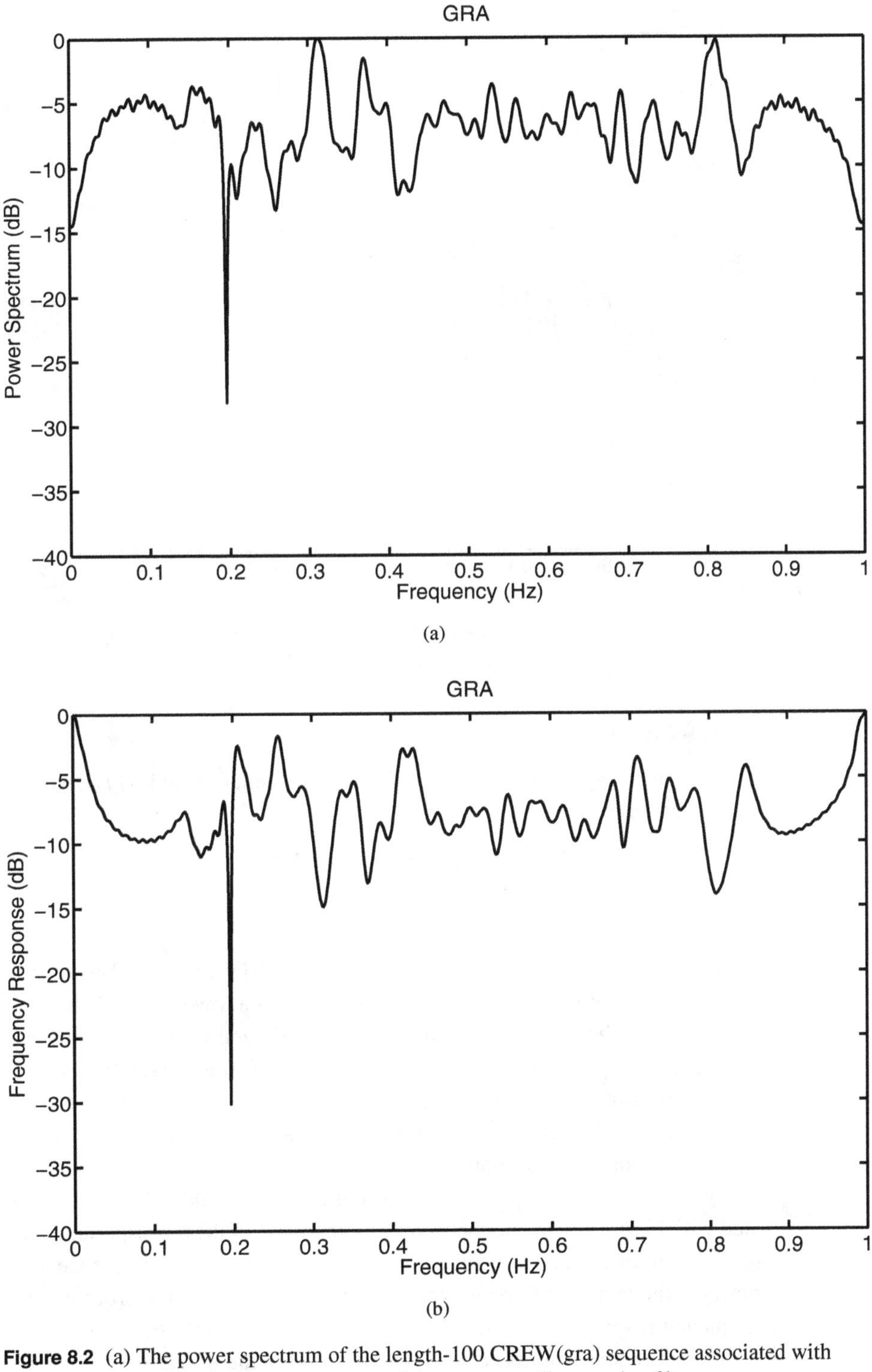

Figure 8.2 (a) The power spectrum of the length-100 CREW(gra) sequence associated with Figure 8.1(a). (b) The frequency response of the corresponding receive filter.

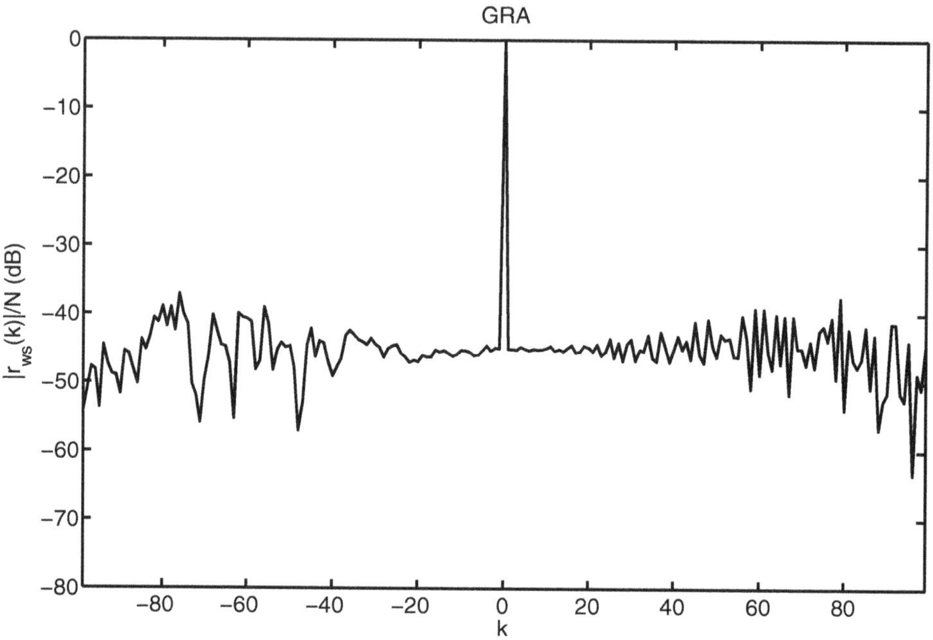

Figure 8.3 The cross-correlation between the length-100 CREW(gra) sequence and the corresponding receive filter associated with Figure 8.1(a).

8.5.2 Barrage jamming

Consider a barrage jamming, located in the frequency band $[f_1, f_2]$, whose power spectrum is given by

$$\eta_p = \begin{cases} 1, & \lfloor (2N-1)f_1 \rfloor \leq p \leq \lfloor (2N-1)f_2 \rfloor, \\ 0, & \text{elsewhere,} \end{cases} \qquad p = 1, \ldots, 2N-1. \tag{8.75}$$

We choose $f_1 = 0.2$ Hz and $f_2 = 0.3$ Hz.

Figure 8.4(a) shows the MSEs of CAN, CAN-IV and CREW(gra) under the unit-modulus constraint. Once again CREW(gra) gives a lower MSE than CAN or CAN-IV. Figure 8.4(b) shows the MSEs of CAN, CAN-IV and CREW(fre) under the constraint PAR ≤ 2. Except for N smaller than 100, CREW(fre) gives the lowest MSE among the three algorithms. Moreover, the performance of CREW(fre) tends to follow the MSE lower bound as N increases, in contrast with CAN-IV's performance that deviates more and more from the lower bound as N increases.

Figure 8.5 shows the power spectrum of the length-300 CREW(fre) sequence associated with Figure 8.4(b) and the frequency response of the corresponding receive filter. In this case, unlike what were shown in Figure 8.2, the probing sequence emits significant power in the jamming frequency band, and the spectrum of the probing sequence and the frequency response of the receive filter are approximately the reciprocal of each other for the entire spectrum. The spectrum of the probing sequence will have a wide notch

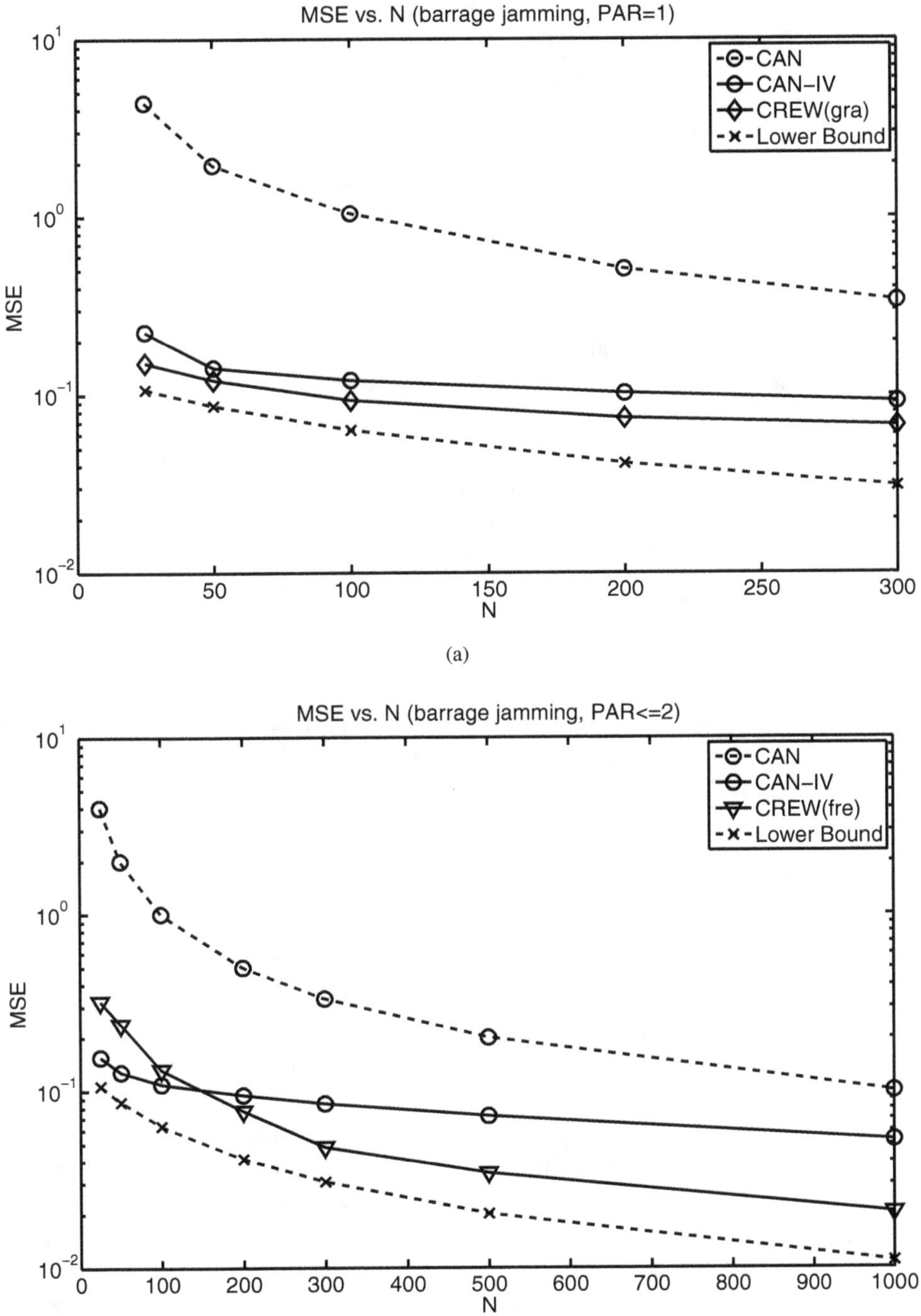

(a)

(b)

Figure 8.4 MSE comparison in the case of barrage jamming. (a) MSE of CAN, CAN-IV and CREW(gra) under the unit-modulus constraint. (b) MSE of CAN, CAN-IV and CREW(fre) under the constraint PAR ≤ 2.

in the jamming frequency band if the jamming power is sufficiently large, as shown in Figure 8.6.

8.5.3 Robust design

Finally we discuss the robustness issue of the aforementioned designs. Consider a strong spot jamming that operates at frequency f_0. In this case the optimum filter in (8.14) will have a notch at frequency f_0. However, if the information about the jamming frequency is not accurate, the receive filter in (8.14) will have a notch at the assumed frequency and so it will pass the jamming interference that in reality operates at a different frequency. To combat such a problem, we can assume a barrage jamming located in a frequency band $[f_1, f_2]$ Hz which is wide enough to include the true frequency of the spot jamming. In this way the receive filter will have a wider stop-band and the jamming interference will be annihilated. Similarly in the case of imprecise information about a barrage jamming, we can consider a frequency band that is wide enough to cover the true jamming band, for the sake of robustness.

As an example, let (8.75) be the true jamming power spectrum with $f_1 = 0.2$ and $f_2 = 0.3$ Hz. We use CREW(fre) to design the transmit sequence and the receive filter. Three scenarios are considered: the exact values of f_1 and f_2 are known, a wrong jamming band $[0.25, 0.35]$ Hz is used, and an imprecise but sufficiently wide jamming band $[0.15, 0.35]$ Hz is used (a case referred to as robust design). Figure 8.7 shows the MSEs (calculated using the true jamming spectrum) for $N = 25, 50, 100, 200, 300, 500, 1000$ under the constraint PAR ≤ 2. It can be observed that the robust design is only weakly affected by the imprecise jamming information.

8.6 Conclusions

We have discussed three algorithms, namely CREW(gra), CREW(fre) and CREW(mat), that can be used for the joint optimization of receive filters and transmit signals. The design goal is to minimize the MSE of a target's scattering coefficient estimate or, equivalently, to maximize the signal-to-clutter-plus-interference ratio. Practical unit-modulus or low-PAR constraints are taken into account in the transmit sequence design. The CREW(gra) algorithm uses a gradient-based approach to minimize the MSE that results from using the optimum IV filter. Usually it yields the best performance for unimodular sequences but it is relatively computationally expensive and cannot be used for the more general PAR constraint. The CREW(fre) algorithm operates in the frequency domain and seeks to obtain an optimum power spectrum for the transmit sequence. The said optimum power spectrum leads to an MSE lower bound. CREW(fre) can be specialized to the matched-filter case, which yields the CREW(mat) algorithm that can be considered as an extension of the previously suggested CAN approach to the case of non-white interference. Both CREW(fre) and CREW(mat) are computationally efficient and can easily handle sequence lengths $\sim 10^3$. Numerical examples have been provided to illustrate the effectiveness of the proposed receive filters and transmit sequences.

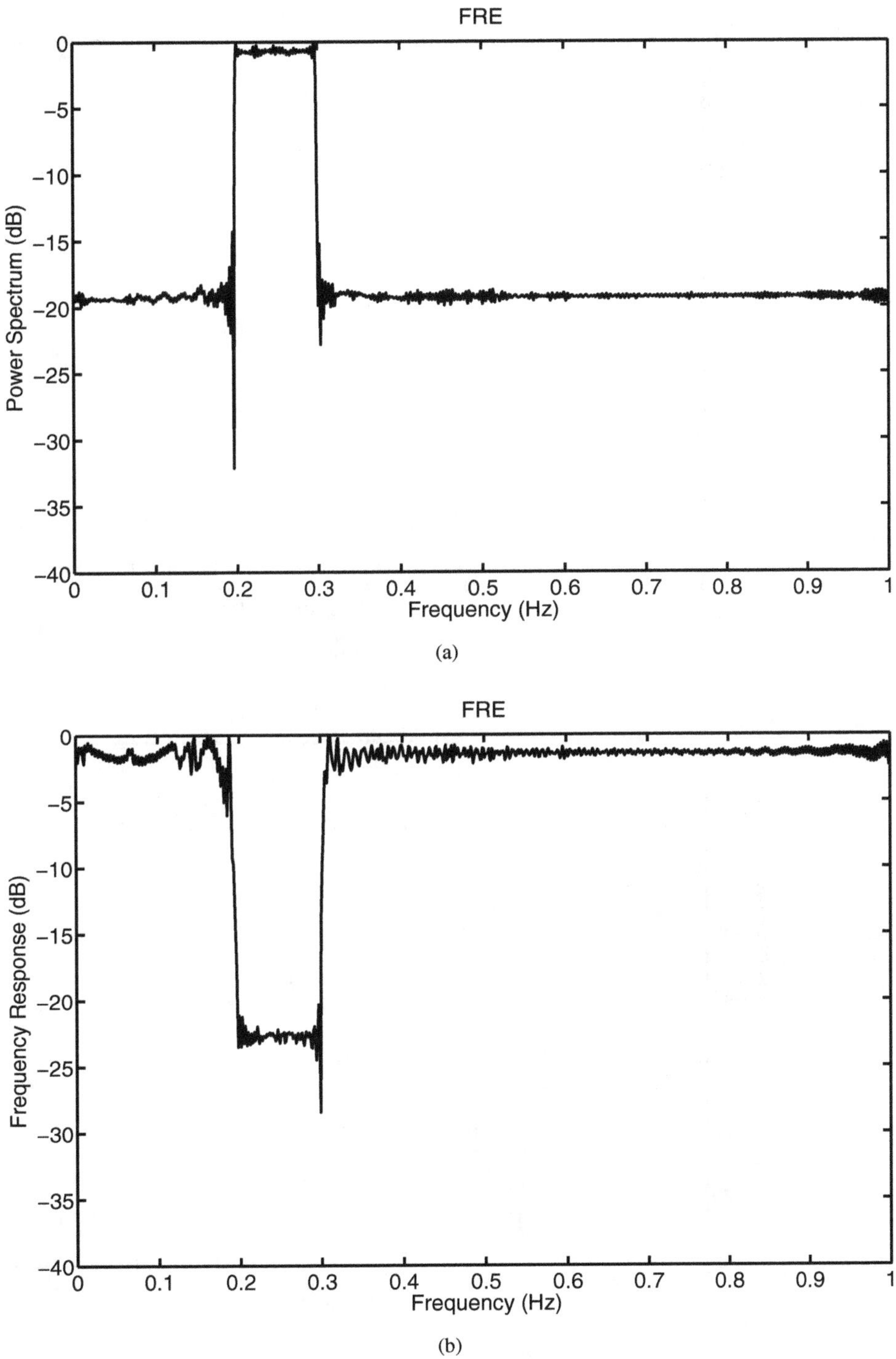

Figure 8.5 (a) The power spectrum of the length-300 CREW(fre) sequence associated with Figure 8.4(b). (b) The frequency response of the corresponding receive filter.

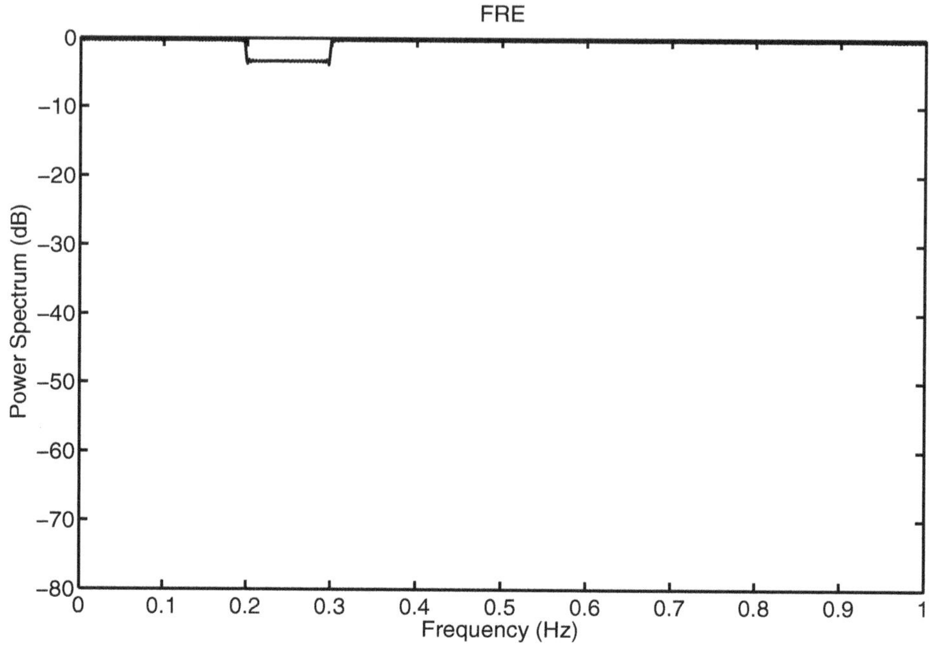

(a)

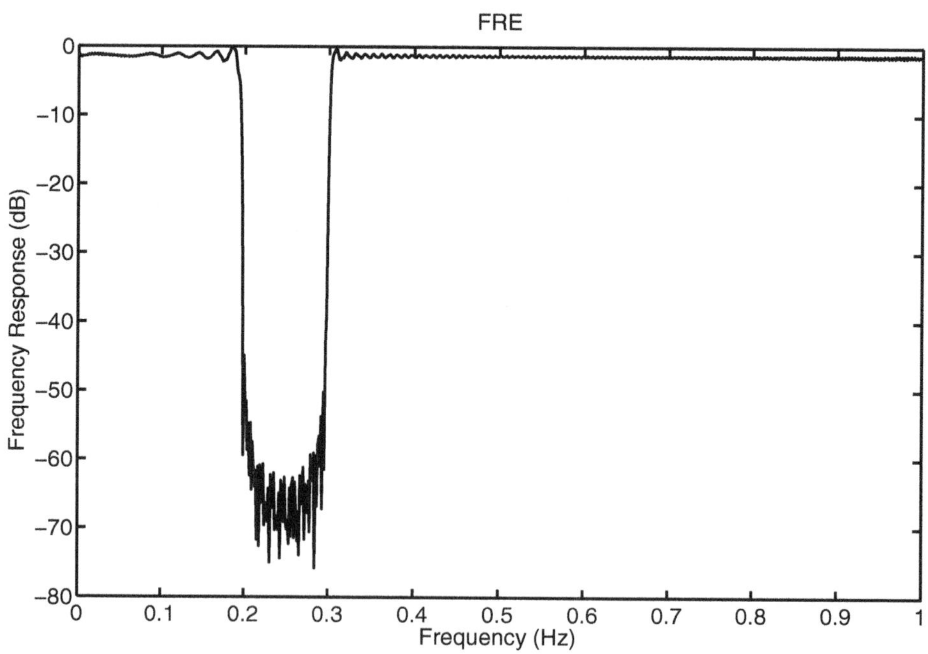

(b)

Figure 8.6 The same as Figure 8.5 except that the jamming power σ_J^2 is changed from 100 to 100000.

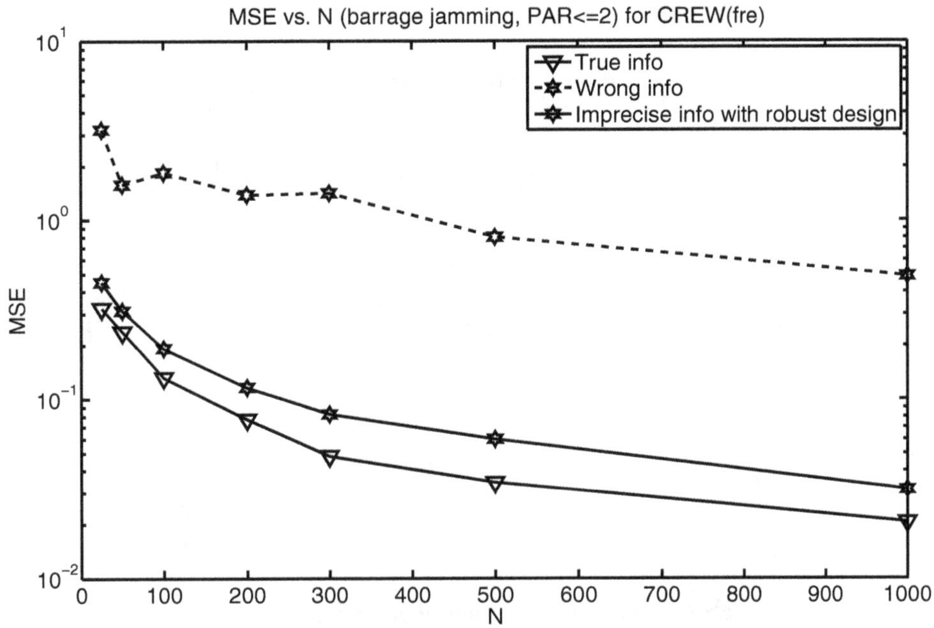

Figure 8.7 MSE comparison of CREW(fre) concerning an imprecise knowledge of the jamming frequency.

Appendix 8A Proof of Equation (8.25)

Observe that the product matrix $\mathbf{A}^H\mathbf{A}$ is not affected by a permutation of the columns in $\mathbf{A}^H$. This implies that

$$\mathbf{A}^H\mathbf{A} = \tilde{\mathbf{A}}^H\tilde{\mathbf{A}} \tag{8.76}$$

where

$$\tilde{\mathbf{A}}^H = \begin{bmatrix} x_N & x_{N-1} & \cdots & x_1 & 0 & \cdots & 0 \\ 0 & x_N & & x_2 & x_1 & & \vdots \\ \vdots & & \ddots & \vdots & & \ddots & \vdots \\ 0 & 0 & & x_N & x_{N-1} & \cdots & x_1 \end{bmatrix}_{N\times(2N-1)} . \tag{8.77}$$

Therefore the (i, j)th element of $\mathbf{A}^H\mathbf{A}$ is given by (assuming that $x_k = 0$ for $k \neq [1, N]$)

$$[\mathbf{A}^H\mathbf{A}]_{ij} = \sum_{p=1}^{2N-1} x_{N+i-p}x^*_{N+j-p} = \sum_{k=1}^{N} x_k x^*_{k-(i-j)} = r_{i-j} \tag{8.78}$$

and this concludes the proof of (8.25).

Appendix 8B Lagrange approach to solving (8.42)

The second-order derivative of the objective function in (8.42) can be shown to be non-positive:

$$\sum_{p=1}^{2N-1} \frac{\partial^2}{\partial^2 z_p}\left(\frac{z_p}{\beta z_p + \Phi_p}\right) = -\sum_{p=1}^{2N-1} \frac{2\beta\Phi_p}{(\beta z_p + \Phi_p)^3} \leq 0 \tag{8.79}$$

since β, $\{z_p\}$ and $\{\Phi_p\}$ are all nonnegative. Hence the objective is a concave function. As the constraints in (8.43) and (8.44) are linear, it follows that (8.42) is a convex optimization problem that can be solved by maximizing the following Lagrange function w.r.t. $\{z_p \geq 0\}$:

$$\begin{aligned} L(\{z_p\}_{p=1}^{2N-1}) &= \sum_{p=1}^{2N-1} \frac{z_p}{\beta z_p + \Phi_p} - \lambda\left(\sum_{p=1}^{2N-1} z_p - N\right) \\ &= \sum_{p=1}^{2N-1}\left(\frac{z_p}{\beta z_p + \Phi_p} - \lambda z_p\right) + \lambda N, \end{aligned} \tag{8.80}$$

where $\lambda > 0$ is the Lagrange multiplier. In fact, owing to the decoupled form of (8.80) we only need to consider maximizing the following function:

$$\tilde{f}(z_p) = \frac{z_p}{\beta z_p + \Phi_p} - \lambda z_p. \tag{8.81}$$

In the same way as for (8.79), it can be shown that $\tilde{f}(z_p)$ is concave; the maximizer of this function can be obtained by setting its first-order derivate to zero:

$$\frac{\partial \tilde{f}(z_p)}{\partial z_p} = \frac{\Phi_p}{(\beta z_p + \Phi_p)^2} - \lambda = 0, \tag{8.82}$$

which leads to the solution

$$\hat{z}_p = \frac{\sqrt{\Phi_p/\lambda} - \Phi_p}{\beta}. \tag{8.83}$$

If $\hat{z}_p \geq 0$, then $\hat{z}_p$ is the sought solution; if $\hat{z}_p < 0$, considering that $\tilde{f}(z_p)$ (a concave function) is increasing for $z_p \leq \hat{z}_p$ and is decreasing for $z_p \geq \hat{z}_p$, $\tilde{f}(z_p)$ must attain its maximum (w.r.t. $z_p \geq 0$) at $z_p = 0$. Therefore, the maximizer of $\tilde{f}(z_p)$, as well as the solution to (8.42), is given by

$$z_p = \max\left\{\frac{\sqrt{\Phi_p/\lambda} - \Phi_p}{\beta}, 0\right\}, \quad p = 1, \ldots, 2N-1. \tag{8.84}$$

The solution as shown in (8.48) is obtained from (8.84) by replacing $\sqrt{1/\lambda}$ with $\lambda\rho$, so that the parameter λ lies in the range [0, 1] (see (8.50)).

Part II

Periodic correlation synthesis

9 Single periodic sequence design

Let $\{x(n)\}_{n=1}^{N}$ be the sequence in question and let $\{\tilde{r}(k)\}$ denote its periodic autocorrelations as defined in (1.16). In most applications the elements of the sequence are constrained in some way; for instance, they may be required to be binary or polyphase, or at least to be unimodular (i.e., to have constant modulus). However we do not make such an assumption in the *theoretical analysis* here, for the sake of generality; therefore we simply let $x(n) \in \mathbb{C}$ $(n = 1, \ldots, N)$ to avoid restraining the validity of the results that follow.

Sequences with impulse-like periodic correlations are required in many applications, particularly in pulse compression radar and in wireless communications [Levanon & Mozeson 2004][Freedman et al. 1995][Stoica et al. 2009*a*][Ling et al. 2010][Torii et al. 2004]. Such sequences can be designed via the minimization of a number of criteria that express, in different ways, the distance between the actual and the desired periodic correlations. In order to describe these criteria, let us assume that $\tilde{r}(0)$ is fixed (i.e., $\tilde{r}(0) =$ constant; for example, in the case of unimodular sequences mentioned above, we have $\tilde{r}(0) = N$ when $|x(n)| = 1$). Also, let $\mathbf{X}$ denote the following $N \times N$ right-circulant matrix:

$$\mathbf{X} = \begin{bmatrix} x(1) & x(2) & \cdots & x(N-1) & x(N) \\ x(N) & x(1) & \cdots & x(N-2) & x(N-1) \\ \vdots & & & & \vdots \\ x(2) & x(3) & \cdots & x(N) & x(1) \end{bmatrix}. \tag{9.1}$$

Making use of $\mathbf{X}$ we can write the $N \times N$ correlation matrix of the sequence in the following form:

$$\begin{bmatrix} \tilde{r}(0) & \tilde{r}(1) & \cdots & \tilde{r}(N-1) \\ \tilde{r}^*(1) & \tilde{r}(0) & \cdots & \tilde{r}(N-2) \\ \vdots & \ddots & \ddots & \vdots \\ \tilde{r}^*(N-1) & \cdots & \tilde{r}^*(1) & \tilde{r}(0) \end{bmatrix} = \mathbf{X}\mathbf{X}^H. \tag{9.2}$$

With (9.2) in mind we can think of designing the desired sequence $\{x(n)\}$ with impulse-like periodic correlations via the minimization of the following criterion:

$$C_1 = \|\mathbf{X}\mathbf{X}^H - \tilde{r}(0)\mathbf{I}_N\|^2 \tag{9.3}$$

It follows readily from (9.2) that

$$C_1 = \sum_{\substack{k=-(N-1)\\k\neq 0}}^{N-1} (N-|k|)|\tilde{r}(k)|^2 = 2\sum_{k=1}^{N-1}(N-k)|\tilde{r}(k)|^2. \tag{9.4}$$

Seemingly the above form of C_1 suggests that the larger correlation lags are less emphasized in this criterion than the smaller correlation lags (for example, the weight of $|\tilde{r}(1)|^2$ in (9.4) is $N-1$, whereas that of $|\tilde{r}(N-1)|^2$ is equal to 1).

The above observation leads to a second, potentially more natural criterion, in which all correlation lags (that are to be minimized) are equally weighted:

$$C_2 = N\sum_{k=1}^{N-1} |\tilde{r}(k)|^2 \tag{9.5}$$

(the factor N in (9.5) will make the comparison with (9.3) neater).

Criteria similar to C_1 and C_2 have been used to design unimodular sequences with impulse-like *aperiodic* correlations in previous chapters (e.g., the criteria (3.2) and (3.41) albeit for multi-sequences). In the aperiodic case, the sequences obtained by minimizing C_1 and respectively C_2 turned out, as expected, to have rather different correlation properties. On the other hand, in the periodic case the two criteria are in fact identical, as can be proved below:

$$\begin{aligned} C_1 &= \sum_{k=1}^{N-1}(N-k)|\tilde{r}(k)|^2 + \sum_{k=(N-1)}^{1} k|\tilde{r}(N-k)|^2 \\ &= \sum_{k=1}^{N-1}\left[(N-k)|\tilde{r}(k)|^2 + k|\tilde{r}^*(k)|^2\right] \\ &= \sum_{k=1}^{N-1} N|\tilde{r}(k)|^2 = C_2, \end{aligned} \tag{9.6}$$

where the periodic and symmetric properties of $\tilde{r}(k)$ (see (1.16)), i.e., $\tilde{r}(N-k) = \tilde{r}(-k) = \tilde{r}^*(-k)$, were used.

In the next section, we will prove the equivalence between C_1 and C_2 in the frequency domain. Doing so leads to two other design criteria, called C_3 and C_4, which are simpler than C_1 or C_2 from the perspective of minimization complexity.

9.1 Design criteria

Let $\mathbf{F}$ denote the $N \times N$ (inverse) DFT matrix, whose (k,p)th element is given by

$$[F]_{kp} = \frac{1}{\sqrt{N}} e^{j\frac{2\pi}{N}kp}, \quad k,p = 1,\ldots,N \tag{9.7}$$

and let

$$X_p = \sum_{n=1}^{N} x(n) e^{-j\frac{2\pi}{N}pn}, \quad p = 1, \ldots, N \tag{9.8}$$

be the discrete Fourier transform of the sequence $\{x(n)\}$. Then it is well known that

$$\mathbf{X} = \mathbf{F}^H \mathbf{D} \mathbf{F}, \tag{9.9}$$

where

$$\mathbf{D} = \begin{bmatrix} X_1 & & \mathbf{0} \\ & \ddots & \\ \mathbf{0} & & X_N \end{bmatrix}. \tag{9.10}$$

For completeness' sake and for readers' convenience we provide a simple proof of (9.9) in Appendix 9A.

Using (9.9) and the fact that $\mathbf{F}$ is unitary (i.e., $\mathbf{F}^H\mathbf{F} = \mathbf{I}$) we can rewrite C_1 as follows:

$$\begin{aligned} C_1 &= \|\mathbf{X}\mathbf{X}^H - \tilde{r}(0)\mathbf{I}\|^2 = \|\mathbf{D}\mathbf{D}^H - \tilde{r}(0)\mathbf{I}\|^2 \\ &= \sum_{p=1}^{N} \left[|X_p|^2 - \tilde{r}(0)\right]^2. \end{aligned} \tag{9.11}$$

Regarding C_2, from (9.2) and (9.9) it follows that

$$\mathbf{F} \begin{bmatrix} \tilde{r}(0) & \tilde{r}(1) & \cdots & \tilde{r}(N-1) \\ \tilde{r}(1)^* & \tilde{r}(0) & \cdots & \tilde{r}(N-1) \\ \vdots & \ddots & \ddots & \vdots \\ \tilde{r}(N-1)^* & \cdots & \tilde{r}(1)^* & \tilde{r}(0) \end{bmatrix} = \mathbf{F}\mathbf{X}\mathbf{X}^H = \mathbf{D}\mathbf{D}^H\mathbf{F} \tag{9.12}$$

and consequently that

$$\mathbf{F} \begin{bmatrix} \tilde{r}(0) \\ \tilde{r}(1)^* \\ \vdots \\ \tilde{r}(N-1)^* \end{bmatrix} = \frac{1}{\sqrt{N}} \begin{bmatrix} |X_1|^2 e^{j\frac{2\pi}{N}} \\ \vdots \\ |X_N|^2 e^{j\frac{2\pi}{N}N} \end{bmatrix}. \tag{9.13}$$

From (9.13) we obtain the following expression for C_2:

$$\begin{aligned} C_2 &= N \sum_{k=1}^{N-1} |\tilde{r}(k)|^2 = N \left(\left\| \mathbf{F} \begin{bmatrix} \tilde{r}(0) \\ \vdots \\ \tilde{r}(N-1)^* \end{bmatrix} \right\|^2 - \tilde{r}(0)^2 \right) \\ &= \sum_{p=1}^{N} \left[|X_p|^4 - \tilde{r}(0)^2 \right]. \end{aligned} \tag{9.14}$$

Comparing (9.11) and (9.14) we see that $C_1 = C_2$ if and only if

$$-2\tilde{r}(0)\sum_{p=1}^{N}|X_p|^2 + N\tilde{r}(0)^2 = -N\tilde{r}(0)^2 \tag{9.15}$$

or, equivalently,

$$\tilde{r}(0) = \frac{1}{N}\sum_{p=1}^{N}|X_p|^2. \tag{9.16}$$

However (9.16) is nothing but a Parseval-type equality; this follows easily from the fact that:

$$\mathbf{F}^H\begin{bmatrix} x(1) \\ \vdots \\ x(N) \end{bmatrix} = \frac{1}{\sqrt{N}}\begin{bmatrix} X_1 \\ \vdots \\ X_N \end{bmatrix}. \tag{9.17}$$

With this observation, the proof of the identity (done in the frequency domain)

$$C_1 = C_2 \tag{9.18}$$

is concluded.

The criterion C_1 as defined in (9.3) is a quartic function of the sequence elements. The following criterion is related to C_1 but it is a simpler quadratic function of the unknown variables:

$$C_3 = \left\|\mathbf{X} - \sqrt{\tilde{r}(0)}\mathbf{U}\right\|^2, \tag{9.19}$$

where $\mathbf{U}$ is an $N \times N$ unitary matrix: $\mathbf{U}^H\mathbf{U} = \mathbf{U}\mathbf{U}^H = \mathbf{I}$. The criterion C_3 is related to C_1 in the sense that C_1 takes on a small value if C_3 does so, and vice versa; in particular, $C_1 = 0$ if and only if $C_3 = 0$. Consequently, an alternative way of designing the sequence is to minimize C_3 (with respect to both $\{x(n)\}$ and $\mathbf{U}$) instead of minimizing C_1.

A similar argument to the above one leads from (9.11) to the following related but simpler criterion in the frequency domain:

$$C_4 = \sum_{p=1}^{N}|X_p - \sqrt{\tilde{r}(0)}e^{j\psi_p}|^2 \tag{9.20}$$

where $\{\psi_p\}$ are auxiliary variables (similar to the $\mathbf{U}$ in (9.19)).

The criteria C_3 and C_4 are related to C_1 in different ways. However, they can be shown to be equivalent, as explained next.

Making use of (9.9) we can rewrite C_3 as follows:

$$C_3 = \left\|\mathbf{D} - \sqrt{\tilde{r}(0)}\mathbf{V}\right\|^2, \tag{9.21}$$

where $\mathbf{V} = \mathbf{F}\mathbf{U}\mathbf{F}^H$ is an $N \times N$ auxiliary unitary matrix. Because $\mathbf{D}$ is diagonal we can expect that the minimizing unitary matrix $\mathbf{V}$ is also diagonal. To show that this is indeed the case, note that the matrices made from the right and left singular vectors of $\mathbf{D}$ are diagonal with unimodular elements on the main diagonal. Then it follows from the result proved in Appendix 3B that the minimizing matrix $\mathbf{V}$ must have the form

$$\mathbf{V} = \begin{bmatrix} e^{j\psi_1} & & \mathbf{0} \\ & \ddots & \\ \mathbf{0} & & e^{j\psi_N} \end{bmatrix}. \tag{9.22}$$

Inserting (9.22) into (9.21) we obtain the criterion C_4 after a simple calculation, a fact that we can informally state as follows:

$$\min_{\{V_{kp}(k \neq p)\}} C_3 = C_4. \tag{9.23}$$

With this observation the proof of the equivalence of C_3 and C_4 is concluded.

9.2 The periodic CAN (PeCAN) algorithm

In the heavily constrained case of binary sequences or of polyphase sequences with a small phase set, perfect sequences (for which $C_1 = C_2 = C_3 = C_4 = 0$), which are also called constant amplitude zero auto-correlation (CAZAC) sequences [Benedetto et al. 2009], are rare occurrences [Torii et al. 2004]. For such cases, minimizing C_1 or C_2 may be the only way available for designing an optimal sequence. Alternatively, we can design a quasi-optimal sequence by minimizing the related but simpler criteria C_3 or C_4 (note that the sequence minimizing C_3 or C_4 is not necessarily the same as the sequence that minimizes C_1 or C_2). We recommend using C_4 for such a design task due to the computational convenience of this criterion. If the performance achieved by minimizing C_4 is not deemed to be satisfactory, then the direct minimization of C_1 or C_2 (which is harder) may be advisable.

In the mildly constrained case of polyphase sequences with large phase sets or of unimodular sequences, on the other hand, perfect sequences exist for any length N [Levanon & Mozeson 2004][Torii et al. 2004]. Furthermore, there are systematic analytical ways (including closed-form expressions) for constructing such perfect sequences. However, even in this situation, designing a perfect sequence by minimizing one of the above criteria (all of which can be made equal to zero in such a case) is of considerable interest for radar and wireless communications in hostile environments (such as covert underwater communications; see Chapter 19). Indeed, in such applications it is important not only to use a perfect sequence to mitigate the multipath interferences, but also to employ one that is hard to guess by the adversary. Perfect sequences given by closed-form expressions or constructed in some other analytical ways are typically easy to guess because they depend on a relatively small number of parameters (such as the sequence length, possible sign changes or phase shifts). On the other hand, perfect sequences generated by minimizing the above criteria (see the algorithm described shortly) depend on the initial sequence used to start the search, and are nearly uncorrelated to one another when obtained using random initializations; and a random initial sequence depends on too many unknowns (between 2^N in the binary case, and infin-

Table 9.1. The PeCAN algorithm

Step 0: Use N independent and uniformly distributed phases in the interval $[0, 2\pi]$ to randomly generate an initial unimodular sequence $\{x(n)\}_{n=1}^{N}$.
Step 1: For given $\{x(n)\}$, compute the FFT of $\{x(n)\}$, i.e., $\{X_p\}_{p=1}^{N}$, and minimize C_4 with respect to $\{\psi_p\}_{p=1}^{N}$. The minimizing $\{\psi_p\}$ are simply given by

$$\psi_p = \arg(X_p), \quad p = 1, \ldots, N. \tag{9.25}$$

Step 2: For given $\{\psi_p\}$, compute the IFFT of $\{e^{j\psi_p}\}$, let us say $\{z_n\}$, and minimize C_4 with respect to $\{x(n)\}$ (subject to $|x(n)| = 1$). The minimizing $\{x(n)\}$ are again easy to compute:

$$x(n) = e^{j \arg(z_n)}, \quad n = 1, \ldots, N. \tag{9.26}$$

Iteration: Repeat steps 1 and 2 until a prespecified stop criterion is satisfied.

---

ity in the unimodular case) to allow exact guessing (even when the sequence's length and the generating algorithm are known). For similar reasons to those mentioned in the previous paragraph, we recommend using C_4 for sequence design in this case as well: indeed C_4 can be efficiently minimized by the algorithm described below, whereas no similarly efficient algorithm is available for minimizing the other criteria.

The criterion C_4 in (9.20) can be written as

$$C_4 = \left\| \mathbf{F}^H \begin{bmatrix} x(1) \\ \vdots \\ x(N) \end{bmatrix} - \begin{bmatrix} \sqrt{\tilde{r}_0} e^{j\psi_1} \\ \vdots \\ \sqrt{\tilde{r}_0} e^{j\psi_N} \end{bmatrix} \right\|^2 = \left\| \begin{bmatrix} x(1) \\ \vdots \\ x(N) \end{bmatrix} - \mathbf{F} \begin{bmatrix} \sqrt{\tilde{r}_0} e^{j\psi_1} \\ \vdots \\ \sqrt{\tilde{r}_0} e^{j\psi_N} \end{bmatrix} \right\|^2. \tag{9.24}$$

The algorithm for minimizing C_4 is called PeCAN (periodic CAN) and is summarized in Table 9.1. Note that PeCAN parallels the CAN algorithm discussed in Chapter 2 which minimizes *aperiodic* correlation sidelobes. Owing to its leverage of FFT operations, PeCAN is computationally efficient and can be used with values of N up to 10^6 on an ordinary PC.

9.3 Numerical examples

We consider the design of a unimodular sequence (with $|x(n)| = 1$) of length $N = 256$. We ran PeCAN with 200 different random initializations. Out of the so-obtained 200 unimodular sequences, the 50 sequences with the lowest integrated sidelobe level, $\sum_{k=1}^{N-1} |\tilde{r}(k)|^2$, were kept. The normalized periodic correlations (in dB) of these 50 sequences, viz.

$$20 \log_{10} \frac{|\tilde{r}(k)|}{N}, \quad k = -255, \ldots, 255, \tag{9.27}$$

are shown in Figure 9.1(a), in a superimposed manner. The values of the periodic correlation sidelobes (corresponding to lags $k \neq 0$) in this figure are around -200 dB (i.e.,

10^{-10}), which can actually be considered zero in practice. The correlations of the corresponding random-phase sequences that are used for initialization are shown in Figure 9.1(b), where it is easy to observe the much higher correlation sidelobes than in Figure 9.1(a). We also computed the normalized periodic cross-correlations of these 50 perfect sequences associated with Figure 9.1(a):

$$\frac{1}{N}\sum_{n=1}^{N} x_{m_1}(n)x_{m_2}^*((n-k) \bmod N) \tag{9.28}$$

$$k = 0, \ldots, 255, \quad m_1, m_2 = 1, \ldots, 50 \quad (m_1 \neq m_2),$$

where $\{x_m(n)\}_{n=1}^N$ denotes the mth perfect sequence. The absolute values of (9.28), for the shown k, m_1 and m_2, turned out to lie in the interval $[2.36 \times 10^{-5}, 0.24]$ (with a median value equal to 0.052), which means that the perfect sequences obtained using PeCAN with random initialization can be considered to be nearly uncorrelated to one another.

9.4 Conclusions

We have studied several criteria that can be used for designing perfect or CAZAC sequences, i.e., sequences whose periodic correlation sidelobes are all zero. The algorithm PeCAN was proposed to minimize one of the criteria that appears relatively easy to handle. PeCAN can be used to efficiently design perfect sequences under the unit-modulus constraint. In addition, the PeCAN algorithm starts from random initializations and can generate many different perfect sequences of the same length.

Appendix 9A Proof of Equation (9.9)

Let $\mathbf{J}$ denote the $N \times N$ shift matrix:

$$\mathbf{J} = \begin{bmatrix} \mathbf{0} & \mathbf{I}_{N-1} \\ 1 & \mathbf{0} \end{bmatrix}. \tag{9.29}$$

Then, we have that

$$\mathbf{J}^T \begin{bmatrix} x(1) \\ \vdots \\ x(N-1) \\ x(N) \end{bmatrix} = \begin{bmatrix} x(N) \\ x(1) \\ \vdots \\ x(N-1) \end{bmatrix} \tag{9.30}$$

and by induction that the kth row of $\mathbf{X}$ is $\mathbf{x}^T\mathbf{J}^k$ (with $\mathbf{x} = \begin{bmatrix} x(1) & \cdots & x(N) \end{bmatrix}^T$).

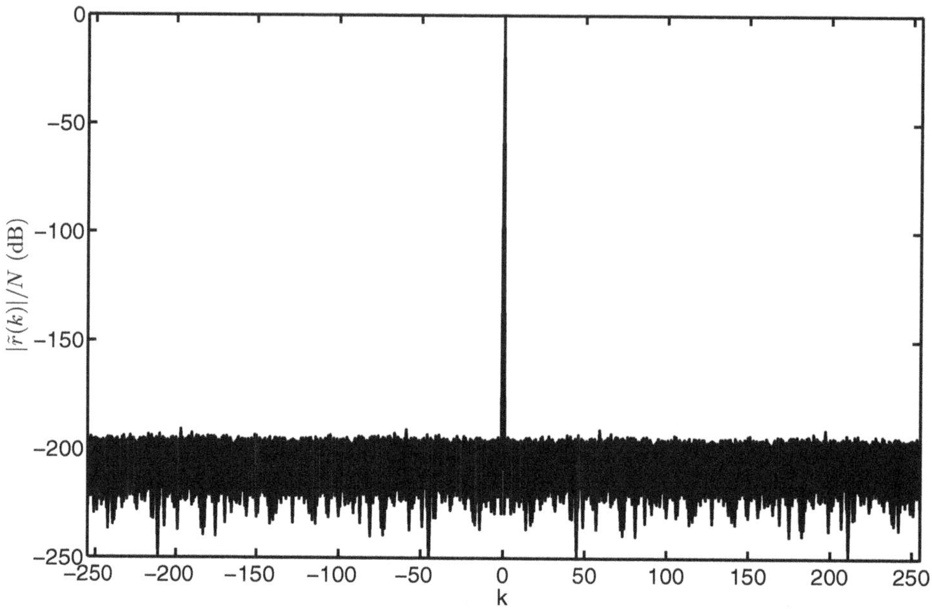

(a)

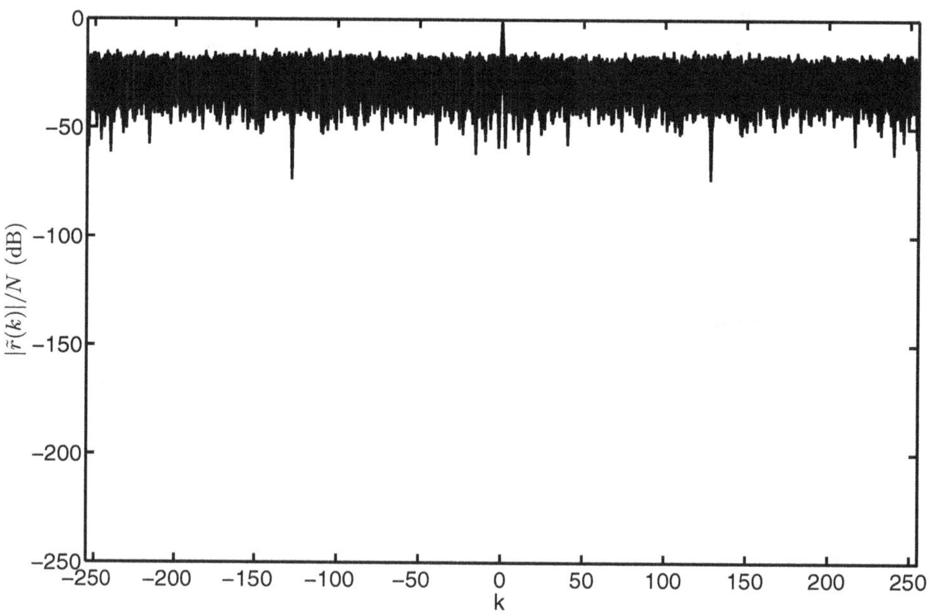

(b)

Figure 9.1 Superimposed normalized periodic auto-correlations (see (9.27)) (a) of 50 perfect sequences (of length $N = 256$) generated using PeCAN with random initialization, and (b) of the corresponding random-phase sequences used for initialization.

Let

$$\mathbf{a}_p = \text{the } p\text{th column of } \mathbf{F}^H \tag{9.31}$$
$$= \left[e^{-j\frac{2\pi}{N}p} \ \dots \ e^{-j\frac{2\pi}{N}Np}\right]^T.$$

Then we have the following equivalences:

$$(9.9) \Longleftrightarrow \mathbf{X}\mathbf{F}^H = \mathbf{F}^H\mathbf{D} \Longleftrightarrow \mathbf{X}\mathbf{a}_p = \mathbf{a}_p X_p$$
$$\Longleftrightarrow \mathbf{x}^T\mathbf{J}^k\mathbf{a}_p = e^{-j\frac{2\pi}{N}kp}X_p, \quad p,k = 1,\dots,N. \tag{9.32}$$

The last equality is obviously true for $k = N$, so let us assume that $k = 1,\dots,N-1$. To show that (9.32) holds in the latter case as well, observe that

$$e^{-j\frac{2\pi}{N}kp}X_p = \sum_{s=1}^{N} x_s e^{-j\frac{2\pi}{N}p(s+k)} \tag{9.33}$$
$$= \mathbf{x}^T \begin{bmatrix} e^{-j\frac{2\pi}{N}p(k+1)} \\ \vdots \\ e^{-j\frac{2\pi}{N}pN} \\ e^{-j\frac{2\pi}{N}p(N+1)} \\ \vdots \\ e^{-j\frac{2\pi}{N}p(N+k)} \end{bmatrix} = \mathbf{x}^T \begin{bmatrix} e^{-j\frac{2\pi}{N}p(k+1)} \\ \vdots \\ e^{-j\frac{2\pi}{N}pN} \\ e^{-j\frac{2\pi}{N}p} \\ \vdots \\ e^{-j\frac{2\pi}{N}pk} \end{bmatrix} = \mathbf{x}^T\mathbf{J}^k\mathbf{a}_p, \quad k = 1,\cdots,N-1,$$

which proves (9.32) and therefore (9.9).

10 Periodic sequence set design

Let $\{x_m(n)\}$ ($m = 1, \ldots, M$ and $n = 1, \ldots, N$) denote a set of M sequences, each of which is of length N. Every element of the sequence set is a complex-valued number that is unimodular: $|x_m(n)| = 1$. The periodic cross-correlation between the m_1th and m_2th sequences at time lag k is defined by

$$\begin{aligned}\tilde{r}_{m_1 m_2}(k) = &\sum_{n=1}^{N} x_{m_1}(n) x^*_{m_2}((n-k) \bmod N) \\ =&\tilde{r}^*_{m_2 m_1}(-k) = \tilde{r}^*_{m_2 m_1}(N-k), \\ &m_1, m_2 = 1, \ldots, M, \quad k = 0, \ldots, N-1.\end{aligned} \tag{10.1}$$

When $m_1 = m_2$, the above correlation becomes the auto-correlation.

Sequence sets having low auto- and cross-correlations are useful in many application areas. In radar range compression, low auto-correlations improve the detection of weak targets [Stimson 1998]; in code division multiple access (CDMA) systems, low auto-correlation helps with synchronization and low cross-correlation reduces interferences from other users [Suehiro 1994]; and the situation is similar in many other applications, such as ultra-sonic imaging [Diaz et al. 1999].

Since the early research on maximum-length sequences (m-sequences) in the 1950s, many sequence families with good correlations have been proposed, such as the Gold sequences [Gold 1967], the Kasami sequences [Kasami 1966] and others [Sarwate & Pursley 1980][Kumar & Moreno 1991][Fan & Darnell 1997][Han & Yang 2009]. Some of these sequence sets asymptotically meet the Welch bound, which is a theoretical lower bound on periodic correlations; see [Welch 1974] as well as Chapter 11. Note that [Tang et al. 2000] extended the Welch bound to situations in which some correlation sidelobes can be made zero for a certain lag range (hereafter correlation sidelobes designate auto-correlation sidelobes and all cross-correlation values); sequences having this property are referred to as the zero-correlation zone (ZCZ) sequences [Fan et al. 1999][Torii et al. 2004][Fan & Mow 2004][Tang & Mow 2008].

In this chapter we describe two cyclic algorithms for the design of unimodular sequence sets with low periodic correlations. The first algorithm can be used to generate sequence sets with almost zero correlation sidelobes in a specified time lag interval; these designed sequence sets are essentially ZCZ sequence sets. The second algorithm aims at achieving good correlations for all time lags; it is based on FFT computations and thus can efficiently generate sequence sets of large size. Unlike most existing sequence construction methods which are algebraic and deterministic in nature, both proposed algorithms start from random initializations and then proceed to cyclically minimize their corresponding metrics. In this way many different sequence sets with good correlation properties can be generated; these randomly distributed sequences are especially useful in applications such as covert underwater acoustic communications (see Chapter 19), countering coherent repeater jamming in radar systems [Skolnik 2008][Deng 2004], etc.

10.1 The Multi-PeCAO algorithm

Let $\mathbf{X}_m$ denote the right circulant matrix associated with the mth sequence:

$$\mathbf{X}_m = \begin{bmatrix} x_m(1) & x_m(2) \cdots & x_m(N) \\ x_m(N) & x_m(1) \cdots & x_m(N-1) \\ \vdots & & \vdots \\ x_m(N-P+2) & \cdots & x_m(N-P+1) \end{bmatrix}_{P\times N}, \tag{10.2}$$

where $P-1$ is the maximum lag of interest (see (10.5) below). Clearly every row of $\mathbf{X}_m$ is a cyclically shifted version of the sequence $\{x_m(n)\}_{n=1}^N$ and $\mathbf{X}_m$ has P rows ($0 < P \leq N$). We stack all the $\{\mathbf{X}_m\}$ together:

$$\mathbf{X} = \begin{bmatrix} \mathbf{X}_1 \\ \vdots \\ \mathbf{X}_M \end{bmatrix}_{MP\times N} \tag{10.3}$$

and note that

$$\mathbf{X}\mathbf{X}^H = \begin{bmatrix} \mathbf{R}_{11} & \mathbf{R}_{12} & \cdots & \mathbf{R}_{1M} \\ \mathbf{R}_{21} & \mathbf{R}_{22} & \cdots & \mathbf{R}_{2M} \\ \vdots & & & \vdots \\ \mathbf{R}_{M1} & \mathbf{R}_{M2} & \cdots & \mathbf{R}_{MM}, \end{bmatrix}_{MP\times MP} \tag{10.4}$$

where

$$\mathbf{R}_{m_1 m_2} = \begin{bmatrix} \tilde{r}_{m_1 m_2}(0) & \tilde{r}_{m_1 m_2}(1) \cdots & \tilde{r}_{m_1 m_2}(P-1) \\ \tilde{r}_{m_1 m_2}(-1) & \ddots \quad \ddots & \vdots \\ \vdots & & \tilde{r}_{m_1 m_2}(1) \\ \tilde{r}_{m_1 m_2}(-P+1) & \cdots & \tilde{r}_{m_1 m_2}(0) \end{bmatrix}. \tag{10.5}$$

Observe that $\tilde{r}_{m_1 m_2}(k)$ ($k = -P+1, \ldots, P-1$) appears $P - |k|$ times in the above Toeplitz matrix, so more emphasis will be given to the correlations with smaller time lags (see below). Because all diagonal elements in the $MP \times MP$ matrix in (10.4) are the in-phase auto-correlations that are equal to N, we can minimize all cross-correlations and out-of-phase auto-correlations for the considered P time lags by minimizing the following

Table 10.1. The Multi-PeCAO algorithm

Step 0: Set the matrix $\mathbf{X}$ to an initial value.
Step 1: Fix $\{x_m(n)\}$ (i.e., $\mathbf{X}$ is given) and compute $\mathbf{U}$ as

$$\mathbf{U} = \mathbf{U}_1\mathbf{U}_2^H \tag{10.8}$$

where $\mathbf{U}_1$ and $\mathbf{U}_2$ are unitary matrices obtained from the SVD of $\mathbf{X}$, i.e., $\mathbf{X} = \mathbf{U}_1\mathbf{S}\mathbf{U}_2^H$.
Step 2: Fix $\mathbf{U}$ and compute $\{x_m(n)\}$ as

$$x = \exp\left[j \arg\left(\sum_{k=1}^{P} \mu_k\right)\right], \tag{10.9}$$

where x denotes a generic element of $\{x_m(n)\}$ and $\{\mu_k\}_{k=1}^P$ are the P elements of $\mathbf{U}$ whose positions are the same as the positions of x in $\mathbf{X}$ (note that each element of $\{x_m(n)\}$ appears P times in $\mathbf{X}$).
Iteration: Repeat steps 1 and 2 until a prespecified stop criterion is satisfied.

criterion:

$$C_P = \left\|\mathbf{X}\mathbf{X}^H - N\mathbf{I}_{MP}\right\|^2. \tag{10.6}$$

An analysis of the criterion C_P shows that only for $P \leq N/M$ is it possible to minimize C_P to zero. Indeed, if $MP > N$, then the matrix $\mathbf{X}$ in (10.3) will be "tall" and thus the rank of $\mathbf{X}\mathbf{X}^H$ will be at most N. In this case the rank of $\mathbf{X}\mathbf{X}^H$ is always smaller than the rank of $\mathbf{I}_{MP}$ no matter what sequences we choose, which means that C_P cannot be made zero. Note that the condition $P \leq N/M$ coincides with the theoretical bound in [Tang et al. 2000].

When $P \leq N/M$, it is easy to observe that C_P becomes zero if $\mathbf{X}$ is a semi-unitary matrix. Therefore, instead of minimizing C_P directly, we can consider the following minimization problem:

$$\begin{aligned} \min_{\{x_m(n)\},\mathbf{U}} \quad & \left\|\mathbf{X} - \sqrt{N}\mathbf{U}\right\|^2 \\ \text{s.t.} \quad & |x_m(n)| = 1, \quad m = 1, \ldots, M, \quad n = 1, \ldots, N \\ & \mathbf{U}\mathbf{U}^H = \mathbf{I} \end{aligned} \tag{10.7}$$

where $\mathbf{U}$ is an $MP \times N$ semi-unitary matrix that serves as an auxiliary variable.

The criterion in (10.7) can be minimized in the following cyclic way. The matrix $\mathbf{X}$ is first initialized by a set of randomly generated unimodular sequences. Then (10.7) is iteratively minimized by fixing $\mathbf{X}$ to compute $\mathbf{U}$, then fixing $\mathbf{U}$ to compute $\mathbf{X}$ and so on, until a given stop criterion is satisfied. During this iterative process, both $\mathbf{U}$ and $\mathbf{X}$ have closed-form updating formulae which are easily derived from the Multi-CAO algorithm steps discussed in Section 3.3. The so-obtained algorithm is termed Multi-PeCAO (multi-sequence periodic CAO) and is summarized in Table 10.1.

10.2 The Multi-PeCAN algorithm

Although it is not possible to make all off-diagonal elements of $\mathbf{XX}^H$ zero for $P = N$, we can at least try to make them small without any emphasis on certain time lags, as might be required in certain applications. Specifically, we will try to minimize the criterion in (10.6) with P replaced by N:

$$C_N = \left\| \mathbf{XX}^H - N\mathbf{I}_{MN} \right\|^2. \tag{10.10}$$

Note that in this section the matrix $\mathbf{X}_m$ in (10.2) is $N \times N$ and $\mathbf{X}$ is $MN \times N$, and thus the periodic correlations of all time lags are incorporated in $\mathbf{XX}^H$. By using the periodicity of periodic correlations, i.e., $\tilde{r}_{ks}(-k) = \tilde{r}_{ks}(-k+N)$ (see (10.1)), it is not difficult to observe that each $\tilde{r}_{ks}(l)$ appears N times in the matrix $\mathbf{XX}^H$ and that

$$C_N = N\,\widetilde{\mathrm{ISL}}, \tag{10.11}$$

where $\widetilde{\mathrm{ISL}}$ is defined in (11.2) and is used in Section 10.3 too.

It is well known that the right circulant matrix $\mathbf{X}_m$ can be diagonalized by the DFT matrix (see Appendix 9A):

$$\mathbf{X}_m = \mathbf{F}^H \mathbf{D}_m \mathbf{F}, \tag{10.12}$$

where

$$[\mathbf{F}]_{kl} = \frac{1}{\sqrt{N}} e^{j\frac{2\pi}{N}(k-1)(l-1)}, \quad k, l = 1, \ldots, N, \tag{10.13}$$

$$\mathbf{D}_m = \begin{bmatrix} y_m(1) & & \mathbf{0} \\ & \ddots & \\ \mathbf{0} & & y_m(N) \end{bmatrix},$$

$$y_m(k) = \sum_{n=1}^{N} x_m(n) e^{-j\frac{2\pi}{N}(k-1)(n-1)}, \quad k = 1, \ldots, N.$$

To simplify the notation, define

$$\tilde{\mathbf{F}} = \begin{bmatrix} \mathbf{F} & & \mathbf{0} \\ & \ddots & \\ \mathbf{0} & & \mathbf{F} \end{bmatrix}_{MN \times MN}, \quad \mathbf{D} = \begin{bmatrix} \mathbf{D}_1 \\ \vdots \\ \mathbf{D}_M \end{bmatrix}_{MN \times N}, \quad \mathbf{y}_p = \begin{bmatrix} y_1(p) \\ \vdots \\ y_M(p) \end{bmatrix}_{M \times 1}, \quad p = 1, \ldots, N. \tag{10.14}$$

Then we can write the criterion C_N in (10.10) as follows:

$$\begin{aligned} C_N &= \left\| (\tilde{\mathbf{F}}^H \mathbf{D}\mathbf{F})(\tilde{\mathbf{F}}^H \mathbf{D}\mathbf{F})^H - N\mathbf{I}_{MN} \right\|^2 \\ &= \left\| \tilde{\mathbf{F}}^H \mathbf{D}\mathbf{D}^H \tilde{\mathbf{F}} - N\mathbf{I} \right\|^2 = \left\| \mathbf{D}\mathbf{D}^H - N\mathbf{I} \right\|^2 \\ &= \sum_{p=1}^{N} \left\| \mathbf{y}_p \mathbf{y}_p^H - N\mathbf{I}_M \right\|^2, \end{aligned} \tag{10.15}$$

Table 10.2. The Multi-PeCAN algorithm

Step 0: Initialize $\{x_m(n)\}$.

Step 1: For fixed $\{x_m(n)\}$, the minimizer $\{\boldsymbol{\alpha}_p\}$ is given by

$$\boldsymbol{\alpha}_p = \frac{\mathbf{c}_p}{\|\mathbf{c}_p\|}, \quad p = 1, \ldots, N \tag{10.18}$$

where

$$\mathbf{c}_p^T \text{ is the } p\text{th row of } \mathbf{F}^H[\mathbf{x}_1 \ \cdots \ \mathbf{x}_M] \tag{10.19}$$

and $\mathbf{x}_m = [x_m(1) \ \cdots \ x_m(N)]^T$.

Step 2: For fixed $\{\boldsymbol{\alpha}_p\}$, the minimizer $\{x_m(n)\}$ ($m = 1, \ldots, M$ and $n = 1, \ldots, N$) is given by

$$x_m(n) = \exp\left[\arg\left(\text{the } (n,m)^{\text{th}} \text{ element of } \mathbf{F}[\boldsymbol{\alpha}_1 \ \cdots \ \boldsymbol{\alpha}_N]^T\right)\right]. \tag{10.20}$$

Iteration: Repeat steps 1 and 2 until a prespecified stop criterion is satisfied.

which can be further rewritten as

$$\begin{aligned} C_N &= \sum_{p=1}^{N} \operatorname{tr}\left[\left(\mathbf{y}_p\mathbf{y}_p^H - N\mathbf{I}\right)\left(\mathbf{y}_p\mathbf{y}_p^H - N\mathbf{I}\right)^H\right] \\ &= \sum_{p=1}^{N} \left(\|\mathbf{y}_p\|^4 - 2N\|\mathbf{y}_p\|^2 + N^2M\right) \\ &= N^2 \sum_{p=1}^{N} \left(\left\|\frac{\mathbf{y}_p}{\sqrt{N}}\right\|^2 - 1\right)^2 + N^3(M-1). \end{aligned} \tag{10.16}$$

Similarly to the case of (10.6) and (10.7), we consider the following related minimization problem instead of minimizing C_N directly:

$$\begin{aligned} \min_{\{x_m(n)\},\{\boldsymbol{\alpha}_p\}} \ & \sum_{p=1}^{N} \left\|\frac{1}{\sqrt{N}}\mathbf{y}_p - \boldsymbol{\alpha}_p\right\|^2 \\ \text{s.t. } & |x_m(n)| = 1, \quad m = 1, \ldots, M \text{ and } n = 1, \ldots, N \\ & \|\boldsymbol{\alpha}_p\|^2 = 1, \quad p = 1, \ldots, N \quad (\boldsymbol{\alpha}_p \text{ is } M \times 1) \end{aligned} \tag{10.17}$$

where $\{\boldsymbol{\alpha}_p\}$ are auxiliary vectors.

The above problem has the same structure as the minimization problem for aperiodic correlations in (3.16). Thus the Multi-CAN algorithm there can be adapted for solving (10.17). The so-obtained algorithm is called Multi-PeCAN (multi-sequence periodic CAN) and is summarized in Table 10.2. Note that Multi-PeCAN is computationally efficient and can be used to design sequence sets with the size $NM \sim 10^5$ on an ordinary PC. Also note that Multi-PeCAN is an extension of the PeCAN algorithm described in Chapter 9.

10.3 Numerical examples

Making use of

$$\tilde{\mathbf{X}} = \begin{bmatrix} x_1(1) & \cdots & x_M(1) \\ x_1(2) & & x_M(2) \\ \vdots & & \vdots \\ x_1(N) & \cdots & x_M(N) \end{bmatrix}_{N\times M}, \quad \mathbf{J} = \begin{bmatrix} \mathbf{0} & \mathbf{I}_{N-1} \\ 1 & \mathbf{0} \end{bmatrix}_{N\times N}, \tag{10.21}$$

the auto- and cross-correlations of $\{x_m(n)\}_{m,n=1}^{M,N}$ at lag k can be expressed in a matrix form as

$$\mathbf{P}(k) \triangleq \tilde{\mathbf{X}}^H \mathbf{J}^k \tilde{\mathbf{X}} = \begin{bmatrix} \tilde{r}_{11}(k) & \cdots & \tilde{r}_{M1}(k) \\ \vdots & & \vdots \\ \tilde{r}_{1M}(k) & \cdots & \tilde{r}_{MM}(k) \end{bmatrix} = \mathbf{P}^H(-k), \tag{10.22}$$
$$k = 0, \ldots, N-1, \quad \text{(assuming that } \mathbf{J}^0 = \mathbf{I}_N).$$

Correspondingly the "correlation level" at lag k is defined as

$$\text{periodic correlation level} = 20 \log_{10} \frac{\|\mathbf{P}(k) - N\mathbf{I}_M \delta_k\|}{\sqrt{MN^2}}, \tag{10.23}$$
$$k = -N+1, \ldots, 0, \ldots, N-1.$$

The above normalization factor $\sqrt{MN^2}$ is the Frobenius norm of $\mathbf{P}(0)$ when all sequences in the set are orthogonal to each other (i.e., $\mathbf{P}(0) = N\mathbf{I}_M$).

10.3.1 Multi-PeCAO

Suppose that there are $M = 4$ transmitters, that each transmit sequence is of length $N = 512$ and that we are interested mainly in the correlations at the first $P = 60$ time lags. The Multi-PeCAO algorithm is used to design such a set of sequences. Figure 10.1 shows the correlation levels of the so-generated sequence set, from which we observe that the correlation levels at the first P time lags (i.e., from 0 to $P-1$) are below −60 dB. The corresponding criterion value C_P is 0.23, while the C_P of a randomly generated unimodular sequence set of the same size is usually on the order of 10^3.

Note that in Figure 10.1, the correlations within the region of interest go up as the time lag's absolute value increases. This is to be expected; see the discussion on "implicit weighting" after (10.5).

An alternative way to the minimization of C_P is a "shifting" approach that can be outlined as follows. We first generate a single so-called "perfect" sequence of length N whose auto-correlations are zero for all time lags, such as a Frank or Chu sequence (see Chapter 1). Actually the PeCAN algorithm discussed in Chapter 9 can be used to generate many such perfect sequences of a given length. Denote this length-N perfect sequence as $\mathbf{x}$. Then the desired sequence set is constructed as $\mathcal{K} = \{\mathbf{x}, T^P(\mathbf{x}), \ldots, T^{(M-1)P}(\mathbf{x})\}$ (assuming $MP \leq N$), where the operator $T^k(\mathbf{x})$ denotes

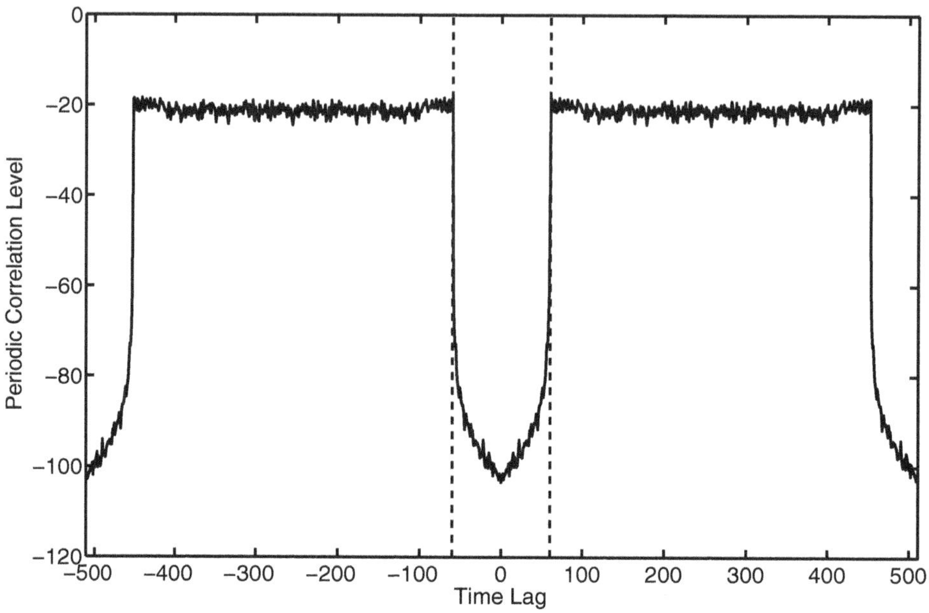

Figure 10.1 The periodic correlation levels of a Multi-PeCAO sequence set. $N = 512$, $M = 4$ and the goal is to minimize the correlations at the first $P = 60$ time lags. (The dashed vertical lines signify the boundaries of the time lag zone under consideration.)

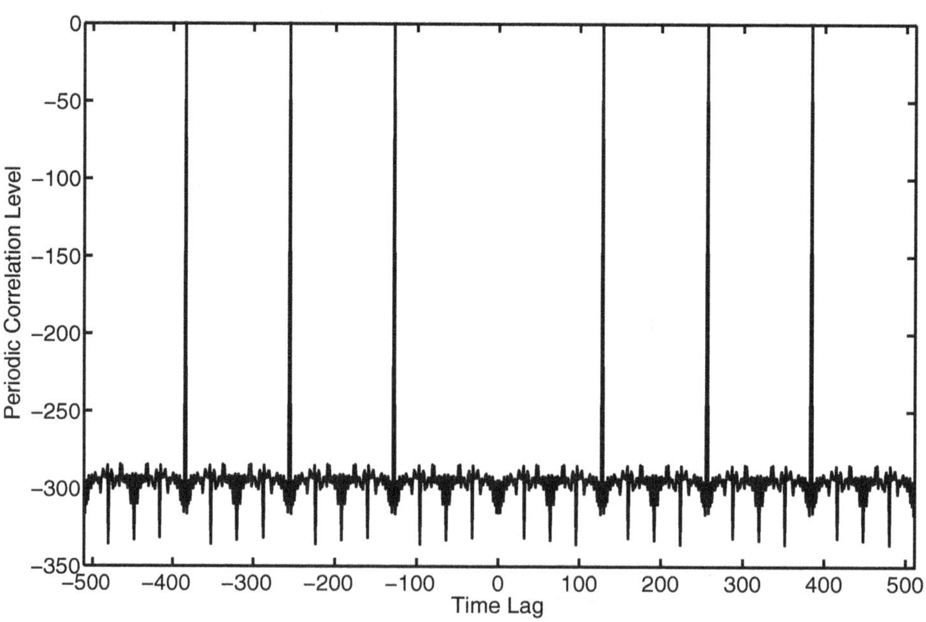

Figure 10.2 The periodic correlation levels of an optimal ZCZ sequence set of $N = 512$ and $M = 4$ ("optimal" means N/M zero-correlation lags), generated using the "shifting" approach.

the k-element right cyclic shift of the sequence $\mathbf{x}$ (considered as a row vector). In this way, the correlations of the sequences in $\mathcal{K}$ at the first P time lags can take values only from the auto-correlations of $\mathbf{x}$ at the first MP time lags, which are all zero. As an example, we choose $N = 512$, $M = 4$ and $P = 128$. For $\mathbf{x}$ equal to the P4 sequence, the correlation levels of the sequence set in $\mathcal{K}$ are shown in Figure 10.2. We observe that the correlations within the region of interest (time lags less than P) are all zero. The trade-off is the high sidelobes (in fact, as high as the in-phase auto-correlation) outside the region of interest. Such a sequence set that has zero correlation sidelobes at the first N/M lags is called an optimal ZCZ sequence set [Torii et al. 2004][Tang & Mow 2008]; here N/M is assumed to be an integer.

10.3.2 Multi-PeCAN

The integrated sidelobe level ($\widetilde{\text{ISL}}$) and the peak sidelobe level ($\widetilde{\text{PSL}}$) metrics (see (11.2) and (11.3)) will be used to measure the goodness of sequences. The Welch bound (see Chapter 11) states that for a set of sequences $\{x_m(n)\}_{m,n=1}^{M,N}$, the maximum correlation sidelobe satisfies the following inequality:

$$\widetilde{\text{PSL}} \geq B = N\sqrt{\frac{M-1}{NM-1}} \tag{10.24}$$

where we have kept the convention of using a tilde to denote *periodic* correlation-related metrics. Usually one considers the asymptotic bound: $B_{\text{asymp}} = \sqrt{N}$, which is close to B when $N \gg 1$. A set of sequences is called optimal if its $\widetilde{\text{PSL}}$ asymptotically attains B_{asymp}, i.e., $\lim_{N\to\infty} \widetilde{\text{PSL}} = \sqrt{N}$. Interestingly, there exist many optimal sequence sets [Kumar & Moreno 1991][Han & Yang 2009] of which we choose the Kasami sequence set [Kasami 1966] as an example to compare with the Multi-PeCAN sequence set proposed in this chapter. A Kasami sequence set is constructed from m-sequences. It has $M = \sqrt{N+1}$ sequences of length N, where N is restricted to $2^k - 1$ for even k (note that the Multi-PeCAN sequence set can be generated for any length desired). Its maximum correlation sidelobe $\widetilde{\text{PSL}}$ equals $1 + \sqrt{N+1}$.

We choose $N = 1023$ and $M = 4$. We generate the Kasami sequence set and the Multi-PeCAN sequence set. The full Kasami sequence set has 32 sequences out of which we choose the first four sequences; any other four sequences lead to very similar results. Figure 10.3 shows their correlation levels and Table 10.3 shows their $\widetilde{\text{ISL}}$ and $\widetilde{\text{PSL}}$ values. The correlation levels of a random-phase sequence set are shown in Figure 10.4 for the purpose of comparison. Compared with the optimal Kasami sequence set, the Multi-PeCAN sequence set gives higher $\widetilde{\text{PSL}}$ but lower $\widetilde{\text{ISL}}$; this can be expected since $\widetilde{\text{ISL}}$ is the criterion that the Multi-PeCAN algorithm aims to minimize. We also observe that the correlation sidelobes of a random-phase sequence set are not significantly higher than those of Kasami or Multi-PeCAN, which in fact implies the difficulty of minimizing both auto- and cross-correlation sidelobes of a sequence set (as opposed to minimizing only the auto-correlation of a single sequence).

Another point worth mentioning here is that different initializations lead to different Multi-PeCAN sequence sets. According to our experience, all these Multi-PeCAN

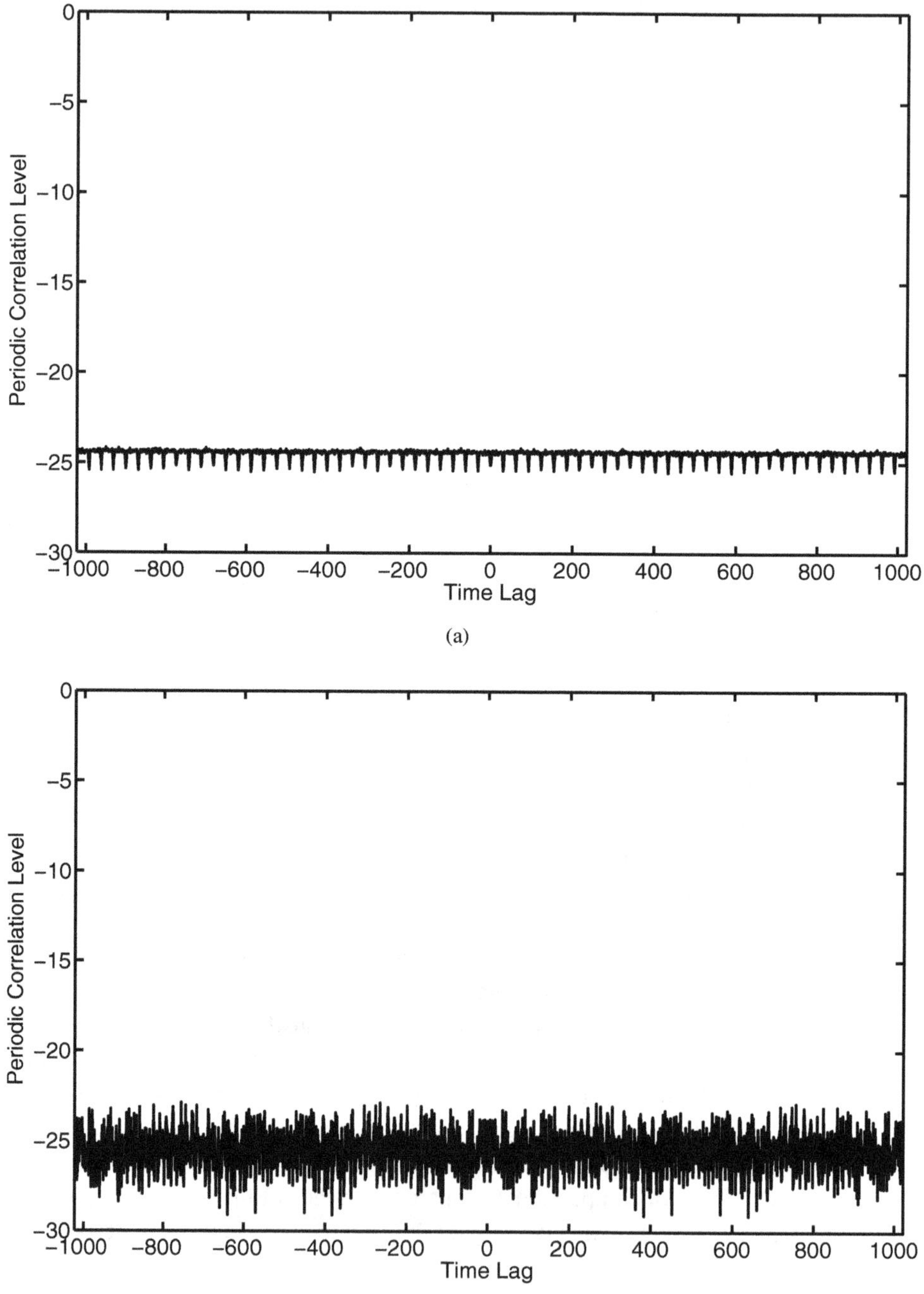

Figure 10.3 The periodic correlation levels of a set of $M = 4$ sequences, each with length $N = 1023$. (a) The Kasami sequence set and (b) the Multi-PeCAN sequence set.

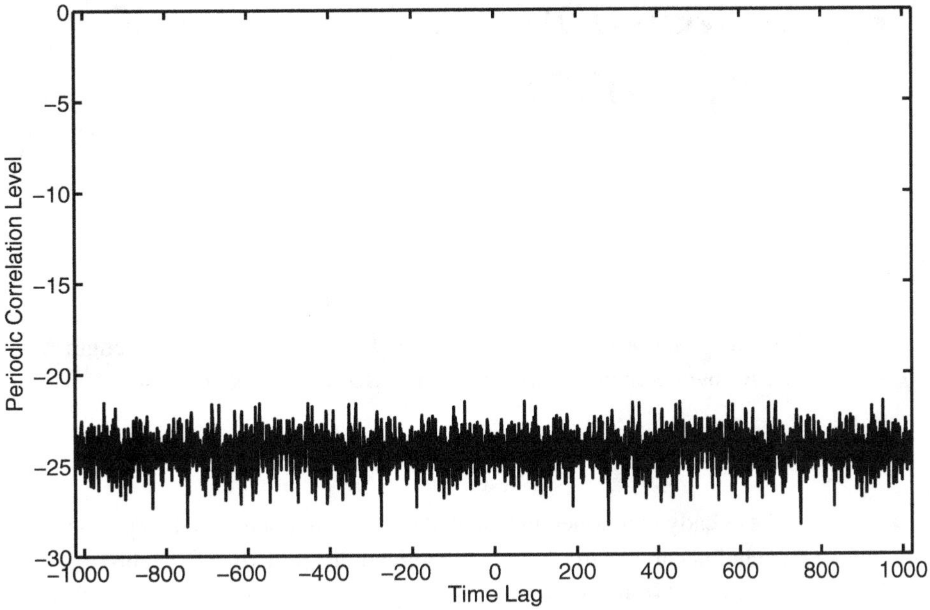

Figure 10.4 The periodic correlation levels of a random-phase sequence set with $N = 1023$ and $M = 4$.

Table 10.3. Values of $\widetilde{\text{ISL}}$ and $\widetilde{\text{PSL}}$ for random-phase, Kasami and Multi-PeCAN sequence sets, $N = 1023, M = 4$

	$\widetilde{\text{ISL}}$	$\widetilde{\text{PSL}}$
random-phase	17050045	94.8
Kasami	15380972	33.0
Multi-PeCAN	12558348	82.2

sequence sets have the same $\widetilde{\text{ISL}}$ value and similar $\widetilde{\text{PSL}}$ values. Therefore the Multi-PeCAN algorithm is able to generate many sequence sets that have similarly low correlation sidelobes.

10.4 Conclusions

In this chapter we have presented two cyclic algorithms that can be used to generate a set of unimodular sequences with low periodic correlation sidelobes (i.e., auto-correlation sidelobes and cross-correlations). The Multi-PeCAO algorithm is able to generate sequence sets that have very low correlation sidelobes within a time lag interval, while the Multi-PeCAN algorithm aims at minimizing the correlation sidelobes at all time lags. The Multi-PeCAN algorithm is computationally efficient and can be used to design large sequence sets.

11 Lower bounds for periodic sequences

Consider a set of M sequences $\{x_k(n)\}$ $(k = 1, \ldots, M)$, each of length N $(n = 1, \ldots, N)$. The following energy constraint is imposed on the sequence set:

$$\sum_{n=1}^{N} |x_k(n)|^2 = N, \quad k = 1, \ldots, M. \tag{11.1}$$

As already mentioned in Chapter 10, sequence sets with low periodic auto- and cross-correlations are desired in many applications, including asynchronous CDMA systems [Suehiro 1994], ultrasonic imaging [Diaz et al. 1999] and radar and sonar [Levanon & Mozeson 2004]. The two metrics that are most commonly used to measure the correlation levels are the integrated sidelobe level ($\widetilde{\text{ISL}}$) and the peak sidelobe level ($\widetilde{\text{PSL}}$):

$$\widetilde{\text{ISL}} = \sum_{k=1}^{M} \sum_{l=1}^{N-1} |\tilde{r}_{kk}(l)|^2 + \sum_{k=1}^{M} \sum_{\substack{s=1 \\ s \neq k}}^{M} \sum_{l=0}^{N-1} |\tilde{r}_{ks}(l)|^2, \tag{11.2}$$

$$\widetilde{\text{PSL}} = \max \{|\tilde{r}_{ks}(l)|\}, \tag{11.3}$$

$$k, s = 1, \ldots, M, \quad l = 0, \ldots, N-1 \ (l \neq 0 \text{ if } k = s).$$

A lower bound on $\widetilde{\text{PSL}}$ (denoted as $\tilde{\text{B}}_{\text{PSL}}$) was derived in [Welch 1974] and several sequence sets have been proposed that asymptotically (in N) achieve the Welch bound, e.g., [Sarwate 1979][Kumar & Moreno 1991][Han & Yang 2009]. In comparison with $\tilde{\text{B}}_{\text{PSL}}$, the corresponding $\widetilde{\text{ISL}}$ lower bound (denoted as $\tilde{\text{B}}_{\text{ISL}}$) has been less discussed in the literature (with [Sarwate 1979][Sarwate 1999] being notable exceptions) and there appears to be hardly any detailed discussion on sequence sets attaining $\tilde{\text{B}}_{\text{ISL}}$. In this chapter we focus on $\widetilde{\text{ISL}}$-related aspects. We use the Multi-PeCAN framework of Chapter 10 to obtain $\tilde{\text{B}}_{\text{ISL}}$ and $\tilde{\text{B}}_{\text{PSL}}$. We also use this framework to derive closed-form expressions for *all* energy constrained periodic sequence sets that attain the $\widetilde{\text{ISL}}$ lower bound.

11.1 Bound derivation

It follows from the discussions in Section 10.2 that the $\widetilde{\text{ISL}}$ metric (for periodic correlations) can be written as follows:

$$\widetilde{\text{ISL}} = \frac{1}{N}\sum_{p=1}^{N}\left\|\mathbf{y}_p\mathbf{y}_p^H - N\mathbf{I}_M\right\|^2 = \frac{1}{N}\sum_{p=1}^{N}\left(\|\mathbf{y}_p\|^2 - N\right)^2 + N^2(M-1) \tag{11.4}$$

where

$$\mathbf{y}_p = \begin{bmatrix} y_1(p) \\ \vdots \\ y_M(p) \end{bmatrix}, \quad y_k(p) = \sum_{n=1}^{N} x_k(n)e^{-j\frac{2\pi}{N}(p-1)(n-1)}. \tag{11.5}$$

Our problem is basically to minimize the $\widetilde{\text{ISL}}$ in (11.4) under the energy constraint (11.1). It follows from Parseval's equality that this energy constraint can be written as

$$\sum_{p=1}^{N}|y_k(p)|^2 = N\sum_{n=1}^{N}|x_k(n)|^2 = N^2, \quad k = 1,\ldots,M. \tag{11.6}$$

Let

$$z_{kp} = |y_k(p)|^2, \quad k = 1,\ldots,M, \quad p = 1,\ldots,N \tag{11.7}$$

and observe that we can minimize (11.4) by solving the following problem:

$$\min_{\{z_{kp}\}} f = \frac{1}{N}\sum_{p=1}^{N}\left(\sum_{k=1}^{M} z_{kp} - N\right)^2 \tag{11.8}$$

$$\text{s.t.} \sum_{p=1}^{N} z_{kp} = N^2, \quad k = 1,\ldots,M \tag{11.9}$$

$$z_{kp} \geq 0, \quad \forall k, p.$$

Using (11.9) leads to a simpler expression for f:

$$\begin{aligned} f &= \frac{1}{N}\sum_{p=1}^{N}\left[\left(\sum_{k=1}^{M} z_{kp}\right)^2 - 2N\sum_{k=1}^{M} z_{kp} + N^2\right] \\ &= g + N^2 - 2MN^2, \end{aligned} \tag{11.10}$$

where

$$g = \frac{1}{N}\sum_{p=1}^{N}\sum_{k=1}^{M}\sum_{s=1}^{M} z_{kp}z_{sp}. \tag{11.11}$$

Define an $M \times N$ matrix $\mathbf{Z}$ whose (k, p)th element is z_{kp}. Then g can be written as

$$\begin{aligned} g &= \frac{1}{N}\sum_{k=1}^{M}\sum_{s=1}^{M}\left(\sum_{p=1}^{N} Z_{kp}(Z^T)_{ps}\right) = \frac{1}{N}\sum_{k=1}^{M}\sum_{s=1}^{M}(ZZ^T)_{ks} \\ &= \frac{1}{N}\mathbf{u}^T\mathbf{Z}\mathbf{Z}^T\mathbf{u}, \end{aligned} \tag{11.12}$$

where

$$\mathbf{u} = \begin{bmatrix} 1 \ 1 \cdots 1 \end{bmatrix}^T \quad (M \times 1). \tag{11.13}$$

Furthermore, using $\mathbf{Z}$ we can rewrite the constraint $\sum_{p=1}^{N} z_{kp} = N^2$ in (11.9) as

$$\mathbf{Zv} = N^2\mathbf{u}, \tag{11.14}$$

where

$$\mathbf{v} = \begin{bmatrix} 1 \ 1 \cdots 1 \end{bmatrix}^T \quad (N \times 1). \tag{11.15}$$

Next note that

$$\mathbf{u}^T\mathbf{Z}\mathbf{\Pi}\mathbf{Z}^T\mathbf{u} \geq 0, \tag{11.16}$$

where

$$\mathbf{\Pi} = \mathbf{I} - \frac{1}{N}\mathbf{v}\mathbf{v}^T \tag{11.17}$$

is a positive semidefinite matrix (since it is the orthogonal projector onto the null space of $\mathbf{v}$). It follows from (11.12), (11.14), (11.16) and (11.17) that

$$\begin{aligned} g =& \frac{1}{N}\mathbf{u}^T\mathbf{Z}\mathbf{Z}^T\mathbf{u} \geq \frac{1}{N}\mathbf{u}^T\mathbf{Z}\frac{\mathbf{v}\mathbf{v}^T}{N}\mathbf{Z}^T\mathbf{u} \\ =& N^2\|\mathbf{u}\|^2\|\mathbf{u}\|^2 = M^2N^2. \end{aligned} \tag{11.18}$$

From (11.4), (11.8), (11.10) and (11.18) we obtain the equations

$$\begin{aligned} \widetilde{\mathrm{ISL}} &= f + N^2(M-1), \\ f &= g + N^2 - 2MN^2, \quad g \geq M^2N^2, \end{aligned} \tag{11.19}$$

which lead to the following lower bound on $\widetilde{\mathrm{ISL}}$:

$$\widetilde{\mathrm{ISL}} \geq N^2M(M-1) \triangleq \tilde{\mathrm{B}}_{\mathrm{ISL}}. \tag{11.20}$$

Also note from (11.2) that

$$\widetilde{\mathrm{ISL}} \leq M(N-1)\,\widetilde{\mathrm{PSL}}^2 + M(M-1)N\,\widetilde{\mathrm{PSL}}^2, \tag{11.21}$$

which, together with (11.20), leads to the following lower bound on $\widetilde{\mathrm{PSL}}$:

$$\widetilde{\mathrm{PSL}} \geq N\sqrt{\frac{M-1}{NM-1}} \triangleq \tilde{\mathrm{B}}_{\mathrm{PSL}}. \tag{11.22}$$

The above $\tilde{\mathrm{B}}_{\mathrm{PSL}}$ coincides with the $\widetilde{\mathrm{PSL}}$ bound introduced in [Welch 1974] and $\tilde{\mathrm{B}}_{\mathrm{ISL}}$ agrees with the $\widetilde{\mathrm{ISL}}$ bound discussed implicitly in [Sarwate 1999].

11.2 Optimal $\widetilde{\text{ISL}}$ sequence sets

A byproduct of the previous discussion is the following result:

$$\widetilde{\text{ISL}} = g - MN^2, \quad g = \frac{1}{N}\sum_{p=1}^{N}\sum_{k=1}^{M}\sum_{s=1}^{M} z_{kp}z_{sp} \geq M^2N^2 \tag{11.23}$$

from which it is easy to see that

$$z_{kp} = N, \quad k = 1,\ldots,M, \quad p = 1,\ldots,N, \tag{11.24}$$

which satisfies the constraint in (11.9), is a particular minimizer of $\widetilde{\text{ISL}}$. Because $z_{kp} = |y_k(p)|^2$, the $\widetilde{\text{ISL}}$ criterion depends only on $|y_k(p)|$ which is the absolute value of the DFT of the kth sequence at the frequency $(p-1)/N$. In other words, the phases of $\{y_k(p)\}$ can be chosen arbitrarily . As an example, if we choose zero phases for all $\{y_k(p)\}$ then (11.24) leads to

$$y_k(p) = \sqrt{N}, \quad k = 1,\ldots,M, \quad p = 1,\ldots,N \tag{11.25}$$

which corresponds to the following sequence set:

$$\begin{aligned} x_k(n) &= \frac{1}{N}\sum_{p=1}^{N} y_k(p)e^{j\frac{2\pi}{N}(n-1)(p-1)} \\ &= \begin{cases} \sqrt{N}, & n = 1, \\ 0, & n = 2,\ldots,N, \end{cases} \qquad k = 1,\ldots,M. \end{aligned} \tag{11.26}$$

Evidently the $\widetilde{\text{ISL}}$ of the sequence set $\{x_k(n)\}$ given by (11.26) is $M(M-1)N^2$, which achieves the $\widetilde{\text{ISL}}$ lower bound in (11.20). Yet such a sequence set is not too useful practically because the cross-correlation at lag zero between any two sequences is as high as the in-phase auto-correlation and the peak-to-average power ratio (PAR) for any sequence in this set is the largest possible.

Next we derive *all* solutions to the minimization problem (11.8), where we note once again that only the energy (not the PAR) of each sequence is constrained. Let $\mathbf{z}_0$ denote an $MN \times 1$ vector whose elements are all N (corresponding to the particular solution in (11.24)) and let $\mathbf{z} = \text{vec}(\mathbf{Z}^T)$ which is an $MN \times 1$ vector that contains the columns of $\mathbf{Z}^T$ stacked on top of each other. Then we can write the vectors $\mathbf{Z}^T\mathbf{u}$ and $\mathbf{Zv}$ (which appear in (11.12) and (11.14), respectively) in the following way:

$$\begin{aligned} \mathbf{Z}^T\mathbf{u} &= \begin{bmatrix} \mathbf{I}_N & \cdots & \mathbf{I}_N \end{bmatrix}_{N\times MN} \mathbf{z} = \mathbf{Az} \quad (\mathbf{A} \triangleq \mathbf{u}^T \otimes \mathbf{I}_N), \\ \mathbf{Zv} &= \begin{bmatrix} \mathbf{v}^T & & \mathbf{0} \\ & \ddots & \\ \mathbf{0} & & \mathbf{v}^T \end{bmatrix}_{M\times MN} \mathbf{z} = \mathbf{Bz} \quad (\mathbf{B} \triangleq \mathbf{I}_M \otimes \mathbf{v}^T). \end{aligned} \tag{11.27}$$

If an $MN \times 1$ vector $\boldsymbol{\delta}$ satisfies

$$\begin{bmatrix} \mathbf{A} \\ \mathbf{B} \end{bmatrix}_{(M+N)\times MN} \boldsymbol{\delta} = 0, \tag{11.28}$$

then $\mathbf{z} = \boldsymbol{\delta}$ leads to $g = 0$ in (11.12) and $\mathbf{Zv} = 0$ in (11.14). From this observation and the fact that $\mathbf{z}_0$ is a particular solution to (11.8), it follows that all solutions to (11.8) are given by

$$\mathbf{z} = \mathbf{z}_0 + \rho\boldsymbol{\delta} \tag{11.29}$$

where ρ is such that $\mathbf{z} > 0$ (element-wise). Note also that $\mathbf{z}_0$ belongs to the range space of $\mathbf{A}^T$, which implies that $\mathbf{z}_0^T\boldsymbol{\delta} = 0$ and thus that $\mathbf{z}_0$ is the minimum-norm solution of (11.8) (indeed, from (11.29) we have that $\|\mathbf{z}\|^2 = \|\mathbf{z}_0\|^2 + \rho^2\|\boldsymbol{\delta}\|^2 \geq \|\mathbf{z}_0\|^2$).

The problem that remains is to determine $\boldsymbol{\delta}$, or equivalently, the null space of $[\mathbf{A}^T\ \mathbf{B}^T]^T$. We assume that $M \geq 2$ and $N \geq 3$ so that $[\mathbf{A}^T\ \mathbf{B}^T]^T$ is a "wide" matrix that has a non-trivial null space. Owing to the special structure of $\mathbf{A}$ and $\mathbf{B}$, a closed-form basis for the null space of $[\mathbf{A}^T\ \mathbf{B}^T]^T$ can be obtained as follows. Define

$$\mathbf{G} = \begin{bmatrix} 1 & & \mathbf{0} \\ -1 & \ddots & \\ & \ddots & 1 \\ \mathbf{0} & & -1 \end{bmatrix}_{N\times(N-1)}, \quad \mathbf{H} = \begin{bmatrix} \mathbf{G} & & \mathbf{0} \\ & \ddots & \\ \mathbf{0} & & \mathbf{G} \\ -\mathbf{G} & \cdots & -\mathbf{G} \end{bmatrix} \tag{11.30}$$

(the dimension of $\mathbf{H}$ is $NM \times (N-1)(M-1)$) and then verify that

$$\mathbf{AH} = \mathbf{0} \ \text{ and } \ \mathbf{BH} = \mathbf{0}. \tag{11.31}$$

Because $\text{rank}(\mathbf{H}) = (N-1)(M-1)$, $\mathbf{H}$ spans a subspace of dimension $(N-1)(M-1)$ that belongs to the null space of $[\mathbf{A}^T\ \mathbf{B}^T]^T$. To show that $\mathbf{H}$ spans the entire null space of $[\mathbf{A}^T\ \mathbf{B}^T]^T$, note that

$$\begin{bmatrix}\mathbf{A}^T\ \mathbf{B}^T\end{bmatrix} = \begin{bmatrix} \mathbf{I}_N & \mathbf{v} & & \mathbf{0} \\ \mathbf{I}_N & & \mathbf{v} & \\ \vdots & & & \ddots \\ \mathbf{I}_N & \mathbf{0} & & \mathbf{v} \end{bmatrix}_{MN\times(M+N)}. \tag{11.32}$$

The first $M + N - 1$ columns of this matrix are obviously linearly independent but not all of its $M + N$ columns are so; indeed, we have that

$$\begin{bmatrix}\mathbf{A}^T\ \mathbf{B}^T\end{bmatrix} \begin{bmatrix} -\mathbf{v} \\ \mathbf{1}_{N\times1} \\ \vdots \\ \mathbf{1}_{N\times1} \end{bmatrix} = \mathbf{0}. \tag{11.33}$$

Hence $\text{rank}([\mathbf{A}^T\ \mathbf{B}^T]^T) = M + N - 1$ and consequently the dimension of the null space of this matrix is $MN - (M + N - 1) = (N-1)(M-1) = \text{rank}(\mathbf{H})$. The proof that $\mathbf{H}$ spans the entire null space of $[\mathbf{A}^T\ \mathbf{B}^T]^T$ is thus concluded.

In summary, the vector $\boldsymbol{\delta}$ can be written as $\boldsymbol{\delta} = \mathbf{Hw}$ where the $(N-1)(M-1) \times 1$ vector $\mathbf{w}$ is arbitrary. Then it follows from (11.29) that all solutions to (11.8) are given by

$$\mathbf{z} = \mathbf{z}_0 + \rho\mathbf{Hw}. \tag{11.34}$$

$\mathbf{x}_1$	$\mathbf{x}_2$
$-0.6535 + 0.7569j$	$0.7905 + 0.6125j$
$-0.3837 - 0.9235j$	$0.6590 + 0.7521j$
$-0.7717 - 0.6360j$	$-0.1753 - 0.9845j$
$-0.7989 - 0.6015j$	$-0.3145 - 0.9493j$
$0.9106 - 0.4133j$	$0.8426 - 0.5385j$
$-0.0453 + 0.9990j$	$0.9004 + 0.4351j$
$0.2875 - 0.9578j$	$-0.6840 - 0.7295j$
$0.1281 - 0.9918j$	$-0.4809 + 0.8768j$
$-0.8022 + 0.5971j$	$0.9599 - 0.2804j$
$-0.9137 - 0.4063j$	$-0.4742 + 0.8804j$

Table 11.1. A Multi-PeCAN sequence set with $N = 10, M = 2$

For any given $\mathbf{w}$ and ρ, we compute $\mathbf{z}$ (i.e., $\{z_{kp}\}$) using (11.34) and $y_k(p) = \sqrt{z_{kp}}e^{j\phi_{kp}}$ where the phases $\{\phi_{kp}\}$ can be chosen randomly. Then the IDFT of $\{y_k(p)\}_{p=1}^{N}$ gives the kth sequence ($k = 1, \ldots, M$). We call such a sequence set (attaining the $\widetilde{\text{ISL}}$ lower bound) an $\widetilde{\text{ISL}}$-optimal sequence set.

Note that in the single-sequence case (i.e., $M = 1$), $[\mathbf{A}^T\ \mathbf{B}^T]^T$ has a trivial null space, and thus there is only one solution to (11.8): $\mathbf{z} = \mathbf{z}_0$. This observation agrees with the known fact that for any single $\widetilde{\text{ISL}}$-optimal sequence (such as a Frank or Chu sequence), the magnitude of its DFT is the same at all frequency points (equal to $\sqrt{N}$ if the sequence energy is constrained to be N). In the case of $M > 1$, a similar characterization can be shown to hold: a sequence set is $\widetilde{\text{ISL}}$-optimal if and only if $\|\mathbf{y}_1\| = \cdots = \|\mathbf{y}_N\|$ (see (11.5) for the definition of $\mathbf{y}_p$).

Finally we remark on the fact that if a set of M sequences achieves $\tilde{\text{B}}_{\text{ISL}}$, then these M sequences are complementary sequences [Sarwate 1999], i.e., $\sum_{k=1}^{M} \tilde{r}_{kk}(l) = 0$ for $l = 1, \ldots, N-1$. Hence the construction method outlined in this section for $\widetilde{\text{ISL}}$-optimal sequence sets can also be used to obtain M complementary sequences.

11.3 Numerical examples

We use the Multi-PeCAN algorithm in Section 10.2 and the closed-form construction in Section 11.2 to generate $\widetilde{\text{ISL}}$-optimal sequence sets. Table 11.1 shows a Multi-PeCAN sequence set with $N = 10$, $M = 2$ while Table 11.2 shows a sequence set of the same size constructed from (11.34). Both sequence sets achieve the $\widetilde{\text{ISL}}$ lower bound: $\tilde{\text{B}}_{\text{ISL}} = 200$. The Multi-PeCAN sequence set is unimodular and its $\widetilde{\text{PSL}}$ is 3.53; the other sequence set has $\text{PAR}_1 = 2.28$ and $\text{PAR}_2 = 2.78$, and its $\widetilde{\text{PSL}}$ is 3.51.

Note that the Multi-PeCAN algorithm uses a random initialization (see Chapter 10) and that different initializations lead to different unimodular sequence sets that are all expected to meet the $\widetilde{\text{ISL}}$ lower bound. Furthermore, because the Multi-PeCAN algorithm relies on FFT operations, it can generate large sequence sets up to $NM \sim 10^5$ on an ordinary PC.

$\mathbf{x}_1$	$\mathbf{x}_2$
$0.3767 - 1.4607j$	$0.7378 + 0.1074j$
$-0.4307 - 0.0888j$	$1.4517 + 0.4993j$
$0.4292 - 0.9076j$	$0.4929 - 0.5050j$
$0.2901 + 1.1674j$	$-0.8559 + 0.2080j$
$0.9873 + 0.9642j$	$-0.1185 - 0.8191j$
$0.9064 + 0.4486j$	$1.6596 - 0.1516j$
$-0.1659 + 0.1825j$	$-0.0035 - 0.6168j$
$0.8787 - 0.7447j$	$-0.5303 + 0.5207j$
$-0.2686 + 0.4341j$	$0.9002 - 0.4568j$
$-0.5939 + 0.3849j$	$-0.0312 + 0.6313j$

Table 11.2. A sequence set constructed from (11.34) with $N = 10, M = 2$

In comparison with the Multi-PeCAN algorithm, which is iterative, the construction method in Section 11.2 has a closed form and thus there is essentially no length limit on the generated sequence sets. Moreover because $\rho, \mathbf{w}$ and $\{\phi_{kp}\}$ (see (11.34)) can be randomly chosen, this method can also generate many different sequence sets that all meet $\tilde{\mathrm{B}}_{\mathrm{ISL}}$. As a matter of fact the Multi-PeCAN sequence sets are just special cases (for which the sequences are unimodular) of (11.34).

11.4 Conclusions

By using the Multi-PeCAN framework introduced in the previous chapter, we proved two lower bounds on $\widetilde{\mathrm{ISL}}$ and $\widetilde{\mathrm{PSL}}$, respectively, for periodic correlations. An analytical formula for generating sequence sets (under the energy rather than the PAR constraint) of any size that attain the $\widetilde{\mathrm{ISL}}$ lower bound was derived. The Multi-PeCAN algorithm, however, can be used to computationally generate *unimodular* sequence sets that attain the $\widetilde{\mathrm{ISL}}$ lower bound.

12 Periodic ambiguity function (PAF)

Paralleling the ambiguity function (AF) definition in (6.1), the *periodic* ambiguity function (PAF) is given by [Freedman & Levanon 1994][Getz & Levanon 1995]

$$\chi_T(\tau, f) = \frac{1}{T}\int_0^T u(t)u^*(t-\tau)e^{-j2\pi f(t-\tau)}dt, \tag{12.1}$$

where τ and f denote, as usual, the time delay and Doppler frequency shift, respectively, and $u(t)$ is a periodic signal with period T:

$$u(t) = u(t - nT), \quad -\infty < t < \infty, \tag{12.2}$$

for any integer n. The PAF defined in (12.1) shows the response of a receiver that is matched to one period of the returned signal, which is a time-delayed and Doppler-shifted replica of the transmitted periodic signal. The receiver can also be matched to more than one period, in which case the definition of PAF becomes

$$\chi_{MT}(\tau, f) = \frac{1}{MT}\int_0^{MT} u(t)u^*(t-\tau)e^{-j2\pi f(t-\tau)}dt, \tag{12.3}$$

where M is an integer larger than 1. As can be expected, (12.1) and (12.3) are closely related to each other:

$$\begin{aligned}\chi_{MT}(\tau, f) &= \frac{1}{MT}\sum_{m=0}^{M-1}\int_{mT}^{(m+1)T} u(t)u^*(t-\tau)e^{-j2\pi f(t-\tau)}dt \\ &= \frac{1}{MT}\sum_{m=0}^{M-1}\int_0^T u(t+mT)u^*(t+mT-\tau)e^{-j2\pi f(t+mT-\tau)}dt \\ &= \frac{1}{M}\left(\sum_{m=0}^{M-1} e^{-j2\pi fmT}\right)\chi_T(\tau, f) = \frac{1-e^{-j2\pi fMT}}{M(1-e^{-j2\pi fT})}\chi_T(\tau, f).\end{aligned} \tag{12.4}$$

Taking the absolute value of each side of (12.4) we obtain the following relationship:

$$|\chi_{MT}(\tau, f)| = |\chi_T(\tau, f)|\left|\frac{\sin(\pi fMT)}{M\sin(\pi fT)}\right|. \tag{12.5}$$

The function $|\sin(\pi fMT)/M\sin(\pi fT)|$ is periodic with respect to f. Its period is $1/T$ and it reaches its highest value of 1 at points that are integer multiples of $1/T$; see Figure 12.1 for its plot when $M = 3$ and $T = 1$. It is readily deduced that the "multi-period" PAF $|\chi_{MT}(\tau, f)|$ has lower sidelobes than $|\chi_T(\tau, f)|$ due to the mask function

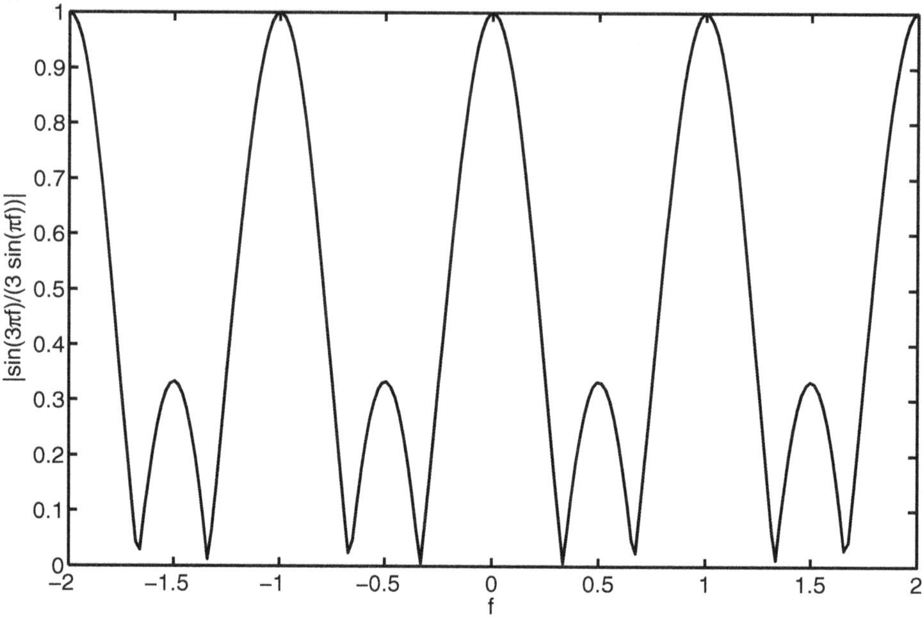

Figure 12.1 The function $\left|\frac{\sin(\pi f MT)}{M \sin(\pi f T)}\right|$ (when $M = 3$ and $T = 1$) that connects two definitions of PAF; see (12.5).

$|\sin(\pi f MT)/M \sin(\pi f T)|$. In this chapter we focus our attention on the PAF $\chi_T(\tau, f)$ as defined in (12.1).

12.1 PAF properties

Consider the periodic waveform

$$u(t) = \sum_{n=-\infty}^{\infty} x(n \bmod N) p_n(t), \qquad -\infty < t < \infty, \tag{12.6}$$

where $\{x(n)\}_{n=1}^{N}$ is the code sequence that is to be designed, $p_n(t)$ is the rectangular shaping pulse of duration t_p as defined in (1.2) and the mod operator is as defined in (1.17).

Figure 12.2 shows the absolute value of the PAF of $u(t)$ when $\{x(n)\}_{n=1}^{N}$ is a length-50 Golomb sequence (see (1.21)). (In what follows we will simply refer to the PAF of a sequence.) In the figure the delay τ is normalized by T (which equals Nt_p), the Doppler shift f is normalized by $1/T$, and the (absolute) value of PAF is normalized in such a way that the peak value at the origin is 1. In contrast with the AF plot of the same Golomb sequence in Figure 6.2, Figure 12.2 exhibits a clear periodic pattern along the time-delay axis.

Figure 12.3 shows the PAF of a length-50 Chu sequence (see (1.22)). It exhibits a Doppler-tolerant property (as indicated by the diagonal ridges) similar to Figure 12.2.

In addition, the volume of the PAF in Figure 12.3 is more concentrated in the ridges than that in Figure 12.2, which leads to larger low-sidelobe regions.

Figure 12.4 shows the PAF of a length-50 random-phase sequence. Different from the PAFs illustrated in Figure 12.2 or 12.3, the PAF of a random-phase sequence has a thumbtack shape and therefore indicates a Doppler-sensitive property. We use this random-phase sequence to initialize the PeCAN algorithm (see Chapter 9) and the PAF of the so-obtained PeCAN sequence is shown in Figure 12.5. Besides the thumbtack shape, we also observe a white horizontal stripe at zero Doppler delay, which results from the almost zero periodic auto-correlation sidelobes of the PeCAN sequence.

Some useful properties of PAF (which apply to any periodic signal $u(t)$) are listed in the following. The associated proofs are omitted since these properties are either self-explanatory or very similar to those of the AF (see Section 6.1):

- Periodicity w.r.t. the time delay:

$$\chi_T(\tau, f) = \chi_T(\tau + nT, f) \tag{12.7}$$

  for any integer n.
- Maximum value at $\{\tau = nT, f = 0\}$:

$$|\chi_T(\tau, f)|^2 \leq \frac{E_T^2}{T^2}, \tag{12.8}$$

  where E_T denotes the energy of $u(t)$ within one period

$$E_T = \int_0^T |u(t)|^2 dt. \tag{12.9}$$

- Symmetry:

$$|\chi_T(\tau, f)| = |\chi_T(-\tau, -f)|. \tag{12.10}$$

- Constant volume within one period:

$$V_T = \int_0^T \int_{-\infty}^{\infty} |\chi_T(\tau, f)|^2 d\tau df = \frac{E_T^2}{T^2}. \tag{12.11}$$

12.2 Discrete-PAF

In Section 6.2, the discrete-AF was derived from the AF definition of the signal model in (6.10). Likewise, we will derive the discrete-PAF definition. It follows from (12.6) that the PAF in (12.1) can be written as

$$\chi_T(\tau, f) = \frac{1}{T} \int_0^T \left(\sum_{n=-\infty}^{\infty} x(n \bmod N) p_n(t) \right) \left(\sum_{m=-\infty}^{\infty} x^*(m \bmod N) p_m(t - \tau) \right) e^{-j2\pi f(t-\tau)} dt. \tag{12.12}$$

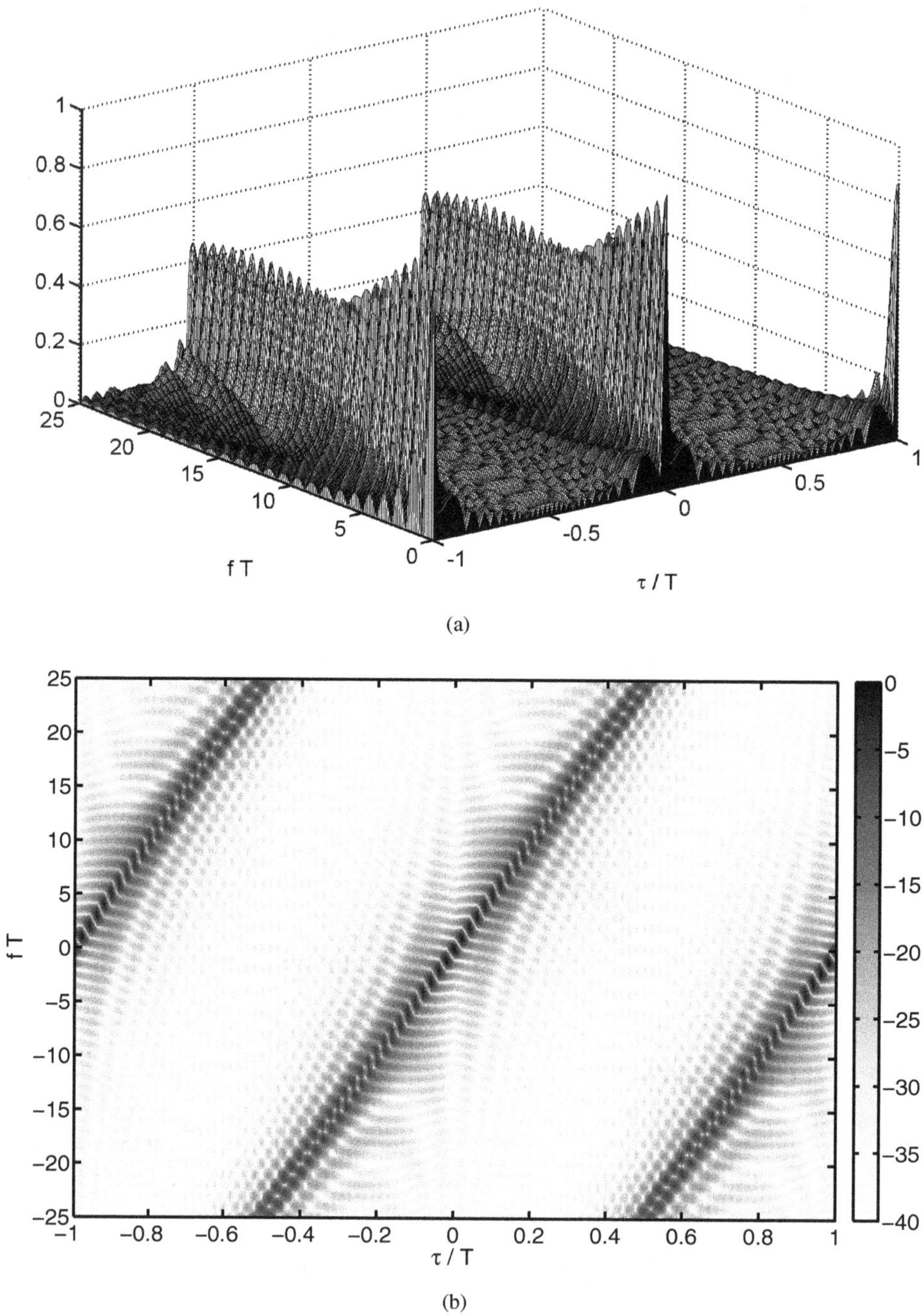

Figure 12.2 The PAF of a length-50 Golomb sequence. (a) 3D plot of the positive Doppler plane and (b) 2D plot of the entire plane.

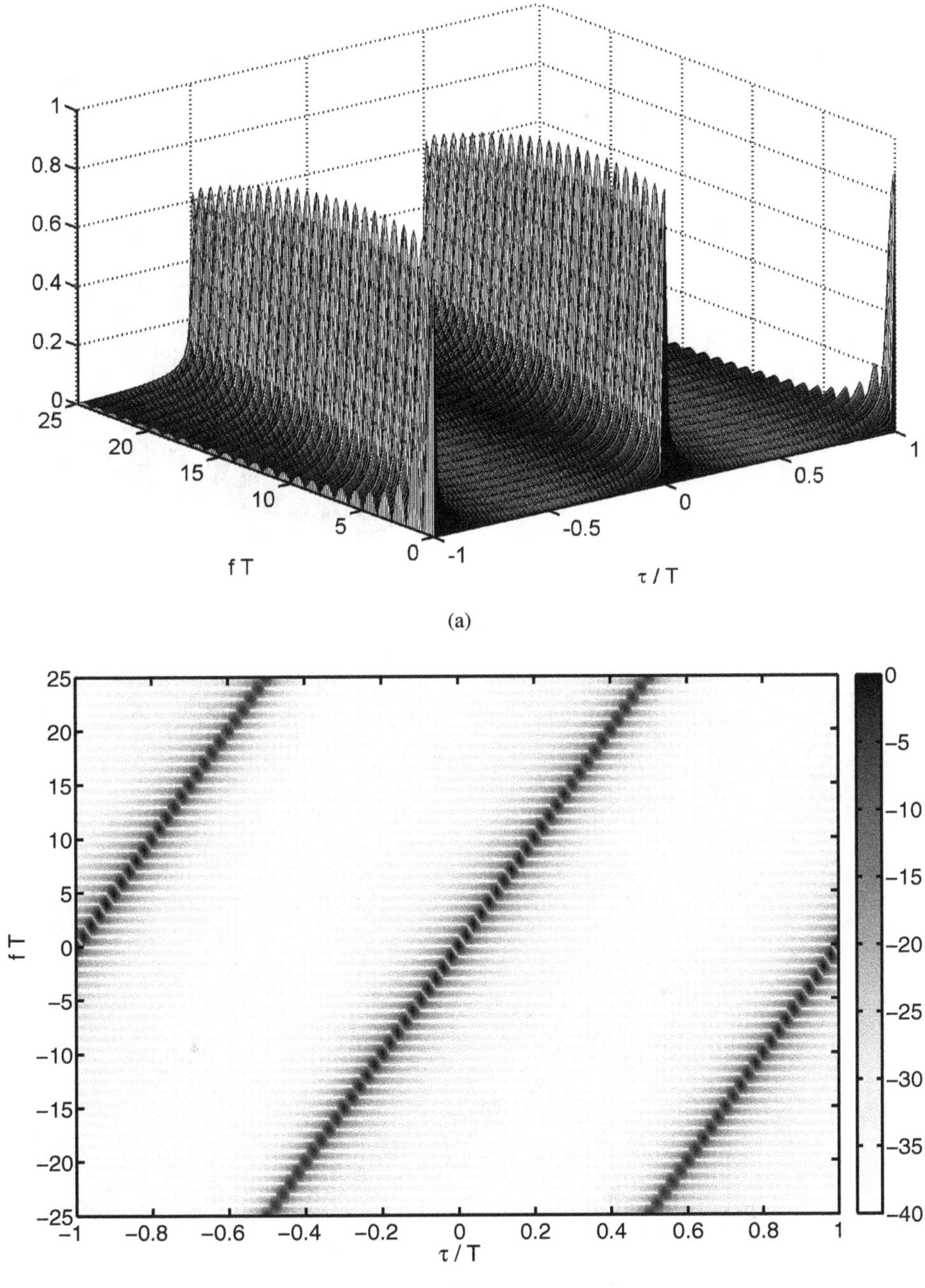

Figure 12.3 The PAF of a length-50 Chu sequence. (a) 3D plot of the positive Doppler plane and (b) 2D plot of the entire plane.

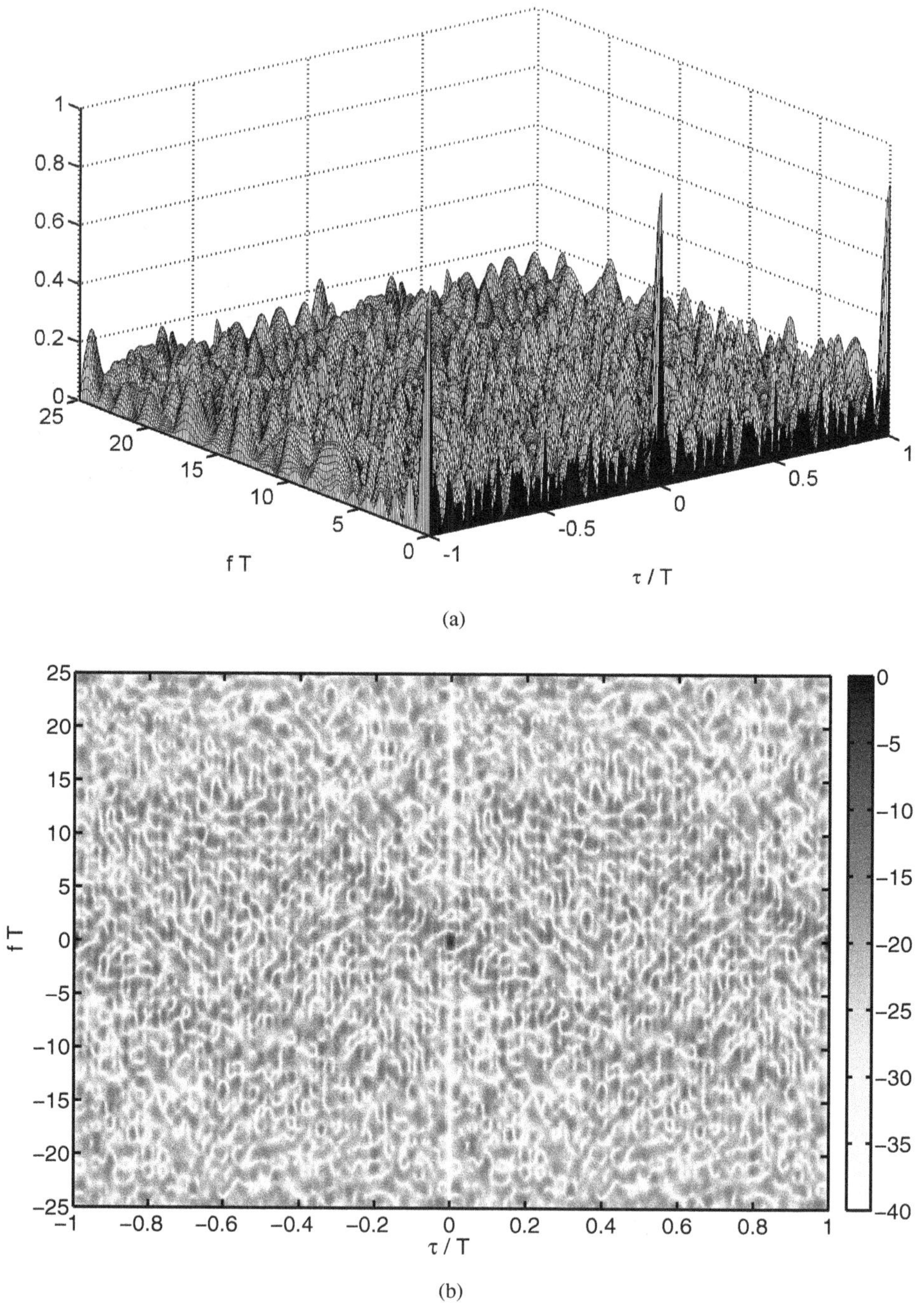

Figure 12.4 The PAF of a length-50 random-phase sequence. (a) 3D plot of the positive Doppler plane and (b) 2D plot of the entire plane.

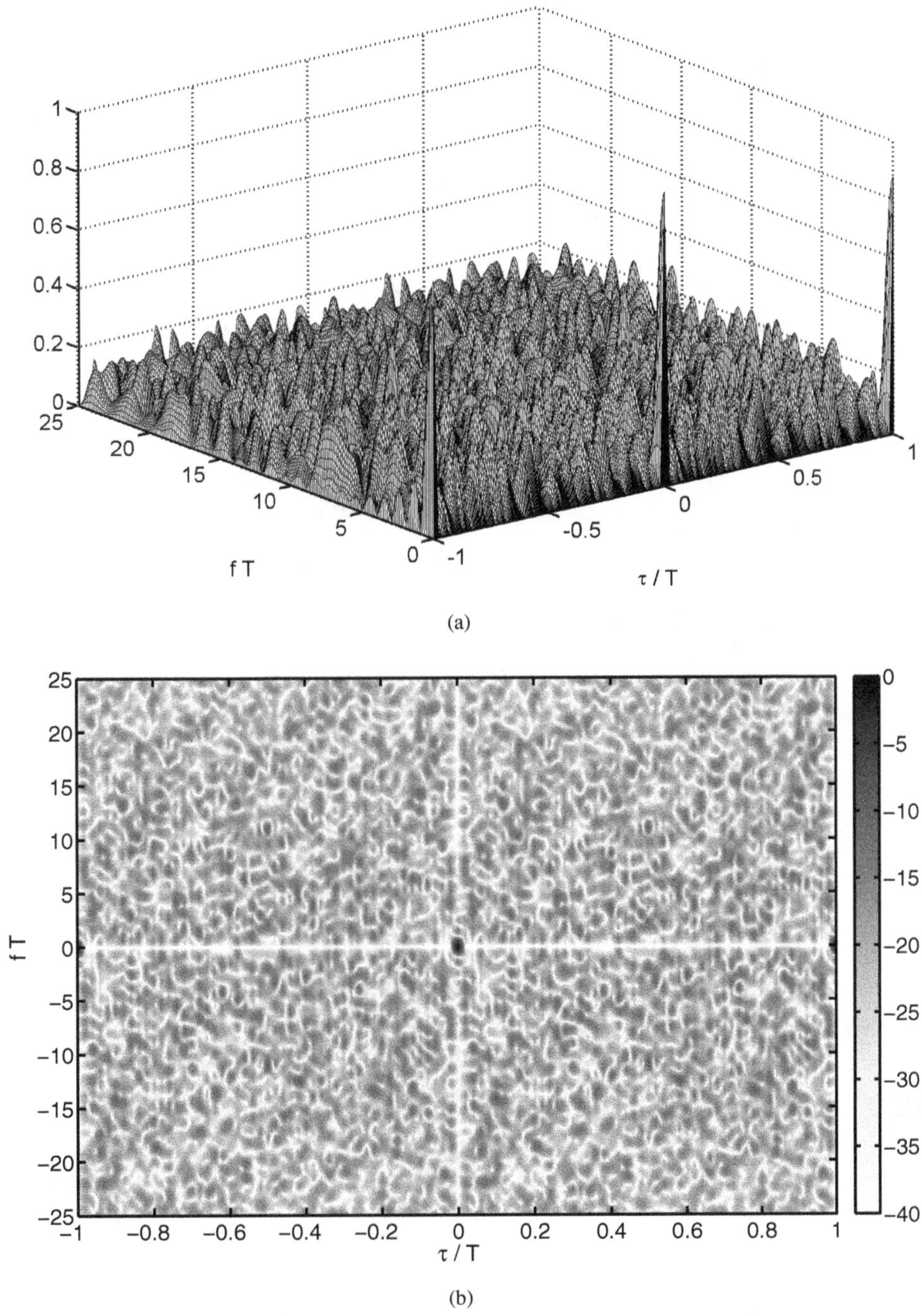

Figure 12.5 The PAF of a length-50 PeCAN sequence initialized by the random-phase sequence associated with Figure 12.4. (a) 3D plot of the positive Doppler plane and (b) 2D plot of the entire plane.

Since $\chi_T(\tau, f)$ is periodic w.r.t. τ, we can restrict our attention to just one period $0 \le \tau \le T$ $(T = Nt_p)$ and then obtain, noting that the duration of $p_n(t)$ is from $(n-1)t_p$ to nt_p,

$$\begin{aligned}\chi_T(\tau, f) =& \frac{1}{T}\int_0^T \left(\sum_{n=1}^{N} x(n)p_n(t)\right)\left(\sum_{m=-N+1}^{N} x^*(m \bmod N)p_m(t-\tau)\right)e^{-j2\pi f(t-\tau)}dt \quad (12.13)\\ =& \frac{1}{T}\sum_{m=-N+1}^{N}\sum_{n=1}^{N} x^*(m \bmod N)\left(\int_0^T p_n(t)p_m(t-\tau)e^{-j2\pi f(t-\tau)}dt\right)x(n).\end{aligned}$$

Using the derivation from (6.13) to (6.16) plus some slight modifications, we get the PAF value at the grid points $\{\tau = kt_p, f = p/Nt_p\}$:

$$\chi_T\left(kt_p, \frac{p}{Nt_p}\right) = \frac{1}{T}e^{j\pi\frac{p}{N}}\operatorname{sinc}\left(\pi\frac{p}{N}\right)\tilde{r}_T(k, p) \tag{12.14}$$

where $\tilde{r}_T(k, p)$ is what we name the discrete-PAF:

$$\begin{aligned}\tilde{r}_T(k, p) =& \sum_{n=1}^{N} x(n)x^*((n-k) \bmod N)e^{-j2\pi\frac{(n-k)p}{N}}, \qquad (12.15)\\ & k = \dots, -2, -1, 0, 1, 2, \dots, \quad p = -\frac{N}{2}, \dots, \frac{N}{2} - 1.\end{aligned}$$

It is easy to observe that $\tilde{r}_T(k, p)$ is periodic w.r.t. k:

$$\tilde{r}_T(k, p) = \tilde{r}_T(k + nN, p) \tag{12.16}$$

for any integer n. In addition, the discrete-PAF is symmetric,

$$|\tilde{r}_T(k, p)| = |\tilde{r}_T(-k, -p)|, \tag{12.17}$$

and has a constant volume within one period,

$$\sum_{k=0}^{N-1}\sum_{p=-N/2}^{N/2-1} |\tilde{r}_T(k, p)|^2 = N^3, \tag{12.18}$$

where we have assumed the energy constraint $\sum_{n=1}^{N} |x(n)|^2 = N$. The associated proofs are omitted once again due to likeness to those for discrete-AF properties.

12.3 Minimizing the discrete-PAF sidelobes

Similarly to minimizing the discrete-AF sidelobes in Section 6.3, we aim to minimize the discrete-PAF sidelobes in a rectangular region:

$$\min_{\{x(n)\}} \sum_{k\in\mathcal{K}}\sum_{p\in\mathcal{P}} |\tilde{r}(k, p)|^2 \tag{12.19}$$

$$\text{s.t. } |x(n)| = 1, \quad n = 1, \dots, N, \tag{12.20}$$

where $\mathcal{K}$ denotes the time-delay set of interest $\mathcal{K} = \{0, \pm1, \dots, \pm(K-1)\}$ $(K < N)$ and $\mathcal{P}$ denotes the Doppler frequency set of interest $\mathcal{P} = \{0, \pm1, \dots, \pm(P-1)\}$ $(P < N/2)$.

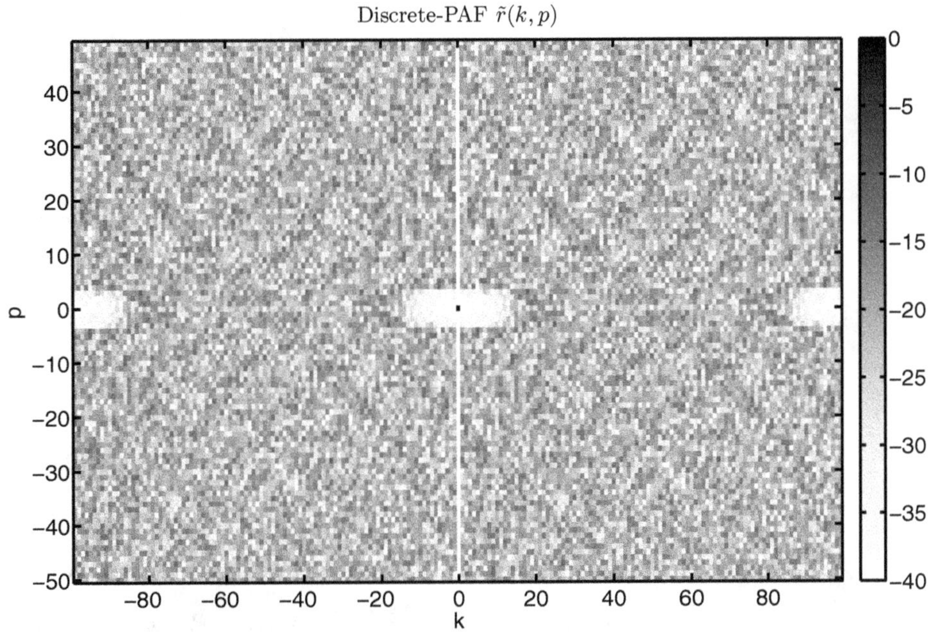

Figure 12.6 A synthesized discrete-PAF $|\tilde{r}(k, p)|$.

Define a set of P sequences $\{x_m(n)\}_{m=1}^{P}$ as in (6.23). Let $\{\tilde{r}_{ml}(k)\}$ denote the *periodic* correlation (see (10.1)) between $\{x_m(n)\}$ and $\{x_l(n)\}$:

$$\begin{aligned}\tilde{r}_{ml}(k) &= \sum_{n=1}^{N} x_m(n)x_l^*((n-k) \bmod N) \\ &= \exp\left\{j2\pi\frac{(m-1)(k \bmod N)}{N}\right\} \cdot \\ &\quad \sum_{n=1}^{N} x(n)x^*((n-k) \bmod N) \exp\left\{-j2\pi\frac{[(n-k) \bmod N](l-m)}{N}\right\}, \quad k \in \mathcal{K}, \\ &\quad m, l = 1, \ldots, P.\end{aligned} \tag{12.21}$$

Noticing the fact that the $\tilde{r}_T(k, p)$ in (12.15) is unchanged when we replace $(n-k)$ by $((n-k) \bmod N)$ in the exponent, it is not difficult to verify that all values of $\{|\tilde{r}(k, p)|\}$ ($k \in \mathcal{K}$ and $p \in \mathcal{P}$) are contained in the set $\{|\tilde{r}_{ml}(k)|\}$ ($k \in \mathcal{K}$ and $m, l = 1, \ldots, P$).

Therefore, in order to minimize the criterion in (12.19), we can equivalently aim to minimize the periodic correlations in (12.21). Define

$$\tilde{\mathbf{X}} = \left[\tilde{\mathbf{X}}_1 \ \cdots \ \tilde{\mathbf{X}}_P\right]_{N \times KP}, \tag{12.22}$$

where

$$\tilde{\mathbf{X}}_m = \begin{bmatrix} x_m(1) & x_m(N) & & x_m(N-K+2) \\ x_m(2) & x_m(1) & \cdots & x_m(N-K+3) \\ \vdots & \vdots & & \vdots \\ x_m(N) & x_m(N-1) & & x_m(N-K+1) \end{bmatrix}_{N\times K}, \quad m = 1, \ldots, P \tag{12.23}$$

(the kth column results from circularly shifting the first column downward by $k-1$ elements). It is not difficult to see that all $\{\tilde{r}_{ml}(k)\}$ ($k \in \mathcal{K}$ and $m, l = 1, \ldots, P$) appear in the matrix $\tilde{\mathbf{X}}^H\tilde{\mathbf{X}}$. Also note that the diagonal elements of $\tilde{\mathbf{X}}^H\tilde{\mathbf{X}}$ are equal to N because each element of $\{x(n)\}$ has unit modulus. Therefore, the criterion in (12.19) can be made small through minimizing the criterion

$$\left\|\tilde{\mathbf{X}}^H\tilde{\mathbf{X}} - N\mathbf{I}_{KP}\right\|^2, \tag{12.24}$$

which, in turn, leads to the following minimization problem (see (6.27) and (6.28)):

$$\begin{aligned} &\min_{\{x(n)\},\mathbf{U}} \|\tilde{\mathbf{X}} - \sqrt{N}\mathbf{U}\|^2, \\ &\text{s.t. } |x(n)| = 1, \quad n = 1, \ldots, N \\ &\quad x_m(n) = x(n)e^{j2\pi\frac{n(m-1)}{N}}, \quad m = 1, \ldots, P, \quad n = 1, \ldots, N, \\ &\quad \mathbf{U}^H\mathbf{U} = \mathbf{I} \quad (\mathbf{U} \text{ is } N \times KP). \end{aligned} \tag{12.25}$$

Observing the structural similarity between the two matrices in (6.26) and (12.23), the cyclic algorithm outlined in Table 6.1 can be used *mutatis mutandis* to solve (12.25). Figure 12.6 shows a synthesized discrete-PAF with parameters $N = 100$, $K = 15$ and $P = 4$. The "white" area in the center of the figure, which is similar to that in Figure 6.7, indicates that the sidelobes near the origin have been successfully suppressed.

12.4 Conclusions

We introduced the periodic ambiguity function (PAF) in this chapter. It is an extension of the periodic correlation function in the presence of a Doppler frequency shift. In much the same way as the AF was discussed in Chapter 6, we examined the properties of PAF, introduced the concept of the discrete-PAF and investigated the minimization of discrete-PAF sidelobes. As a final remark, the "cross-periodic-AF" could be introduced and analyzed parallelling Chapter 7 without any essential difficulty. However, we will not pursue this topic further here.

Part III

Transmit beampattern synthesis

13 Narrowband beampattern to covariance matrix

Antenna array beampattern design has been a well-studied topic and there is a considerable literature available from classic analytical design [Mailloux 1982][Dolph 1946][Elliott 1975][Ward et al. 1996][Van Trees 2002] to more recent works that resort to numerical optimization [Lebret & Boyd 1997][Scholnik & Coleman 2000][Cardone et al. 2002][San Antonio & Fuhrmann 2005][Li, Xie, Stoica, Zheng & Ward 2007]. The predominant problem considered in the literature refers to the *receive* beampattern design, which is concerned with designing weights for the received signal so that the signal component impinging from a particular direction is reinforced while those from other directions are attenuated, a way in which certain signal properties (e.g., the signal power or direction-of-arrival) can be estimated. Such a problem usually boils down to the design of an FIR (finite-impulse-response) filter in the narrowband case or a set of FIR filters in the wideband case.

The *transmit* beampattern design, on the other hand, is concerned with designing the probing signals to approximate a desired transmit beampattern (i.e., an energy distribution in space and frequency). It has been often stated that the receive and transmit beampattern designs are essentially equivalent, which is partly true in the sense that the two scenarios bear similar problem formulations and that the FIR filter taps obtained via receive pattern design can be used theoretically as the probing signal to achieve an identical transmit pattern. In practice, however, the transmit beampattern design problem appears to be much harder because of the energy and peak-to-average power ratio (PAR) constraints on the transmit waveforms. In particular, a digital-to-analog converter scales the signal by the maximum allowable amplitude and a saturated power amplifier works well only when the signal is constant-modulus [Skolnik 2008][Patton & Rigling 2008]. If the transmitted signals have wildly varying magnitudes, we risk energy loss or even signal distortion. As a result, the transmit beampattern design must be subject to the constraint that the transmit waveforms have unit modulus or low PARs. On the contrary in the receive beampattern design, the FIR taps can take on any values, although certain easy-to-meet constraints (e.g., the symmetry of the filter coefficients) are usually imposed. Therefore, except in a few simple situations such as the phased array case, the transmit beampattern design should be treated differently from the more prevalent receive beampattern design.

There have been works such as [Fuhrmann & San Antonio 2004][Forsythe & Bliss 2005] on probing signal design in the MIMO (multi-input multi-output) radar field. Unlike a phased array which transmits scaled versions of a single waveform, a MIMO radar can transmit probing waveforms that may be chosen at will [Fishler, Haimovich, Blum, Chizhik, Cimini & Valenzuela 2004][Fishler, Haimovich, Blum, Cimini, Chizhik & Valenzuela 2004][Robey et al. 2004][Xu et al. 2006][Fishler et al. 2006][Li & Stoica 2009]. In this chapter, the criterion used in [Fuhrmann & San Antonio 2004] will be modified to include the cross-correlation between the signals bounced back to the radar from the targets of interest. An efficient semidefinite quadratic programming (SQP) algorithm will be used for solving the signal design problem in polynomial time. Note that the formulation used in this chapter is for narrowband signals. Our discussions focus on the MIMO radar application, but the results are useful in many other applications (see, e.g., Chapter 17). The problem of wideband beampattern synthesis is to be dealt with in Chapter 15.

13.1 Problem formulation

Consider M transmit antennas and let $x_m(n)$ denote the discrete-time baseband signal transmitted by the mth antenna. Also, let θ denote the location parameter(s) of a generic focal point, for example, its azimuth angle and its range. Then, under the assumption that the transmitted probing signals are narrowband and that the propagation is nondispersive, the baseband signal at location θ can be described by the expression [Fuhrmann & San Antonio 2004][Stoica & Moses 2005]

$$\sum_{m=1}^{M} e^{-j2\pi f_0 \tau_m(\theta)} x_m(n) \triangleq \mathbf{a}^H(\theta)\mathbf{x}(n), \quad n = 1, \ldots, N, \tag{13.1}$$

where f_0 is the carrier frequency, $\tau_m(\theta)$ is the time needed by the signal emitted via the mth transmit antenna to arrive at the target, N denotes the number of samples of each transmitted signal pulse,

$$\mathbf{x}(n) = \left[x_1(n) \; x_2(n) \; \cdots \; x_M(n) \right]^T, \tag{13.2}$$

and

$$\mathbf{a}(\theta) = \left[e^{j2\pi f_0 \tau_1(\theta)} \; e^{j2\pi f_0 \tau_2(\theta)} \; \cdots \; e^{j2\pi f_0 \tau_M(\theta)} \right]^T. \tag{13.3}$$

Assuming that the transmit array is calibrated, $\mathbf{a}(\theta)$ is a known function of θ.

It follows from (13.1) that the power of the probing signal at a generic focal point with location θ is given by

$$P(\theta) = \mathbf{a}^H(\theta)\mathbf{R}\mathbf{a}(\theta), \tag{13.4}$$

where $\mathbf{R}$ is the covariance matrix of $\mathbf{x}(n)$, i.e.,

$$\mathbf{R} = E\{\mathbf{x}(n)\mathbf{x}^H(n)\}. \tag{13.5}$$

The "spatial spectrum" in (13.4), as a function of θ, will be called the *transmit beampattern*.

One of our problems consists of choosing **R**, under a uniform elemental power constraint,

$$R_{mm} = \frac{c}{M}, \quad m = 1, \ldots, M; \quad \text{with } c \text{ given}, \tag{13.6}$$

where R_{mm} denotes the (m, m)th element of **R**, to achieve the following goals:

(a) Maximize the total spatial power at a number of given target locations, or more generally, match a desired transmit beampattern.
(b) Minimize the cross-correlation between the probing signals at a number of given target locations; note from (13.1) that the cross-correlation between the probing signals at locations θ and $\bar{\theta}$ is given by $\mathbf{a}^H(\theta)\mathbf{R}\mathbf{a}(\bar{\theta})$.

According to (a) above, we would like to choose **R** such that the available transmit power is used to maximize the probing signal power at the locations of the targets of interest and to minimize it anywhere else. This is a natural goal that needs no additional comments. Regarding (b), we note from [Xu et al. 2006] and its references that the statistical performance of any adaptive MIMO radar technique depends heavily on the cross-correlation (beam)pattern $\mathbf{a}^H(\theta)\mathbf{R}\mathbf{a}(\bar{\theta})$ (for $\theta \neq \bar{\theta}$): the said performance degrades rapidly as the cross-correlation increases (to emphasize the importance of this fact, we note that in the phased array radar case, the probing signals at any two (different) target locations are fully correlated/coherent and, as a consequence, the standard adaptive techniques are not applicable). We will illustrate the above fact numerically in Section 13.3, where we apply the adaptive techniques of [Xu et al. 2006] to the data collected by a simulated MIMO radar with identically located transmit and receive antennas. Such data, under the simplifying assumption of point targets, can be described by the following equation [Xu et al. 2006, Fishler et al. 2006]:

$$\mathbf{y}(n) = \sum_{k=1}^{K} \beta_k \mathbf{a}^*(\theta_k)\mathbf{a}^H(\theta_k)\mathbf{x}(n) + \boldsymbol{\epsilon}(n), \tag{13.7}$$

where K is the number of targets that reflect the signals back to the radar receiver, $\{\beta_k\}$ are the complex amplitudes proportional to the radar cross sections (RCS's) of those targets, $\{\theta_k\}$ are their location parameters and $\boldsymbol{\epsilon}(n)$ denotes the interference-plus-noise term.

Another beampattern design problem we consider is how to choose **R**, under the uniform elemental power constraint in (13.6), to achieve the following goals:

(c) Minimize the sidelobe level in a prescribed region.
(d) Achieve a predetermined 3 dB mainbeam width.

In the next section, we will show how to formulate mathematically the goals in (a) and (b) or in (c) and (d) above, and how to solve the so-obtained design problems for **R**.

Remark: The optimal designs presented in the next section, particularly those in Sections 13.2.3 and 13.2.5, can be modified in a straightforward manner to accommodate other transmit power constraints, for example, a nonuniform elemental power constraint or a total transmit power constraint, $\operatorname{tr}(\mathbf{R}) = c$, where $\operatorname{tr}(\cdot)$ denotes the matrix trace. However, the constraint in (13.6) appears to be the most practically relevant one. □

Once $\mathbf{R}$ has been determined, a signal sequence $\{\mathbf{x}(n)\}$ that has $\mathbf{R}$ as its covariance matrix can be synthesized in a number of ways. Herein we simply set $\mathbf{x}(n) = \mathbf{R}^{1/2}\mathbf{w}(n)$, where $\{\mathbf{w}(n)\}$ is a sequence of i.i.d. random vectors with mean zero and covariance matrix $\mathbf{I}$, and $\mathbf{R}^{1/2}$ denotes a square root of $\mathbf{R}$. However, we note that such a synthesizing procedure may not give a signal that satisfies all practical requirements of a real-world sensing system (e.g., the above signal does not have a constant modulus). The topic of synthesizing practical probing radar signals with a given covariance matrix is discussed in Chapter 14.

13.2 Optimal designs

We consider four MIMO design problems in this section, that rely on either no or some prior information and which employ different criteria to formulate mathematically the goals (a) and (b) or goals (c) and (d) in the previous section. The phased array counterparts of several of the MIMO designs will be discussed as well.

To begin with, we assume that the radar has no prior information about the scene of interest.

13.2.1 Maximum power design for unknown target locations

Let us assume that there are $\tilde{K}$ ($\tilde{K} \leq K$) targets of interest. Without loss of generality, they are assumed to be at locations $\{\theta_k\}_{k=1}^{\tilde{K}}$. Then the cumulated power of the probing signals at the target locations is given by

$$\sum_{k=1}^{\tilde{K}} \mathbf{a}^H(\theta_k)\mathbf{R}\mathbf{a}(\theta_k) \triangleq \operatorname{tr}(\mathbf{RB}), \tag{13.8}$$

where

$$\mathbf{B} = \sum_{k=1}^{\tilde{K}} \mathbf{a}(\theta_k)\mathbf{a}^H(\theta_k). \tag{13.9}$$

In this subsection, we assume that the radar has no prior knowledge on $\mathbf{B}$. As a consequence, we can think of choosing $\mathbf{R}$ such that it maximizes (13.8) in the worst-case scenario:

$$\begin{aligned} &\max_{\mathbf{R}} \min_{\mathbf{B}} \operatorname{tr}(\mathbf{RB}) \\ &\quad \text{s.t. } R_{mm} = \frac{c}{M}, \quad m = 1, \ldots, M, \\ &\quad \mathbf{R} \geq 0 \\ &\quad \mathbf{B} \geq 0; \quad \mathbf{B} \neq 0, \end{aligned} \tag{13.10}$$

where the notation $\mathbf{R} \geq 0$ means that $\mathbf{R}$ is a positive semidefinite matrix, and the constraint $\mathbf{B} \neq 0$ is required to eliminate the trivial "solution" $\mathbf{B} = 0$ to the inner minimization.

The solution to a maximin design problem similar to (13.10), but with the uniform elemental power constraint $R_{mm} = c/M$, $m = 1, \ldots, M$, replaced by a less stringent total power constraint $\text{tr}(\mathbf{R}) = c$ was shown in [Stoica & Ganesan 2002] to be

$$\mathbf{R} = \frac{c}{M}\mathbf{I}. \tag{13.11}$$

Because (13.11) also satisfies the uniform elemental power constraint, it is the solution to the maximin design problem in (13.10) as well. This solution is easy to understand intuitively: without prior information as to where the focal points of interest are located, the MIMO radar will transmit a spatially white probing signal, which gives a constant power at any location θ, namely $(c/M)\|\mathbf{a}(\theta)\|^2 = c$ (note from (13.3) that $\|\mathbf{a}(\theta)\|^2 = M$).

Next we consider three design problems which assume that information about the (approximate) locations of the targets of interest is available. We will explain in due course how the said information can be obtained.

13.2.2 Maximum power design for known target locations

Assume that an estimate $\hat{\mathbf{B}}$ of $\mathbf{B}$ is available. Then the inner minimization in (13.10) can be omitted, and the problem becomes one of maximizing the total power at the locations of the targets of interest, under the uniform elemental power constraint. While this problem is a semidefinite program (SDP) and can, therefore, be efficiently solved numerically, it does not appear to admit a closed-form solution, unlike (13.10). For this reason, in the following we consider the said problem but with a total power constraint instead of the elemental power one, namely,

$$\begin{aligned} &\max_{\mathbf{R}} \ \text{tr}(\mathbf{R}\hat{\mathbf{B}}) \\ &\text{s.t.} \ \ \text{tr}(\mathbf{R}) = c, \\ &\qquad \mathbf{R} \geq 0. \end{aligned} \tag{13.12}$$

By a well-known inequality in matrix theory, we have

$$\text{tr}(\mathbf{R}\hat{\mathbf{B}}) \leq \lambda_{\max}(\hat{\mathbf{B}})\text{tr}(\mathbf{R}) = c\lambda_{\max}(\hat{\mathbf{B}}), \tag{13.13}$$

where $\lambda_{\max}(\hat{\mathbf{B}})$ denotes the largest eigenvalue of $\hat{\mathbf{B}}$, and where the last equality follows from the constraint $\text{tr}(\mathbf{R}) = c$. The upper bound in (13.13) is evidently achieved for

$$\mathbf{R} = c\mathbf{u}\mathbf{u}^H, \tag{13.14}$$

where $\mathbf{u}$ is the (unit-norm) eigenvector of $\hat{\mathbf{B}}$ associated with $\lambda_{\max}(\hat{\mathbf{B}})$ [Stoica & Ganesan 2002].

Remark: For $\tilde{K} = 1$, (13.14) reduces to

$$\mathbf{R} = c\frac{\mathbf{a}(\hat{\theta})\mathbf{a}^H(\hat{\theta})}{\|\mathbf{a}(\hat{\theta})\|^2}, \tag{13.15}$$

whose use leads to the delay-and-sum transmit beamformer commonly employed in phased array radar systems. □

The maximum power design in (13.14) is quite simple to compute and use; in particular, the covariance matrix in (13.14) can be synthesized using a constant-modulus scalar signal pre-multiplied by **u**. However, the design in (13.14) has a number of drawbacks:

1. The elemental transmit powers corresponding to (13.14) might vary widely.
2. While the design (13.14) maximizes the total power at the locations of the targets of interest, the way this power is distributed per each individual target is not controlled; consequently, the resulting powers at the target locations can be rather different from one another and from some possible desired relative levels.
3. The design (13.14) does not control the cross-correlation (beam)pattern either. The result is that for (13.14), and in fact for any rank-one design, the normalized magnitude of the pattern is given by (for $\theta \neq \bar{\theta}$):

$$\frac{\left|\mathbf{a}^H(\theta)\mathbf{R}\mathbf{a}(\bar{\theta})\right|}{\left[\mathbf{a}^H(\theta)\mathbf{R}\mathbf{a}(\theta)\right]^{1/2}\left[\mathbf{a}^H(\bar{\theta})\mathbf{R}\mathbf{a}(\bar{\theta})\right]^{1/2}} = \frac{\left|\mathbf{a}^H(\theta)\mathbf{u}\right|\left|\mathbf{u}^H\mathbf{a}(\bar{\theta})\right|}{\left|\mathbf{a}^H(\theta)\mathbf{u}\right|\left|\mathbf{a}^H(\bar{\theta})\mathbf{u}\right|} = 1. \tag{13.16}$$

The signals backscattered to the radar by any two targets are therefore fully coherent, which in particular makes the adaptive localization techniques inapplicable.

The next design replaces the maximum power criterion with a beampattern matching one that accommodates the uniform elemental transmit power constraint and allows the (approximate) control of the power at each target location; the new criterion also includes a term that penalizes large values of the cross-correlation (beam)pattern.

Remark: Maximizing the signal-to-interference-plus-noise ratio (SINR) at the receiver leads to a problem that has precisely the form in (13.10) or (13.12), but with a different matrix **B**. To see this, note from (13.7) that maximizing the receiver's SINR with respect to **R** is equivalent to maximizing the following criterion:

$$\operatorname{tr}\left(\sum_{k=1}^{K}\sum_{p=1}^{K}\beta_k\beta_p^H\mathbf{a}_k^H\mathbf{a}_k^H\mathbf{R}\mathbf{a}_p\mathbf{a}_p^T\right) \triangleq \operatorname{tr}\left(\mathbf{R}\tilde{\mathbf{B}}\right), \tag{13.17}$$

where $\mathbf{a}_k$ is a short notation for $\mathbf{a}(\theta_k)$, and

$$\tilde{\mathbf{B}} = \sum_{k=1}^{K}\sum_{p=1}^{K}(\beta_k\beta_p^H)\left(\mathbf{a}_p^T\mathbf{a}_k^H\right)\left(\mathbf{a}_p\mathbf{a}_k^H\right) \tag{13.18}$$

(it can be readily checked that $\tilde{\mathbf{B}} \geq 0$). Clearly, the cost functions in (13.10), (13.12) and (13.18) have the same form. Furthermore, for well-separated targets (for which $\mathbf{a}_p^T\mathbf{a}_k^H \approx 0$ for $p \neq k$) with similar β_ks we have $\tilde{\mathbf{B}} \approx \mathbf{B}$ (to within a multiplicative constant).

Maximizing the SINR of the received data is presumably a more justifiable goal than maximizing the signal's power at the target locations, as in (13.12). Nevertheless, we focus on (13.12) herein, because (13.12) is closer than (13.17) to the general framework of transmit beampattern matching design of the next subsection; additionally, the design derived from (13.12), as well as the one introduced in the following, rely only on a model for the transmit beampattern (see (13.1) - (13.5)), whereas (13.17) and the corresponding design would also require the use of a model for the received data (as in (13.7)). □

13.2.3 Beampattern matching design

Let $\phi(\theta)$ denote a desired transmit beampattern, and let $\{\mu_l\}_{l=1}^{L}$ be a fine grid of points that cover the location sectors of interest. We assume that the said grid contains points which are good approximations of the locations $\{\theta_k\}_{k=1}^{\tilde{K}}$ of interest and also, like in the previous subsection, that we dispose of (initial) estimates $\{\hat{\theta}_k\}_{k=1}^{\tilde{K}}$ of $\{\theta_k\}_{k=1}^{\tilde{K}}$. We will explain how to obtain $\phi(\theta)$ and $\{\hat{\theta}_k\}_{k=1}^{\tilde{K}}$ at the end of this subsection and also in the next section.

Our goal here is to choose $\mathbf{R}$ such that the transmit beampattern, $\mathbf{a}^H(\theta)\mathbf{R}\mathbf{a}(\theta)$, matches or rather approximates (in a least squares (LS) sense) the desired transmit beampattern, $\phi(\theta)$, over the sectors of interest, and also such that the cross-correlation (beam)pattern, $\mathbf{a}^H(\theta)\mathbf{R}\mathbf{a}(\bar{\theta})$ (for $\theta \neq \bar{\theta}$), is minimized (once again, in a LS sense) over the set $\{\hat{\theta}_k\}_{k=1}^{\tilde{K}}$. Mathematically, therefore, we want to solve the following problem:

$$\min_{\alpha,\mathbf{R}} \left\{ \frac{1}{L}\sum_{l=1}^{L} w_l \left[\alpha\phi(\mu_l) - \mathbf{a}^H(\mu_l)\mathbf{R}\mathbf{a}(\mu_l)\right]^2 + \frac{2w_c}{\tilde{K}^2-\tilde{K}}\sum_{k=1}^{\tilde{K}-1}\sum_{p=k+1}^{\tilde{K}} \left|\mathbf{a}^H(\hat{\theta}_k)\mathbf{R}\mathbf{a}(\hat{\theta}_p)\right|^2 \right\} \tag{13.19}$$

$$\text{s.t. } R_{mm} = \frac{c}{M}, \quad m = 1, \ldots, M,$$

$$\mathbf{R} \geq 0,$$

where $w_l \geq 0$ ($l = 1, \ldots, L$) is the weight for the lth grid point and $w_c \geq 0$ is the weight for the cross-correlation term. The value of w_l should be larger than that of w_k if the beampattern matching at μ_l is considered to be more important than the matching at μ_k. Note that by choosing $\max_l w_l > w_c$ we can give more weight to the first term in the design criterion above, and vice versa for $\max_l w_l < w_c$.

Note from (13.19) that the scaling factor α is obtained optimally as part of the solution to the LS matching problem. The reason for introducing α in the design problem is that typically $\phi(\theta)$ is given in a "normalized form" (e.g., satisfying $\phi(\theta) \leq 1, \forall\theta$), and our interest lies in approximating an appropriately scaled version of $\phi(\theta)$, not $\phi(\theta)$ itself.

Next we show that the design problem (13.19) can be efficiently solved in polynomial time as a SQP. We also explain how the prior information needed to define (13.19) can be obtained.

To show that the problem (13.19) is a SQP, we need some additional notation. Let vec($\mathbf{R}$) denote the $M^2 \times 1$ vector obtained by stacking the columns of $\mathbf{R}$ on top of each other. Let $\mathbf{r}$ denote the $M^2 \times 1$ real-valued vector made from R_{mm} ($m = 1, \ldots, M$) and the real and imaginary parts of R_{mp}, ($m, p = 1, \ldots, M; p > m$). Then, given the Hermitian symmetry of $\mathbf{R}$, we can write

$$\text{vec}(\mathbf{R}) = \mathbf{J}\mathbf{r} \tag{13.20}$$

for a suitable $M^2 \times M^2$ matrix $\mathbf{J}$ whose elements are easily derived constants $(0, \pm j, \pm 1)$. Making use of (13.20) and of some simple properties of the vec operator, the reader can verify that (the symbol $\otimes$ denotes the Kronecker product operator):

$$\begin{aligned}\mathbf{a}^H(\mu_l)\mathbf{R}\mathbf{a}(\mu_l) &= \text{vec}\left[\mathbf{a}^H(\mu_l)\mathbf{R}\mathbf{a}(\mu_l)\right] \\ &= \left[\mathbf{a}^T(\mu_l) \otimes \mathbf{a}^H(\mu_l)\right]\mathbf{J}\mathbf{r} \\ &\triangleq -\mathbf{g}_l^T\mathbf{r},\end{aligned} \tag{13.21}$$

and

$$\begin{aligned}\mathbf{a}^H(\hat{\theta}_k)\mathbf{R}\mathbf{a}(\hat{\theta}_p) &= \left[\mathbf{a}^T(\hat{\theta}_p) \otimes \mathbf{a}^H(\hat{\theta}_k)\right]\mathbf{J}\mathbf{r} \\ &\triangleq \mathbf{d}_{k,p}^H\mathbf{r}.\end{aligned} \tag{13.22}$$

Inserting (13.21) and (13.22) into (13.19) yields the following more compact form of the design criterion (which shows clearly the quadratic dependence on $\mathbf{r}$ and α):

$$\begin{aligned}&\frac{1}{L}\sum_{l=1}^{L} w_l\left[\alpha\phi(\mu_l) + \mathbf{g}_l^T\mathbf{r}\right]^2 + \frac{2w_c}{\tilde{K}^2 - \tilde{K}}\sum_{k=1}^{\tilde{K}-1}\sum_{p=k+1}^{\tilde{K}} \left|\mathbf{d}_{k,p}^H\mathbf{r}\right|^2 \\ =&\frac{1}{L}\sum_{l=1}^{L} w_l\left\{\left[\,\phi(\mu_l)\ \mathbf{g}_l^T\,\right]\begin{bmatrix}\alpha \\ \mathbf{r}\end{bmatrix}\right\}^2 + \frac{2w_c}{\tilde{K}^2 - \tilde{K}}\sum_{k=1}^{\tilde{K}-1}\sum_{p=k+1}^{\tilde{K}} \left|\left[\,0\ \mathbf{d}_{k,p}^H\,\right]\begin{bmatrix}\alpha \\ \mathbf{r}\end{bmatrix}\right|^2 \\ \triangleq&\boldsymbol{\rho}^T\boldsymbol{\Gamma}\boldsymbol{\rho},\end{aligned} \tag{13.23}$$

where

$$\boldsymbol{\rho} = \begin{bmatrix}\alpha \\ \mathbf{r}\end{bmatrix} \tag{13.24}$$

and

$$\boldsymbol{\Gamma} = \frac{1}{L}\sum_{l=1}^{L} w_l\begin{bmatrix}\phi(\mu_l) \\ \mathbf{g}_l\end{bmatrix}\left[\,\phi(\mu_l)\ \mathbf{g}_l^T\,\right] + \text{Re}\left\{\frac{2w_c}{\tilde{K}^2 - \tilde{K}}\sum_{k=1}^{\tilde{K}-1}\sum_{p=k+1}^{\tilde{K}}\begin{bmatrix}0 \\ \mathbf{d}_{k,p}\end{bmatrix}\left[\,0\ \mathbf{d}_{k,p}^H\,\right]\right\}. \tag{13.25}$$

The matrix $\boldsymbol{\Gamma}$ above is usually rank deficient. For example, in the case of an M-sensor uniform linear array with half-wavelength or smaller inter-element spacing and for $w_c = 0$, one can show that the rank of $\boldsymbol{\Gamma}$ is $2M$. The rank deficiency of $\boldsymbol{\Gamma}$, however, does not pose any serious problem for the SQP solver outlined below.

Making use of the form in (13.23) of the design criterion, we can rewrite (13.19) as the following SQP [Sturm 1999][Boyd & Vandenberghe 2004]:

$$\begin{aligned}&\min_{\delta,\boldsymbol{\varrho}} \delta \\ &\text{s.t. } \|\boldsymbol{\varrho}\| \leq \delta \\ &\quad R_{mm}(\boldsymbol{\varrho}) = \frac{c}{M}, \quad m = 1, \ldots, M \\ &\quad \mathbf{R}(\boldsymbol{\varrho}) \geq 0,\end{aligned} \tag{13.26}$$

where ($\boldsymbol{\Gamma}^{1/2}$ denotes a square root of $\boldsymbol{\Gamma}$)

$$\varrho = \boldsymbol{\Gamma}^{1/2}\rho \tag{13.27}$$

and where we have indicated explicitly the (linear) dependence of $\mathbf{R}$ on ϱ. For practical values of the array size M, the SQP above can be efficiently solved on a PC using public domain software [Sturm 1999].

In some applications, we would like that the synthesized beampattern at some given locations be very close to the desired values. As already mentioned, to a certain extent, this design goal can be achieved by the selection of the weights $\{w_l\}$ of the design criterion in (13.19). However, if we want the beampattern to match the desired values exactly, then selecting the weights $\{w_l\}$ is not enough and we have to modify the design problem as we now explain.

Consider, for instance, that we want the transmit beampattern at a number of points to be equal to certain desired levels. Then the optimization problem we need to solve is (13.19) with the following additional constraints:

$$\mathbf{a}^H(\check{\mu}_l)\mathbf{R}\mathbf{a}(\check{\mu}_l) = \zeta_l, \quad l = 1, \ldots, \check{L}, \tag{13.28}$$

where $\{\zeta_l\}$ are pre-determined levels. A similar modification of (13.19) takes place when the transmit beampattern at a number of points $\{\check{\mu}_l\}_{l=1}^{\check{L}}$ is restricted to be less than or equal to certain desired levels. The extended problems (with additional either equality or inequality constraints) are also SQPs, and therefore, similarly to (13.19), they can be solved efficiently using readily available software [Sturm 1999][Boyd & Vandenberghe 2004].

To conclude this subsection, we explain briefly how the desired transmit beampattern, $\phi(\theta)$, and the (initial) location estimates can be obtained (this aspect is further discussed in the next section). Because at the beginning of the operation, the MIMO radar system is assumed to have no prior knowledge of the scene, we transmit a maximin power optimal signal towards the targets, for which $\mathbf{R} = (c/M)\mathbf{I}$ (see (13.11)). Using the data $\mathbf{y}(n)_{n=1}^{N}$ collected by the receiving array of the system, we then compute the generalized likelihood ratio test (GLRT) function in [Xu et al. 2006], which is given by

$$\tilde{\phi}(\theta) = 1 - \frac{\mathbf{a}^H(\theta)\hat{\mathbf{R}}_{yy}^{-1}\mathbf{a}(\theta)}{\mathbf{a}^H(\theta)\hat{\mathbf{Q}}^{-1}\mathbf{a}(\theta)}, \tag{13.29}$$

where

$$\hat{\mathbf{Q}} = \hat{\mathbf{R}}_{yy} - \frac{\hat{\mathbf{R}}_{yx}\mathbf{a}(\theta)\mathbf{a}^H(\theta)\hat{\mathbf{R}}_{yx}^H}{\mathbf{a}^H(\theta)\hat{\mathbf{R}}_{xx}\mathbf{a}(\theta)}, \tag{13.30}$$

with

$$\hat{\mathbf{R}}_{yx} = \frac{1}{N}\sum_{n=1}^{N}\mathbf{y}(n)\mathbf{x}^H(n), \tag{13.31}$$

and $\hat{\mathbf{R}}_{xx}$ and $\hat{\mathbf{R}}_{yy}$ similarly defined. (Note that, while $\mathbf{R} = (c/M)\mathbf{I}$, the sample matrix $\hat{\mathbf{R}}_{xx}$ will in general be somewhat different from $(c/M)\mathbf{I}$.) The above function $\tilde{\phi}(\theta)$ possesses the following useful properties (see [Xu et al. 2006] for details):

1. It has values close to one in the vicinity of the target locations $\{\theta_k\}_{k=1}^{K}$, and close to zero elsewhere;
2. Unlike the spatial (pseudo) spectra obtained with other methods, (13.29) takes on small values even at the locations of possibly strong jammers (assuming that the jamming signals are uncorrelated with $\mathbf{x}(n)$);
3. The peaks of (13.29) around the target locations have widths that lead to a good compromise between resolution and robustness.

With the above features in mind, we can use the locations of interest of the dominant peaks of $\tilde{\phi}(\theta)$ as estimates of $\{\theta_k\}_{k=1}^{\hat{K}}$ and also to obtain a desired transmit beampattern – see the next section for details. Note that, in view of the features above, the MIMO radar will not waste power by probing either jammer locations (which may have the added bonus of making the radar harder to detect) or locations of uninteresting targets (which allows the radar to transmit spatially more power towards the targets of interest).

13.2.4 Minimum sidelobe beampattern design

In some applications, the beampattern design goal is to minimize the sidelobe level in a certain sector, when pointing the MIMO radar toward θ_0 (let us say). Such a minimum sidelobe beampattern design problem, with the uniform elemental transmit power constraint, can be formulated as follows:

$$\begin{aligned} &\min_{t,\mathbf{R}}(-t) \\ &\text{s.t. } \mathbf{a}^H(\theta_0)\mathbf{R}\mathbf{a}(\theta_0) - \mathbf{a}^H(\mu_l)\mathbf{R}\mathbf{a}(\mu_l) \geq t, \quad \forall \mu_l \in \Omega, \\ &\qquad \mathbf{a}^H(\theta_1)\mathbf{R}\mathbf{a}(\theta_1) = 0.5\mathbf{a}^H(\theta_0)\mathbf{R}\mathbf{a}(\theta_0), \\ &\qquad \mathbf{a}^H(\theta_2)\mathbf{R}\mathbf{a}(\theta_2) = 0.5\mathbf{a}^H(\theta_0)\mathbf{R}\mathbf{a}(\theta_0), \\ &\qquad \mathbf{R} \geq 0, \\ &\qquad R_{mm} = \frac{c}{M}, \quad m = 1, \ldots, M, \end{aligned} \tag{13.32}$$

where $\theta_2 - \theta_1$ (with $\theta_2 > \theta_0$ and $\theta_1 < \theta_0$) determines the 3 dB mainbeam width and Ω denotes the sidelobe region of interest. This is a SDP that can be solved in polynomial time using public domain software [Sturm 1999]. Similarly to the optimal SQP-based design of the previous subsection, if desired, the elemental power constraint can be replaced by a total power constraint. Note that we can relax somewhat the constraints in (13.32) defining the 3 dB mainbeam width; for instance, we can replace them by $(0.5 - \delta)\mathbf{a}^H(\theta_0)\mathbf{R}\mathbf{a}(\theta_0) \leq \mathbf{a}^H(\theta_i)\mathbf{R}\mathbf{a}(\theta_i) \leq (0.5 + \delta)\mathbf{a}^H(\theta_0)\mathbf{R}\mathbf{a}(\theta_0)$, $i = 1, 2$, for some small δ. Such a relaxation leads to a design with lower sidelobes, and to an optimization problem that is feasible more often than (13.32).

We can also introduce some flexibility in the elemental power constraint by allowing the elemental power to be within a certain range around c/M, while still maintaining the same total transmit power of c. Such a relaxation of the design problem allows lower sidelobe levels and smoother beampatterns, as we will show later on via some numerical examples.

13.2.5 Phased-array beampattern design

Finally, we comment on the conventional phased array beampattern design problem in which only the array weight vector can be adjusted and therefore all antennas transmit the same differently scaled waveform. We can readily modify the previously described beampattern matching or minimum sidelobe beampattern designs for the case of phased arrays by adding the constraint

$$\text{rank}(\mathbf{R}) = 1 \tag{13.33}$$

to (13.19) or (13.32), respectively. However, due to the rank-one constraint, both these originally convex optimization problems become non-convex. The lack of convexity makes the rank-one constrained problems much harder to solve than the original convex optimization problems [Orsi et al. 2004]. Semi-definite relaxation (SDR) is often used to obtain approximate solutions to such rank-constrained optimization problems [Boyd & Vandenberghe 2004]. The SDR is obtained by omitting the rank constraint. Hence, interestingly, the MIMO beampattern design problems are the SDR's of the corresponding phased array beampattern design problems.

In the numerical examples of the next section, we have used the Newton-like algorithm presented in [Orsi et al. 2004] to solve the rank-one constrained design problems for phased arrays. The said algorithm uses SDR to obtain an initial solution, which is the exact solution to the corresponding MIMO beampattern design problem. Although the convergence of the Newton-like algorithm is not guaranteed [Orsi et al. 2004], we did not encounter any apparent problem in our numerical simulations. An interesting detail here is that the approach in [Orsi et al. 2004] is for real-valued vectors and matrices; therefore we had to re-write the rank-one constraint in (13.33) in terms of real-valued quantities:

$$\text{rank}(\tilde{\mathbf{R}}) = 2, \tag{13.34}$$

where

$$\tilde{\mathbf{R}} = \begin{bmatrix} \text{Re}(Rbf) & -\text{Im}(\mathbf{R}) \\ \text{Im}(\mathbf{R}) & \text{Re}(\mathbf{R}) \end{bmatrix}, \tag{13.35}$$

The equivalence between (13.33) and (13.34) is proven in Appendix 13A.

13.3 Numerical examples

We present several numerical examples to demonstrate the merits of the proposed probing signal designs for MIMO radar systems. We consider a MIMO radar with a uniform linear array (ULA) comprising $M = 10$ antennas with half-wavelength spacing between adjacent antennas. The said array is used both for transmitting and for receiving. Without loss of generality, the total transmit power is set to $c = 1$.

13.3.1 Beampattern matching design

Consider first a scenario where $K = 3$ targets are located at $\theta_1 = -40°$, $\theta_2 = 0°$, and $\theta_3 = 40°$ with complex amplitudes equal to $\beta_1 = \beta_2 = \beta_3 = 1$. There is a strong jammer at 25° with an unknown waveform (uncorrelated with the transmitted MIMO radar waveforms) with a power equal to 10^6 (60 dB). Each transmitted signal pulse has $N = 256$ samples. The received signal is corrupted by zero-mean circularly symmetric spatially and temporally white Gaussian noise with variance σ^2. We assume that only the targets reflect the transmitted signals. In practice, the background can also reflect the signals. In the latter case, transmitting most of the power towards the targets should generate much less clutter returns than when transmitting power omni-directionally. Therefore, a MIMO radar system with a proper transmit beampattern design might provide even larger performance gains than those demonstrated herein.

Since we do not assume any prior knowledge about the target locations, the initial probing relies on the maximum power beampattern design for unknown target locations, i.e., $\mathbf{R} = (c/M)\mathbf{I}$. The corresponding transmit beampattern is omnidirectional with power equal to $c = 1$ at any θ. Using the data collected as a result of this initial probing, the target locations can be estimated using the GLRT technique, (13.29) - (13.31) , outlined in the previous section. Alternatively, location estimates can be obtained using the Capon technique, as the maximum points of the following spatial spectrum (see [Xu et al. 2006] for details):

$$\frac{\left|\mathbf{a}^H(\theta)\hat{\mathbf{R}}_{yy}^{-1}\hat{\mathbf{R}}_{yx}\mathbf{a}^H(\theta)\right|}{\left[\mathbf{a}^H(\theta)\hat{\mathbf{R}}_{yy}^{-1}\mathbf{a}(\theta)\right]\left[\mathbf{a}^T(\theta)\hat{\mathbf{R}}_{xx}\mathbf{a}^H(\theta)\right]}. \tag{13.36}$$

An example of the Capon spectrum for $\sigma^2 = -10$ dB is shown in Figure 13.1(a), where very narrow peaks occur around the target locations. Note that in Figure 13.1(a), a false peak occurs around $\theta = 25°$ due to the presence of the very strong jammer. The corresponding GLRT pseudo-spectrum as a function of θ is shown in Figure 13.1(b). Note that the GLRT is close to one at the target locations and close to zero at any other locations including the jammer location. Therefore, the GLRT can be used to reject the jammer peak in the Capon spectrum. The remaining peak locations in the Capon spectrum are the estimated target locations. Note that the Capon spectrum has sharper peaks than the GLRT function and hence, if desired, we can use the Capon estimates of the target locations in lieu of the GLRT estimates.

The initial target locations obtained by Capon or by GLRT can be used to compute the maximum power design in (13.14); we will use the GLRT estimates in what follows. An example of the transmit beampattern synthesized using the so-obtained $\mathbf{R}$ is shown in Figure 13.2. Since the rank of $\mathbf{R}$ is equal to one for this design, the MIMO radar operates as a conventional phased array radar in this case. As a consequence, in the presence of multiple targets, no data-adaptive approach can be used to obtain enhanced estimates of the target locations since the signals reflected by the targets are coherent with each other.

The initial target location estimates obtained using Capon or the GLRT can also be used to derive a desired beampattern for the beampattern matching design. In the fol-

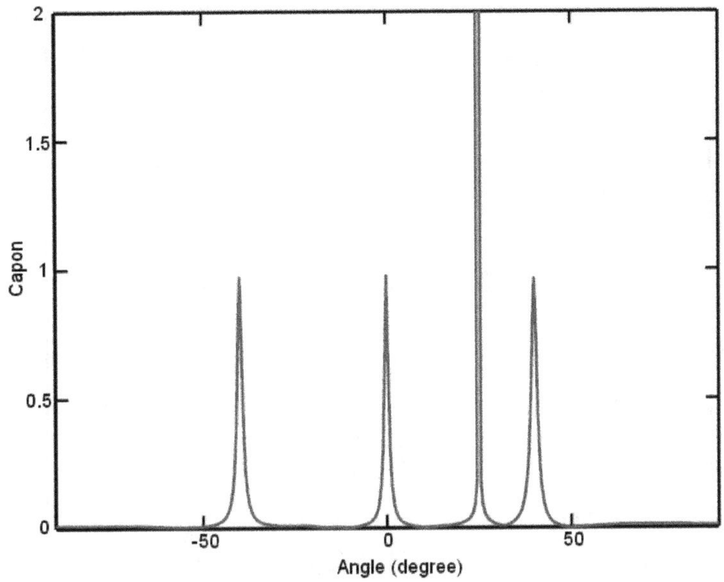

(a)

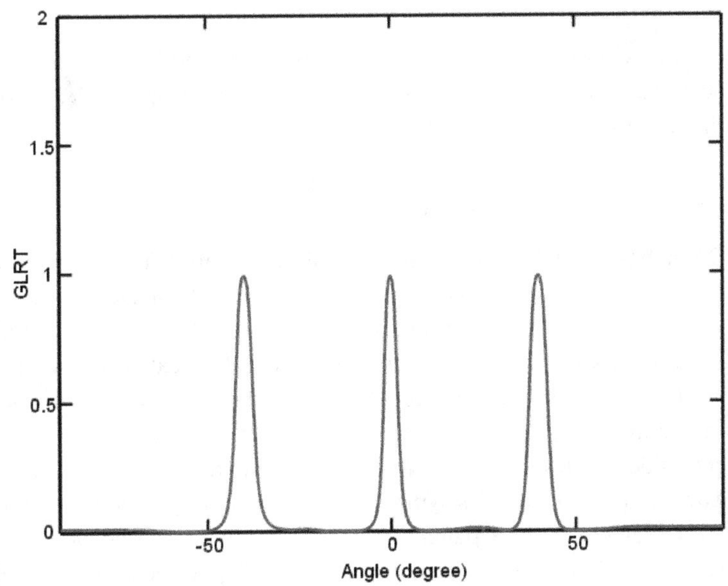

(b)

Figure 13.1 The Capon spatial spectrum and the GLRT pseudo-spectrum as functions of θ, for the initial omnidirectional probing. (a) Capon and (b) GLRT.

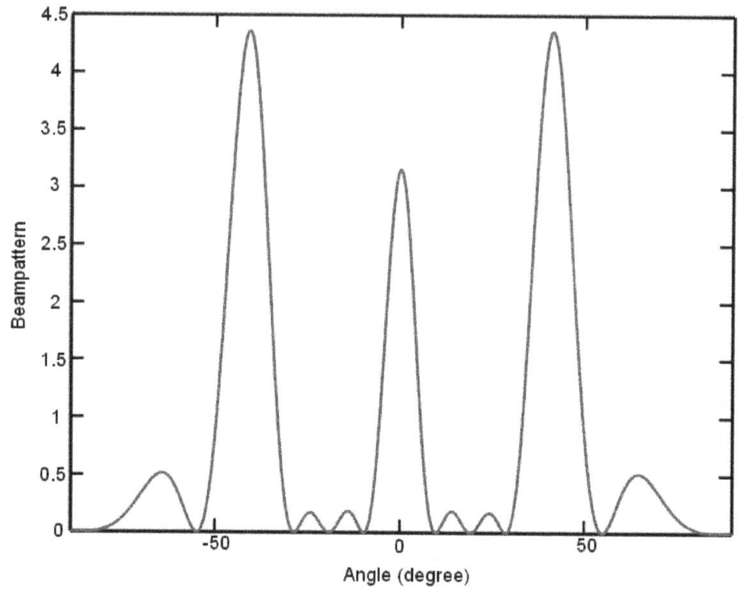

Figure 13.2 A transmit beampattern formed via maximum power design for given target locations (estimated via initial omnidirectional probing).

lowing numerical examples, we form the desired beampattern by using the dominant peak locations of the GLRT pseudo-spectrum, denoted as $\hat{\theta}_1, \ldots, \hat{\theta}_{\hat{K}}$, as follows (with $\hat{K}$ being the resulting estimate of K):

$$\phi(\theta) = \begin{cases} 1, \ \theta \in [\hat{\theta}_k - \Delta, \hat{\theta}_k + \Delta], & k = 1, \ldots, \hat{K}, \\ 0, \ \text{otherwise}, \end{cases} \tag{13.37}$$

where 2Δ is the chosen beamwidth for each target (Δ should be greater than the expected error in $\{\hat{\theta}_k\}$). Figure 13.3(a) is obtained using $\Delta = 10^\circ$ in the beampattern matching design in (13.19) with a mesh grid size of 0.1°, $w_l = 1$, $l = 1, \ldots, L$, and $w_c = 0$. The dashed line shows the desired beampattern in (13.37) scaled by the optimal value of α. Figure 13.3(b) shows the corresponding optimal phased array beampattern (obtained using the additional constraint $\text{rank}(\mathbf{R}) = 1$). Note that the phased array beampattern has higher sidelobe levels than its MIMO counterpart. Also, note that the synthesized MIMO transmit beampattern is symmetric (or nearly so), which is quite natural in view of the fact that the desired pattern is symmetric, whereas the optimal phased array beampattern is asymmetric (generating a symmetric pattern with a phased array would worsen the matching performance significantly). More importantly, in the presence of multiple targets, even though phased arrays can be used to form a transmit beampattern with peaks at the target locations, no data-adaptive approach can be used for localization or detection purposes since the signals reflected by the targets are coherent with each other.

Note that although we used $w_c = 0$ to obtain Figure 13.3(a), we have found out that the signals reflected by the targets exhibit low cross-correlations among them. As Δ is

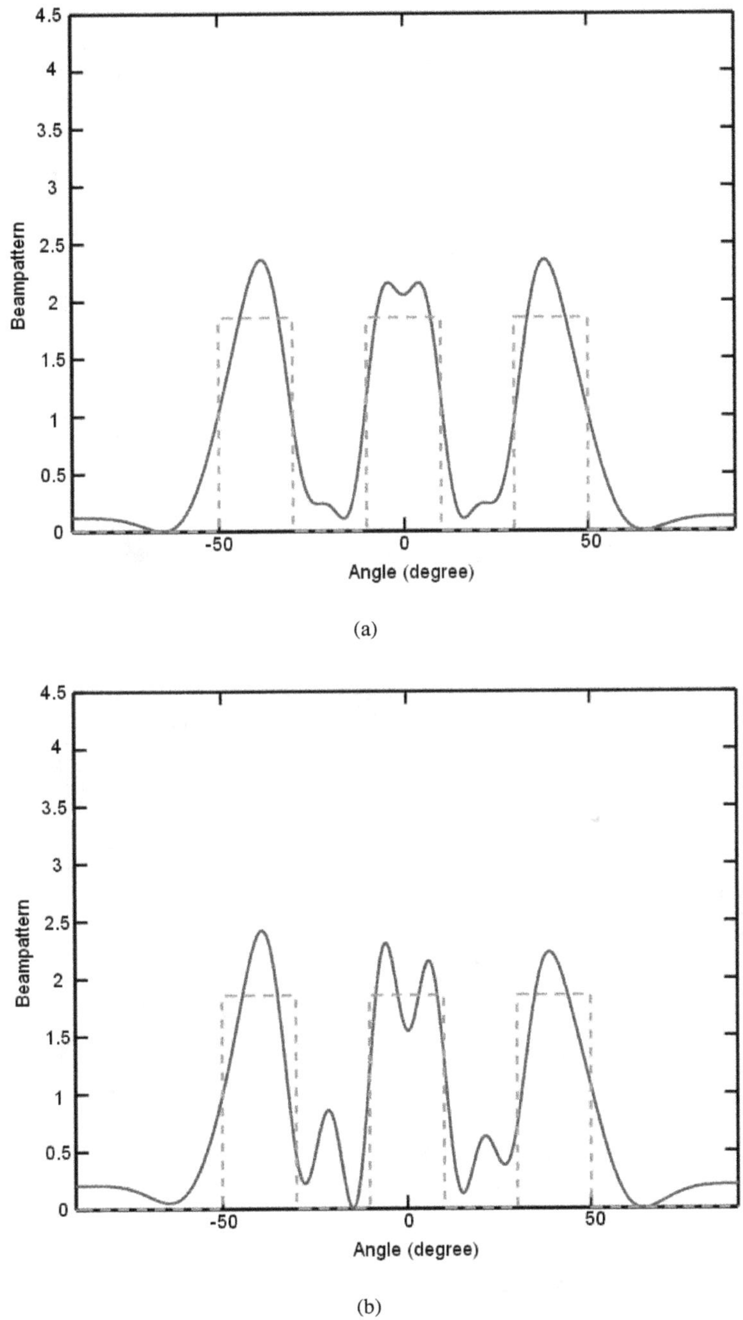

Figure 13.3 Transmit beampatterns formed via (a) MIMO beampattern matching design with $w_c = 0$ under the uniform elemental power constraint when $\Delta = 10°$ and (b) phased array beampattern matching design with $w_c = 0$ under the uniform elemental power constraint when $\Delta = 10°$. The desired beampatterns (scaled by α) are shown by dashed lines.

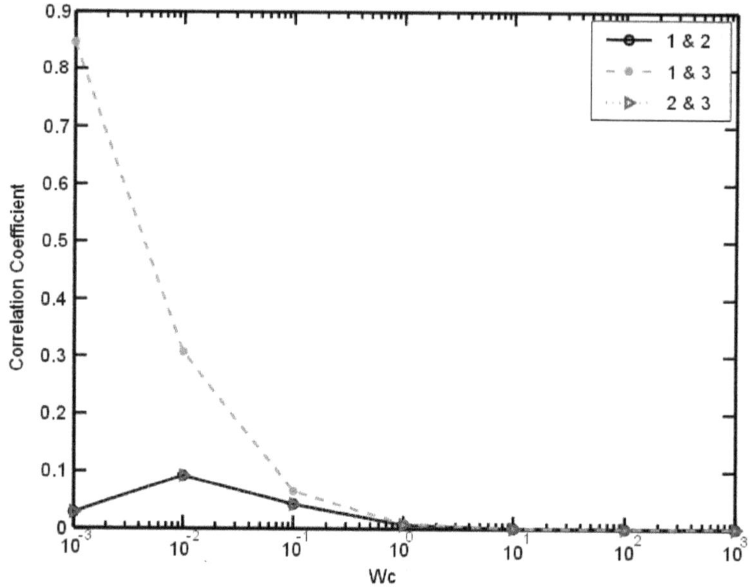

(a)

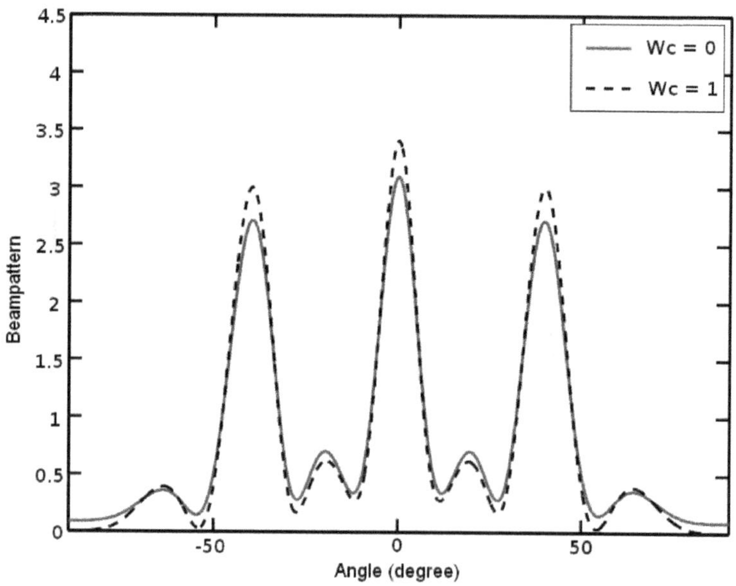

(b)

Figure 13.4 MIMO beampattern matching designs with $\Delta = 5°$ under the uniform elemental power constraint. (a) Cross-correlation coefficients of the three target reflected signals as functions of w_c, and (b) comparison of the beampatterns obtained with $w_c = 0$ and $w_c = 1$.

decreased, however, the cross-correlations become stronger when $w_c = 0$; consequently to achieve low cross-correlations in such a case, we need to increase the weight of the second term of the cost function in (13.19). The normalized magnitudes of the cross-correlation coefficients of the target reflected signals, as functions of w_c, are shown in Figure 13.4(a) for $\Delta = 5°$. We note that when w_c is close to zero, the first and third reflected signals are highly correlated, which can degrade significantly the performance of any adaptive technique. For $w_c = 1$, on the other hand, all cross-correlation coefficients are approximately zero. An example of the beampattern obtained with $w_c = 1$ is shown in Figure 13.4(b), where it is compared with the corresponding beampattern obtained with $w_c = 0$. Note that the designs obtained with $w_c = 1$ and with $w_c = 0$ are similar to one another even though the cross-correlation behavior of the former is much better than that of the latter.

In practice, the theoretical covariance matrix $\mathbf{R}$ of the transmitted signals is realized via the sample covariance matrix $\hat{\mathbf{R}}_{xx} = \frac{1}{N}\sum_{n=1}^{N} \mathbf{x}(n)\mathbf{x}^H(n)$, which may cause the synthesized transmit beampattern to be slightly different from the designed beampattern (unless $\hat{\mathbf{R}}_{xx} = \mathbf{R}$, which holds for instance if $\mathbf{x}(n) = \mathbf{R}^{1/2}\mathbf{w}(n)$ and $\frac{1}{N}\sum_{n=1}^{N} \mathbf{w}(n)\mathbf{w}^H(n) = \mathbf{I}$ exactly; in what follows, however, we assume that $\{\mathbf{w}(n)\}$ is a temporally and spatially white-noise signal from which the last equality holds only approximately in finite samples.) Let $\epsilon(\theta)$ denote the relative difference of the beampatterns obtained by using $\hat{\mathbf{R}}_{xx}$ and $\mathbf{R}$:

$$\epsilon(\theta) = \frac{\mathbf{a}^H(\theta)(\hat{\mathbf{R}}_{xx} - \mathbf{R})\mathbf{a}(\theta)}{\mathbf{a}^H(\theta)\mathbf{R}\mathbf{a}(\theta)}, \quad \theta \in [-90°, 90°]. \tag{13.38}$$

Figure 13.5(a) shows an example of $\epsilon(\theta)$, as a function of θ, for the beampattern design in Figure 13.4(b) with $w_c = 1$ and for $N = 256$. Note that the difference is quite small. We define the mean-squared error (MSE) between the beampatterns obtained by using $\hat{\mathbf{R}}_{xx}$ and $\mathbf{R}$ as the average of the square of (13.38) over all mesh grid points and over the set of Monte-Carlo trials. The MSE as a function of N, obtained from 1000 Monte-Carlo trials, is shown in Figure 13.5(b). As expected, the larger the sample number N, the smaller the MSE.

Next, we consider estimating the complex amplitudes $\{\beta_k\}$ of the reflected signals, (see (13.7)), in addition to estimating their location parameters $\{\theta_k\}$. Consider an example where two of the targets are closely spaced. We assume that there are $K = 3$ targets, located at $\theta_1 = -40°$, $\theta_2 = 0°$, and $\theta_3 = 3°$ with complex amplitudes equal to $\beta_1 = \beta_2 = \beta_3 = 1$. There is a strong jammer at 25° with an unknown waveform, which is uncorrelated with the transmitted MIMO radar waveforms, and with a power equal to 10^6 (60 dB). Each transmitted signal pulse has $N = 256$ samples. The received signal is corrupted by zero-mean circularly symmetric spatially and temporally white Gaussian noise with variance $\sigma^2 = -10$ dB. Figures 13.6(a) and 13.6(b) show the Capon spectrum and the GLRT pseudo-spectrum, respectively, for the initial omnidirectional probing; as can be seen from these figures, the two closely spaced targets cannot be resolved. Using this initial probing result, we derive an optimal beampattern matching design using (13.19) with a mesh grid size of 0.1°, $w_l = 1$, $l = 1, \cdots, L$, and $w_c = 1$. Since the initial probing indicated only two dominant peaks, these two peak locations are used in

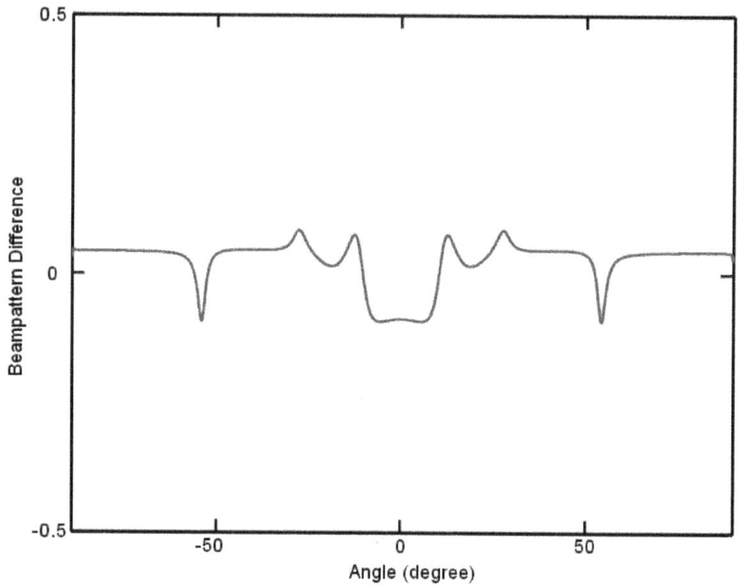

(a)

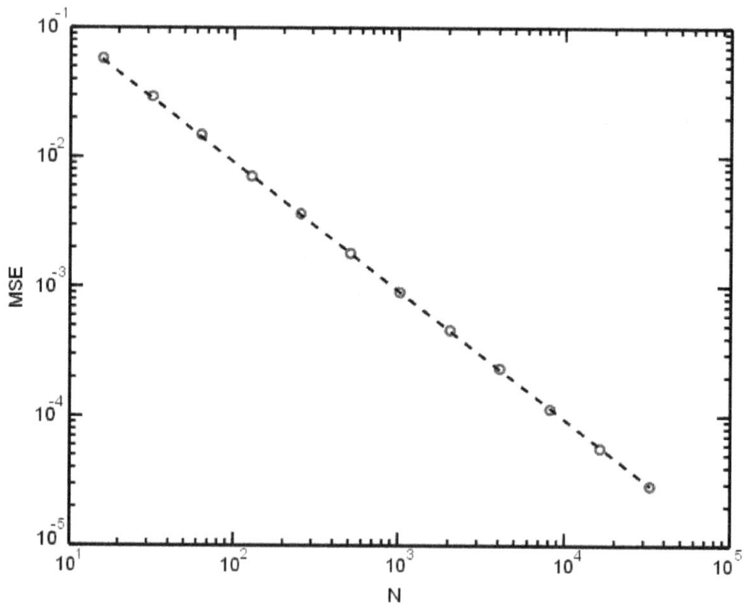

(b)

Figure 13.5 Analysis of the beampattern difference resulting from using $\hat{\mathbf{R}}_{xx}$ in lieu of $\mathbf{R}$. (a) Beampattern difference versus θ when $N = 256$, and (b) average MSE of the beampattern difference as a function of the sample number N.

(13.19). The desired beampattern is given by (13.37) with $\Delta = 10°$ and $\hat{K} = 2$. Figures 13.7(a) and 13.7(b), respectively, show the Capon spectrum and the GLRT pseudo-spectrum for the optimal probing. In principle, the two closely spaced targets are now resolved.

To conclude this subsection, we consider an example where the desired beampattern has only one wide beam centered at 0° with a width of 60°. Figure 13.8(a) shows the result for the beampattern matching design in (13.19) with a mesh grid size of 0.1°, $w_l = 1$, $l = 1, \cdots, L$, and $w_c = 0$. Figure 13.8(b) shows the corresponding phased array beampattern obtained by using the additional constraint of rank($\mathbf{R}$) = 1 in (13.19). We note that, under the elemental power constraint, the number of degrees of freedom (DOF) of the phased array that can be used for beampattern design is equal to only $M - 1$ (real-valued parameters); consequently, it is difficult for the phased array to synthesize a proper wide beam (also called spoiled beam). The MIMO design, however, can be used to achieve a beampattern significantly closer to the desired beampattern due to its much larger number of DOF, viz. $M^2 - M$. Interestingly, we have observed in a number of numerically simulated cases that, under the total power constraint, the optimal MIMO beampattern and the optimal phased array beampattern were quite close to one another. The elemental powers of the phased array design obtained under the total power constraint, however, varied significantly, which may be undesirable in many applications.

13.3.2 Minimum sidelobe beampattern design

Consider the beampattern design problem in (13.32) with the mainbeam centered at $\theta_0 = 0°$ and with a 3 dB width equal to 20° ($\theta_1 = -10°$, $\theta_2 = 10°$). The sidelobe region is $\Omega = [-90°, -20°] \cup [20°, 90°]$. The minimum-sidelobe beampattern design obtained by using (13.32) with a mesh grid size of 0.1° is shown in Figure 13.9(a). Note that the peak sidelobe level achieved by the MIMO design is approximately 18 dB below the mainlobe peak level. Figure 13.9(b) shows the corresponding phased array beampattern obtained by using the additional constraint rank($\mathbf{R}$) = 1 in (13.32). The phased array design fails to provide a proper mainlobe (it suffers from peak splitting) and its peak sidelobe level is about 5 dB higher than that of its MIMO counterpart.

Figure 13.10 is similar to Figure 13.9 except that now we allow the elemental powers to be between 80% and 120% of $c/M = 1/10$, while the total power is still constrained to be $c = 1$. Observe that by allowing such a flexibility in setting the elemental powers, we can bring down the peak sidelobe level of the MIMO beampattern by about 3 dB. The phased array design, on the other hand, does not appear to improve in any significant way.

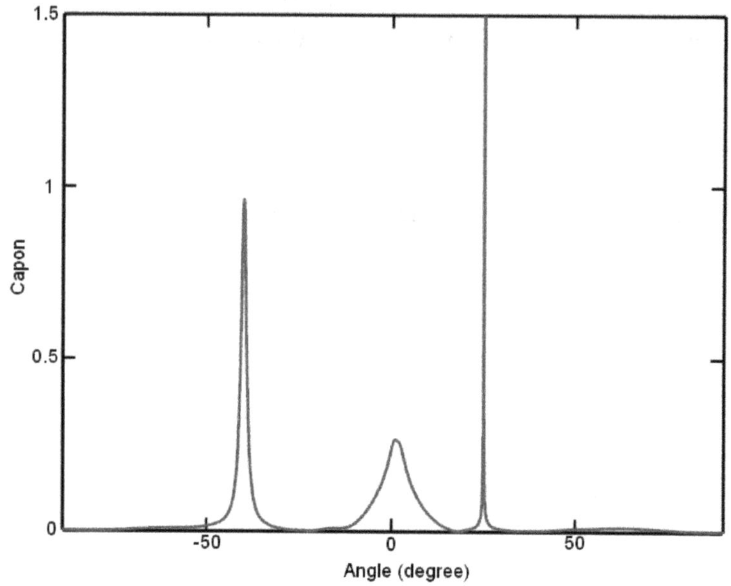

(a)

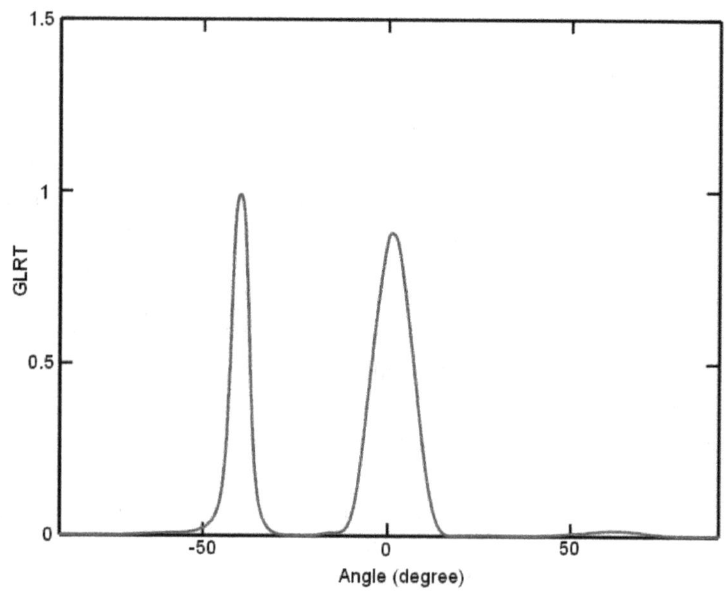

(b)

Figure 13.6 The Capon spatial spectrum and the GLRT pseudo-spectrum as functions of θ. (a) Capon for the initial omnidirectional probing and (b) GLRT for the initial omnidirectional probing.

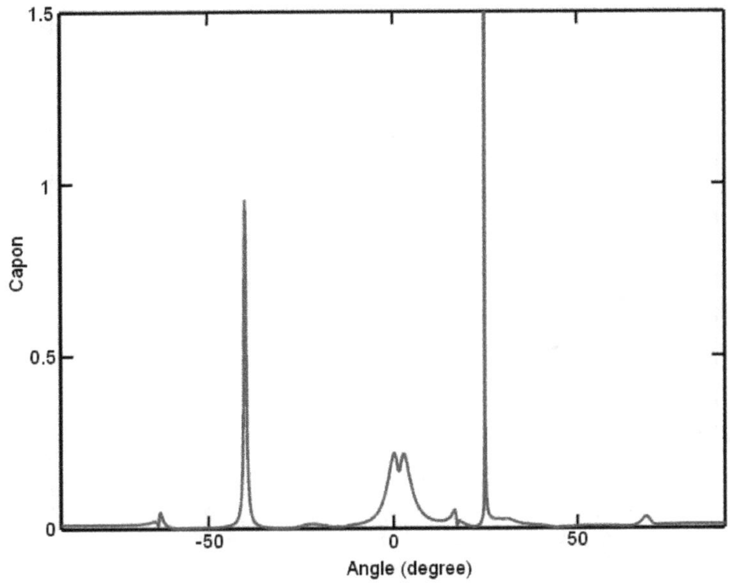

(a)

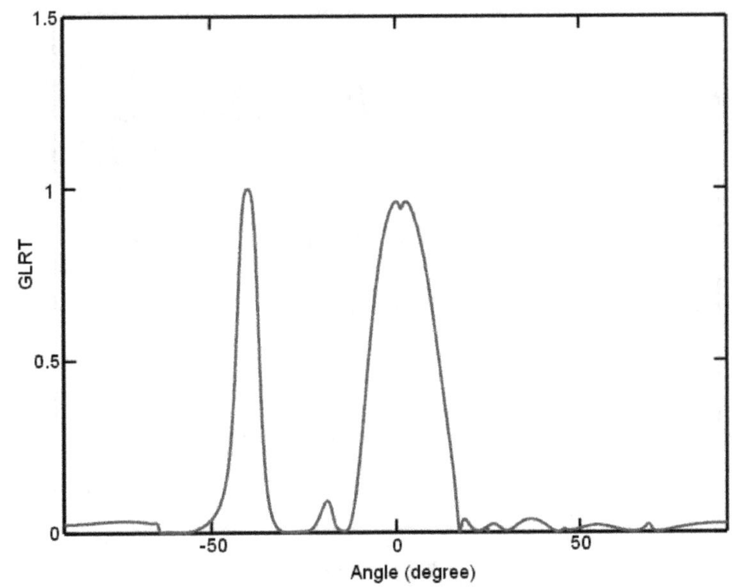

(b)

Figure 13.7 The Capon spatial spectrum and the GLRT pseudo-spectrum as functions of θ. (a) Capon for the optimal probing and (b) GLRT for the optimal probing.

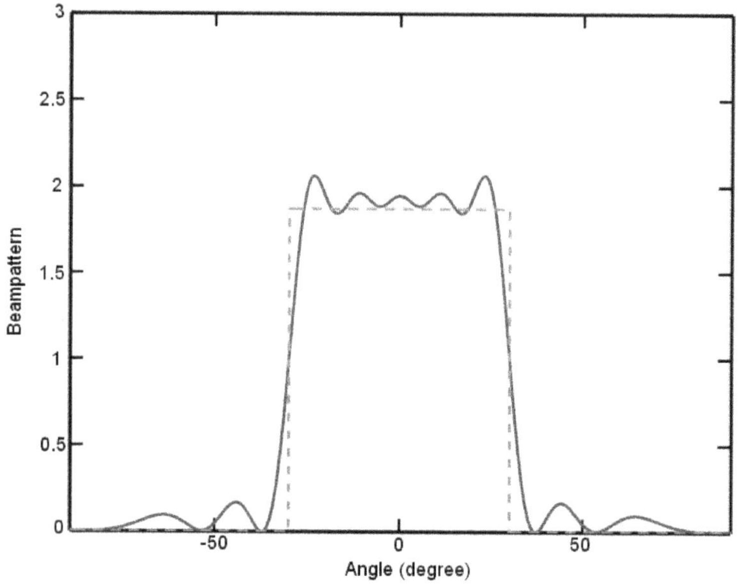

(a)

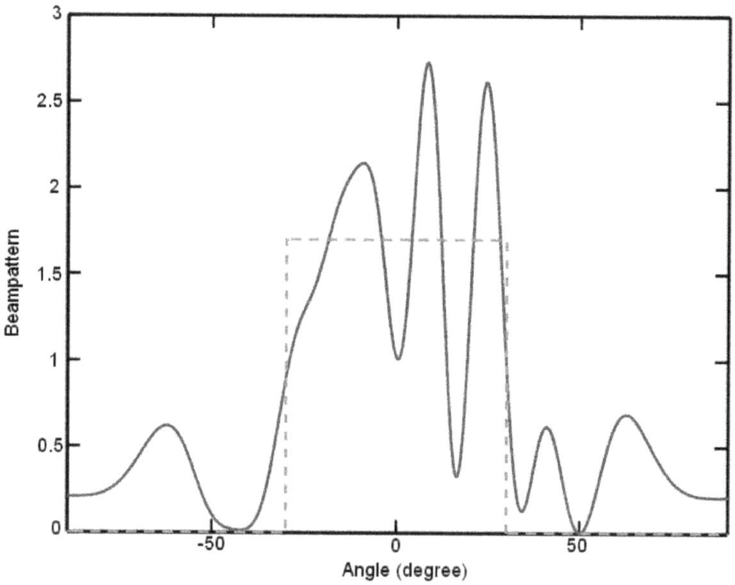

(b)

Figure 13.8 Beampattern matching designs under the uniform elemental power constraint. (a) MIMO and (b) phased array.

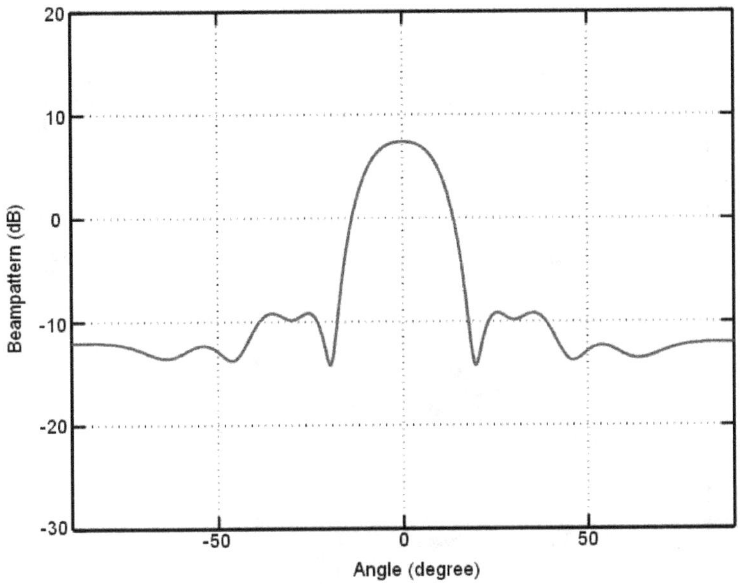

(a)

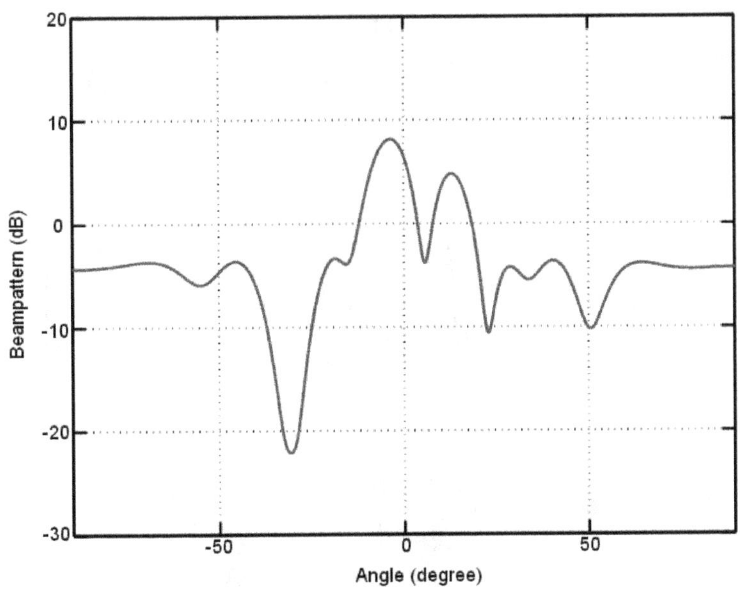

(b)

Figure 13.9 Minimum sidelobe beampattern designs, under the uniform elemental power constraint, when the 3 dB mainbeam width is 20°. (a) MIMO and (b) phased array.

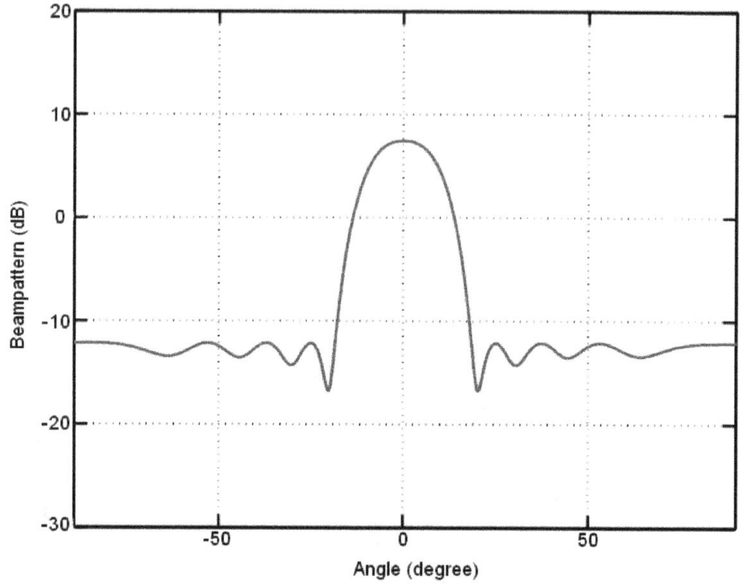

(a)

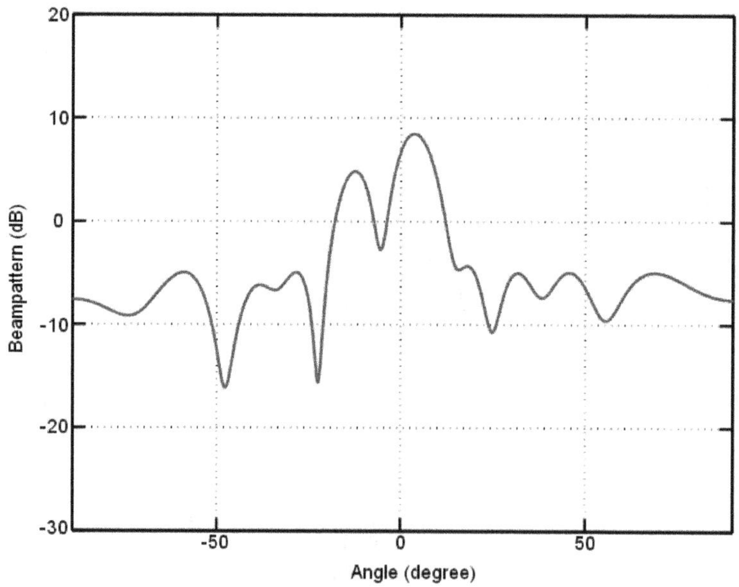

(b)

Figure 13.10 Minimum sidelobe beampattern designs, under a relaxed (±20%) elemental power constraint, when the 3 dB mainbeam width is 20°. (a) MIMO and (b) phased array.

13.4 Conclusions

We considered several transmit beampattern design problems for multi-antenna systems. In a MIMO radar application example, we have shown that beampattern design, by focusing the transmit power around the locations of the targets of interest while minimizing the cross-correlations of the signals reflected back to the radar, can significantly improve the parameter estimation accuracy of the adaptive MIMO radar techniques as well as enhance their resolution. We have also shown that, due to its significantly larger number of degrees of freedom, a multi-antenna system can achieve much better transmit beampatterns under the practical uniform elemental transmit power constraint than its phased array counterpart. Note that in this chapter the beampattern was synthesized by designing the waveform covariance matrix. In the next chapter we will show how to design the waveforms in order to match a given covariance matrix.

Appendix 13A Covariance matrix rank

Lemma: Let $\mathbf{R} \in \mathcal{C}^{M\times M}$, and let $\tilde{\mathbf{R}} \in \mathcal{R}^{2M\times 2M}$ be as defined in (13.35). Then

$$\text{rank}(\mathbf{R}) = M - m \Longleftrightarrow \text{rank}(\tilde{\mathbf{R}}) = 2(M-m), \quad m = 0, \cdots, M. \tag{13.39}$$

Proof: Let $\mathbf{v} \in \mathcal{C}^{M\times 1}$, $\mathbf{v} \neq 0$, be a vector in the null space of $\mathbf{R}$, $\mathcal{N}(\mathbf{R})$, i.e.,

$$\mathbf{R}\mathbf{v} = 0. \tag{13.40}$$

This implies that

$$\tilde{\mathbf{R}} \begin{bmatrix} \text{Re}(\mathbf{v}) \\ \text{Im}(\mathbf{v}) \end{bmatrix} = 0. \tag{13.41}$$

Moreover, since (13.40) also implies $\mathbf{R}(j\mathbf{v}) = 0$, we must also have

$$\tilde{\mathbf{R}} \begin{bmatrix} -\text{Im}(\mathbf{v}) \\ \text{Re}(\mathbf{v}) \end{bmatrix} = 0. \tag{13.42}$$

The vectors appearing in (13.41) and (13.42) are linearly independent of each other. Indeed, if we assume that they were not, then there would exist a nonzero complex-valued scalar, say $\zeta \neq 0$, such that:

$$\begin{bmatrix} \text{Re}(\mathbf{v}) & -\text{Im}(\mathbf{v}) \\ \text{Im}(\mathbf{v}) & \text{Re}(\mathbf{v}) \end{bmatrix} \begin{bmatrix} \text{Re}\{\zeta\} \\ \text{Im}\{\zeta\} \end{bmatrix} = 0 \Longrightarrow \mathbf{v}\zeta = 0 \Longrightarrow \mathbf{v} = 0, \tag{13.43}$$

which is a contradiction to the assumption that $\mathbf{v} \neq 0$.

Thus, we have shown that from each $\mathbf{v} \in \mathcal{N}(\mathbf{R})$ we can obtain (as in (13.41) and (13.42)) two linearly independent vectors in $\mathcal{N}(\tilde{\mathbf{R}})$. Furthermore, we can use an argument similar to (13.43) to show that if the vectors $\mathbf{v}_1, \mathbf{v}_2, \ldots \in \mathcal{N}(\mathbf{R})$ are linearly independent, then so are the corresponding vectors in $\mathcal{N}(\tilde{\mathbf{R}})$. It follows from these observations that

$$\text{rank}(\mathbf{R}) = M - m \Longrightarrow \text{rank}(\tilde{\mathbf{R}}) \leq 2(M-m), \quad m \in [0, M]. \tag{13.44}$$

Conversely, for each $\nu \neq 0$ satisfying $\tilde{\mathbf{R}}\nu = 0$ (i.e., $\nu \in \mathcal{N}(\tilde{\mathbf{R}})$), we can write ν as $\left[\operatorname{Re}(\mathbf{v}^T)\ \operatorname{Im}(\mathbf{v}^T)\right]^T$, and therefore we can build a $\mathbf{v}$ such that $\mathbf{v} \in \mathcal{N}(\mathbf{R})$. Furthermore, due to the structure of $\tilde{\mathbf{R}}$, it follows, as above, that also $\left[-\operatorname{Im}(\mathbf{v}^T)\ \operatorname{Re}(\mathbf{v}^T)\right]^T \in \mathcal{N}(\tilde{\mathbf{R}})$, and that $\left[\operatorname{Re}(\mathbf{v}^T)\ \operatorname{Im}(\mathbf{v}^T)\right]^T$ and $\left[-\operatorname{Im}(\mathbf{v}^T)\ \operatorname{Re}(\mathbf{v}^T)\right]^T$ are linearly independent of each other. Therefore, for any two such linearly independent vectors in $\mathcal{N}(\tilde{\mathbf{R}})$, there is one vector $\mathbf{v} \in \mathcal{N}(\mathbf{R})$. Again, similarly to what was shown above, the linear independence of the vectors in $\mathcal{N}(\tilde{\mathbf{R}})$ implies that of the corresponding vectors in $\mathcal{N}(\mathbf{R})$. Therefore, we have shown that

$$\operatorname{rank}(\tilde{\mathbf{R}}) = 2(M - m) \Longrightarrow \operatorname{rank}(\mathbf{R}) \leq M - m, \quad m \in [0, M]. \tag{13.45}$$

The stated result in (13.39) follows from (13.44) and (13.45). Indeed, if $\operatorname{rank}(\mathbf{R}) = M - m$, then we must have $\operatorname{rank}(\tilde{\mathbf{R}}) = 2(M - m)$ (otherwise (13.44) and (13.45) would imply that $\operatorname{rank}(\tilde{\mathbf{R}}) < 2(M - m)$, by (13.44), and thus that $\operatorname{rank}(\mathbf{R}) < M - m$, by (13.45), which is a contradiction). Similarly, (13.44) and (13.45) can be used to conclude that $\operatorname{rank}(\tilde{\mathbf{R}}) = 2(M - m) \Longrightarrow \operatorname{rank}(\mathbf{R}) = M - m$. □

14 Covariance matrix to waveform

Chapter 13 focused on the optimization of the waveform covariance matrix **R** for transmit beampattern design. While designing **R** is an important step, the final goal is the waveform design and we might therefore think of designing directly the probing signals by optimizing a given performance measure with respect to the matrix **X** of the signal waveforms. However, compared with optimizing the same performance measure with respect to the covariance matrix **R** of the transmitted waveforms, optimizing directly with respect to **X** is a more complicated problem. This is so because **X** has more unknowns than **R** and the dependence of various performance measures on **X** is more intricate than the dependence on **R** (as **R** is a quadratic function of **X**).

In this chapter we consider the following problem: with **R** obtained in a previous (optimization) stage, we want to determine a signal waveform matrix **X** whose covariance matrix is equal or close to **R**, and which also satisfies some practically motivated constraints (such as constant-modulus or low PAR constraints). We present a cyclic algorithm for the synthesis of such an **X**. We also consider the case where the synthesized waveforms are required to have good auto- and cross-correlation properties in time. Several numerical examples are provided to demonstrate the effectiveness of the proposed methodology.

14.1 Problem formulation

Consider an active sensing system equipped with M transmit antennas. Let the columns of $\mathbf{X} \in \mathbb{C}^{N\times M}$ be the transmitted waveforms, where N denotes the number of samples in each waveform.

In many practical systems the need to avoid expensive amplifiers and A/D converters has led to the requirement that the transmitted signals have constant modulus. Let $x_m(n)$ be the (n, m)th-element of **X**. The constant-modulus requirement means that:

$$|x_m(n)| = \sqrt{\gamma} \quad \text{for some given constant } \gamma \text{ and for } n = 1, \ldots, N. \tag{14.1}$$

(For example, we can choose $\gamma = R_{mm}$, where R_{mm} is the mth diagonal element of **R**; we omit the dependence of γ on m for notational simplicity.) However, (14.1) may be too hard a requirement on the signal matrix. In some modern systems this requirement can be replaced by the low PAR constraint (see also (4.14)):

$$\mathrm{PAR}(\mathbf{x}_m) \triangleq \frac{\max_n |x_m(n)|^2}{\frac{1}{N}\sum_{n=1}^{N}|x_m(n)|^2} \leq \rho, \quad \text{for a given } \rho \in [1, N], \tag{14.2}$$

(where, once again, we omit the dependence of ρ on m for notational simplicity). If we add to (14.2) a power constraint, viz.

$$\frac{1}{N}\sum_{n=1}^{N}|x_m(n)|^2 = \gamma \quad (\text{e.g., } \gamma = R_{mm}), \tag{14.3}$$

then the low PAR constraint is described by the following equations:

$$\begin{cases} \frac{1}{N}\sum_{n=1}^{N}|x_m(n)|^2 = \gamma, \\ |x_m(n)|^2 \leq \rho\gamma, \quad n = 1, \ldots, N. \end{cases} \tag{14.4}$$

In what follows, we use C to denote the set of signal matrices $\mathbf{X}$ that satisfy these constraints.

Let

$$\mathbf{R} \triangleq \frac{1}{N}\mathbf{X}^H\mathbf{X} \tag{14.5}$$

be the (sample) covariance matrix of the transmitted waveforms. The class of (unconstrained) signal waveform matrices $\mathbf{X}$ that realize a given covariance matrix $\mathbf{R}$ is given by

$$\frac{1}{\sqrt{N}}\mathbf{X}^H = \mathbf{R}^{1/2}\mathbf{U}^H, \tag{14.6}$$

where $\mathbf{U}^H$ is an arbitrary semi-unitary $M \times N$ matrix ($\mathbf{U}^H\mathbf{U} = \mathbf{I}$), and $\mathbf{R}^{1/2}$ is a Hermitian square root of the matrix $\mathbf{R}$ (we assume hereafter that $N \geq M$). Then a possible mathematical formulation of the problem of synthesizing the probing signal matrix $\mathbf{X}$ (without time correlation considerations) is as follows:

$$\min_{\mathbf{X}\in C;\mathbf{U}} \left\|\mathbf{X} - \sqrt{N}\mathbf{U}\mathbf{R}^{1/2}\right\|^2. \tag{14.7}$$

In many practical applications an important requirement on the synthesized waveforms is that they also have good temporal correlation properties. The synthesis of constant-modulus transmit signals with good auto- and cross-correlation properties is considered in Chapter 3. Here we extend the problem to the synthesis of a waveform matrix that realizes an arbitrary covariance matrix $\mathbf{R}$ under the low PAR constraint, and which also has good correlation properties in time.

Let

$$r_{m\tilde{m}}(p) = \sum_{n=p+1}^{N} x_m(n)x_{\tilde{m}}^*(n-p) = r_{\tilde{m}m}^*(-p), \quad p = 0, 1, \ldots, N-1, \tag{14.8}$$

denote the (cross)-correlation of $x_m(n)$ and $x_{\tilde{m}}(n)$ at lag p (see also (3.1)). To ensure that $\mathbf{X}$ has good auto- and cross-correlation properties in time, we require that

$$\sum_{m=1}^{M}\sum_{\tilde{m}=1}^{M}\sum_{p=-P+1,p\neq 0}^{P-1} |r_{m\tilde{m}}(p)|^2 \text{ is "small"}, \tag{14.9}$$

where P is an integer selected on an application-dependent basis. The above equation means that we want the sidelobes of the auto-correlation functions $\{r_{mm}(p)\}$ $(p \neq 0)$ and also of the cross-correlation functions $\{r_{m\tilde{m}}(p)\}_{m\neq\tilde{m}}$ $(p \neq 0)$ to be "small" at all lags $p = -P+1, \ldots, -1, 1, \ldots, P-1$.

Let $\bar{\mathbf{X}}$ be the block-Toeplitz matrix defined in (3.42). Note that $\bar{\mathbf{X}}^H$ is $MP \times (P + N - 1)$ and in what follows, we assume that $MP < P + N - 1$. The auto- and cross-correlations of $\{x_m(n)\}_{m=1,n=1}^{M,N}$ are the elements of the positive semidefinite matrix $\bar{\mathbf{X}}^H\bar{\mathbf{X}}$. The design problem of synthesizing $\mathbf{R}$ (see (14.5)) as well as of satisfying (14.9) consists thus of achieving:

$$\left\|\bar{\mathbf{X}}^H\bar{\mathbf{X}} - N\tilde{\mathbf{R}}\right\|^2 = \text{"small"}, \tag{14.10}$$

where $\tilde{\mathbf{R}} = \mathbf{R} \otimes \mathbf{I}$ with $\otimes$ denoting the Kronecker matrix product.

Following the idea that led to (14.7), the mathematical formulation of this extended problem of designing the probing signal matrix $\mathbf{X}$ with time correlation considerations is as follows:

$$\min_{\mathbf{X}\in C;\tilde{\mathbf{U}}} \left\|\bar{\mathbf{X}} - \sqrt{N}\tilde{\mathbf{U}}\tilde{\mathbf{R}}^{1/2}\right\|^2, \tag{14.11}$$

where $\tilde{\mathbf{U}}^H$ is an arbitrary semi-unitary $MP \times (P + N - 1)$ matrix ($\tilde{\mathbf{U}}^H\tilde{\mathbf{U}} = \mathbf{I}$).

14.2 Cyclic algorithm for signal synthesis

Under the unit-modulus constraint, the problem in (14.11) can be solved by the Multi-CAO algorithm outlined in Section 3.3. The only difference is that the matrix $\bar{\mathbf{X}}^H$ in (3.46) shall now be replaced by $\tilde{\mathbf{R}}^{1/2}\bar{\mathbf{X}}^H$.

Under the low PAR constraint in (14.4), the generic problem in (3.48) should be reformulated in the present case as

$$\begin{aligned}\sum_{k=1}^{P} |x - \mu_k|^2 &= P|x|^2 - 2\text{Re}\left(x^* \sum_{k=1}^{P}\mu_k\right) + \sum_{k=1}^{P}|\mu_k|^2 \\ &= P\left|x - \frac{\sum_{k=1}^{P}\mu_k}{P}\right|^2 + \text{const},\end{aligned} \tag{14.12}$$

where const denotes a term that does not depend on x. Let $\mathbf{x}_m$ denote the mth column of $\mathbf{X}$. Then given (14.12), Step 2 of the Multi-CAO algorithm (see Table 3.3) reduces to M

independent minimization problems of the following (for $m = 1, \ldots, M$) form:

$$\begin{aligned} &\min_{\mathbf{x}_m} \|\mathbf{x}_m - \mathbf{z}\|^2 \\ &\text{s.t. } \frac{1}{N}\sum_{n=1}^{N} |x_m(n)|^2 = \gamma \\ &|x_m(n)|^2 \le \rho\gamma, \quad n = 1, \ldots, N, \end{aligned} \tag{14.13}$$

where the pth element of $\mathbf{z}$ is defined as $(\sum_{k=1}^{P} \mu_k)/P$. The above minimization problem can be solved by the nearest-vector algorithm discussed in Section 4.2.

The above outlined algorithm will be simply referred to as the CA algorithm in the following numerical examples.

14.3 Numerical examples

The transmitted waveforms have good auto- and cross-correlation properties in time if

$$\left\|\mathbf{X}^H \mathbf{J}_p \mathbf{X}\right\|^2 \tag{14.14}$$

is "small" (ideally, zero) for $p = -P+1, \ldots, -1, 1, \ldots, P-1$, where

$$\mathbf{J}_p = \begin{bmatrix} \overbrace{\mathbf{0} \quad\quad 1}^{p+1} & \mathbf{0} \\ & \ddots \\ & & 1 \\ & & \\ \mathbf{0} & & \end{bmatrix}_{L\times L} = \mathbf{J}_{-p}, \quad p \ge 0, \tag{14.15}$$

is a "shift" matrix. Hence, we use the normalized values of (14.14), which we call "correlation levels", to evaluate the time correlation properties of $\mathbf{X}$.

Assume that we are given a desired beampattern $\phi(\theta)$ defined over a region of interest Ω. Let $\{\mu_l\}_{l=1}^{L}$ be a fine grid of points that covers Ω. Let $\{\theta_k\}_{k=1}^{\tilde{K}}$ be a given set of locations of possible interest. The mathematical problem of beampattern matching through designing the covariance matrix $\mathbf{R}$ has been formulated in (13.19) and is repeated here for convenience:

$$\begin{aligned} &\min_{\alpha,\mathbf{R}} \left\{ \frac{1}{L}\sum_{l=1}^{L} w_l \left[\alpha\phi(\mu_l) - \mathbf{a}^H(\mu_l)\mathbf{R}\mathbf{a}(\mu_l)\right]^2 + \frac{2w_c}{\tilde{K}^2 - \tilde{K}} \sum_{k=1}^{\tilde{K}-1}\sum_{p=k+1}^{\tilde{K}} \left|\mathbf{a}^H(\hat{\theta}_k)\mathbf{R}\mathbf{a}(\hat{\theta}_p)\right|^2 \right\} \\ &\text{s.t. } R_{mm} = \frac{c}{M}, \quad m = 1, \ldots, M, \\ &\quad\ \mathbf{R} \ge 0, \end{aligned} \tag{14.16}$$

where α is a scaling factor, and $w_l \ge 0$ $(l = 1, \ldots, L)$ as well as $w_c \ge 0$ are weights. We have shown in Chapter 13 that this design problem is a semidefinite quadratic program (SQP) that can be efficiently solved in polynomial time.

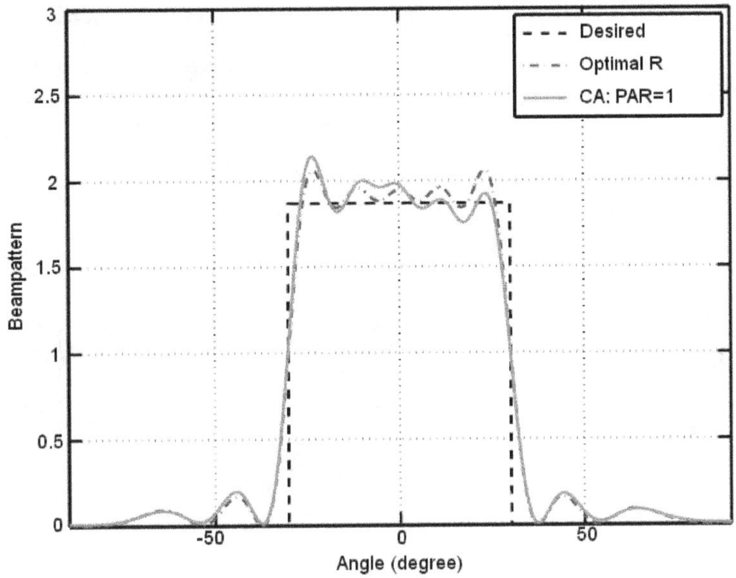

Figure 14.1 Beampattern matching design with the desired main-beam width of 60° and under the uniform elemental power constraint. The probing signals are synthesized for $M = 10$, $N = 256$ and $P = 1$ by using CA under PAR = 1.

In the first numerical example, the transmit array is assumed to be uniform linear and to comprise $M = 10$ sensors with half-wavelength interelement spacing. The sample number N is set to 256. The uniform elemental power constraint with $\gamma = 1$ (see (14.4)) is used in the design of $\mathbf{R}$. For Ω, we choose a mesh grid size of 0.1°. The desired beampattern has one wide main-beam centered at 0° with a width of 60°. The weighting factor w_l in (14.16) is set to 1 and w_c is set to 0. Figures 14.1 and 14.2 show the beampatterns using the CA synthesized waveforms under the constraints of PAR = 1 (constant-modulus), PAR $\leq$ 1.1 and PAR $\leq$ 2 and for $P = 1$. For comparison purposes, we also show the desired beampattern $\phi(\theta)$ scaled by the optimal value of α as well as the beampattern obtained from the optimal $\mathbf{R}$ (i.e., the matrix $\mathbf{R}$ that is the solution to (14.16)). Note that the beampattern obtained using the CA synthesized waveforms is close to the desired one even under the constant-modulus constraint, and it approaches the one obtained from the optimal $\mathbf{R}$ as the PAR value increases.

We next consider the same waveform synthesis problem but with $P = 10$. We use the same initial values in Step 0 of the CA algorithm (the initial value for $\tilde{\mathbf{X}}$ when $P = 10$ is constructed from the initial value for $\mathbf{X}$ when $P = 1$). The resulting beampatterns are similar to those shown in Figures 14.1 and 14.2; see Figures 14.3 and 14.4. Figure 14.5 shows the correlation levels (normalized values of (14.14)) versus p for the CA synthesized waveforms corresponding to Figures 14.1–14.4. The correlation levels decrease as the PAR value increases, as expected. As we can see, the correlation levels of the wave-

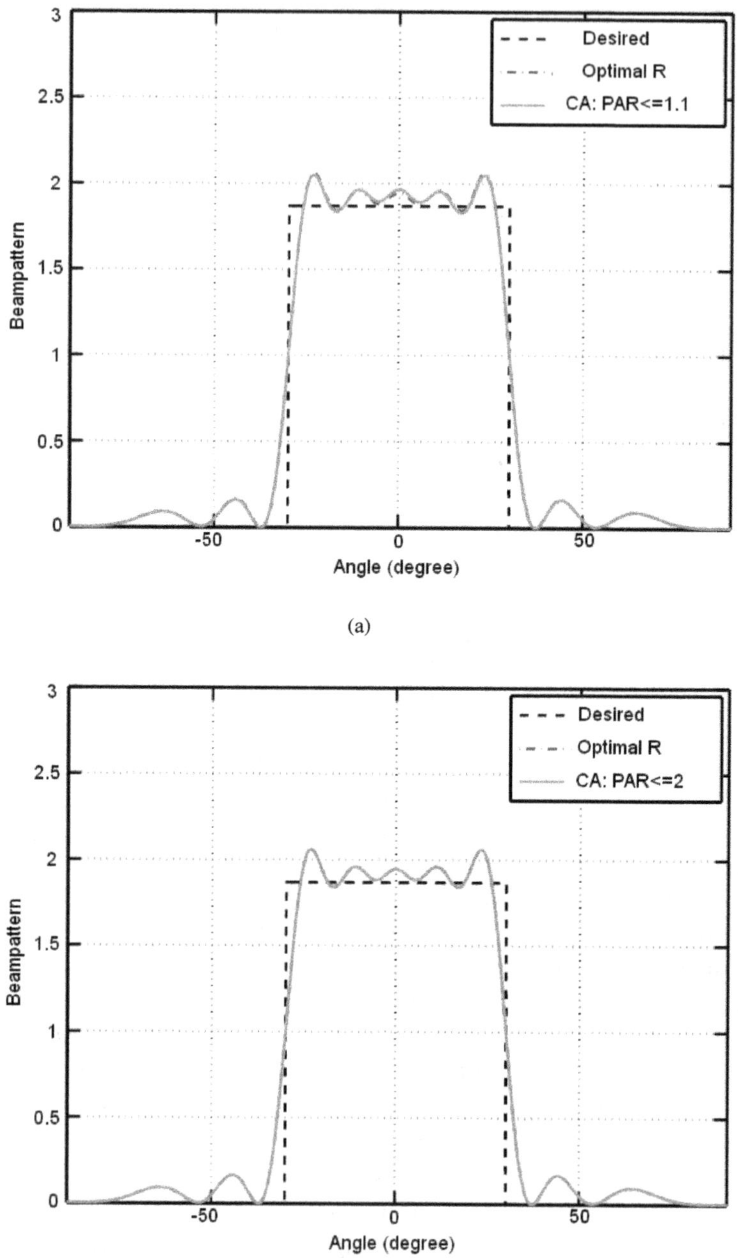

(a)

(b)

Figure 14.2 The same as Figure 14.1 except for the PAR constraint. (a) PAR $\leq$ 1.1 and (b) PAR $\leq$ 2 (the two curves corresponding to "Optimal **R**" and "CA: PAR $\leq$ 2" coincide with each other).

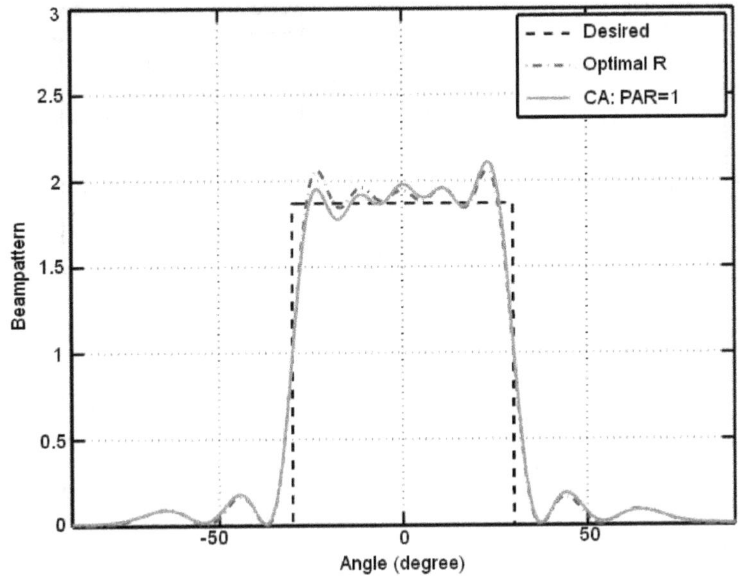

Figure 14.3 Beampattern matching design with the desired main-beam width of 60° and under the uniform elemental power constraint. The probing signals are synthesized for $M = 10$, $N = 256$ and $P = 10$ by using CA under PAR = 1.

forms shown in Figure 14.5(b) are (much) lower than those shown in Figure 14.5(a) due to optimizing the temporal auto- and cross-correlations of the waveforms.

14.4 Conclusions

Transmit waveform synthesis plays an important role in many applications that require beampattern design, such as MIMO radar and MIMO communications. We have presented a computationally attractive cyclic optimization algorithm for the synthesis of the waveform matrix to (approximately) realize a given covariance matrix $\mathbf{R}$ under the constant-modulus constraint or a low PAR constraint. We have also considered the case where the synthesized waveforms are required to have good temporal auto- and cross-correlation properties. A number of numerical examples have been provided to demonstrate the effectiveness of the proposed algorithm.

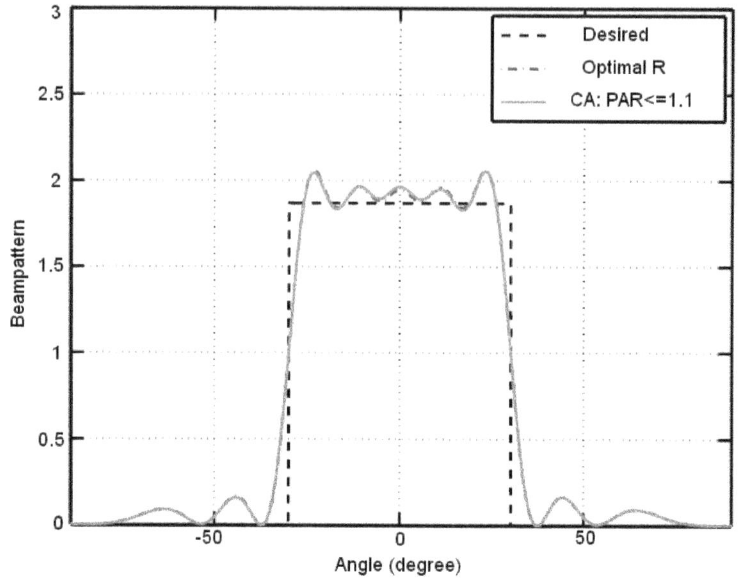

(a)

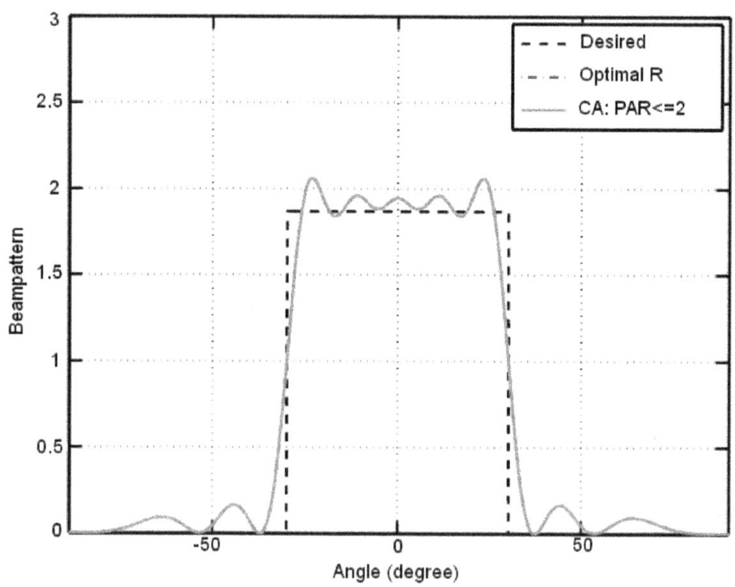

(b)

Figure 14.4 The same as Figure 14.3 except for the PAR constraint. (a) PAR ≤ 1.1 and (b) PAR ≤ 2 (the two curves corresponding to "Optimal $\mathbf{R}$" and "CA: PAR ≤ 2" coincide with each other).

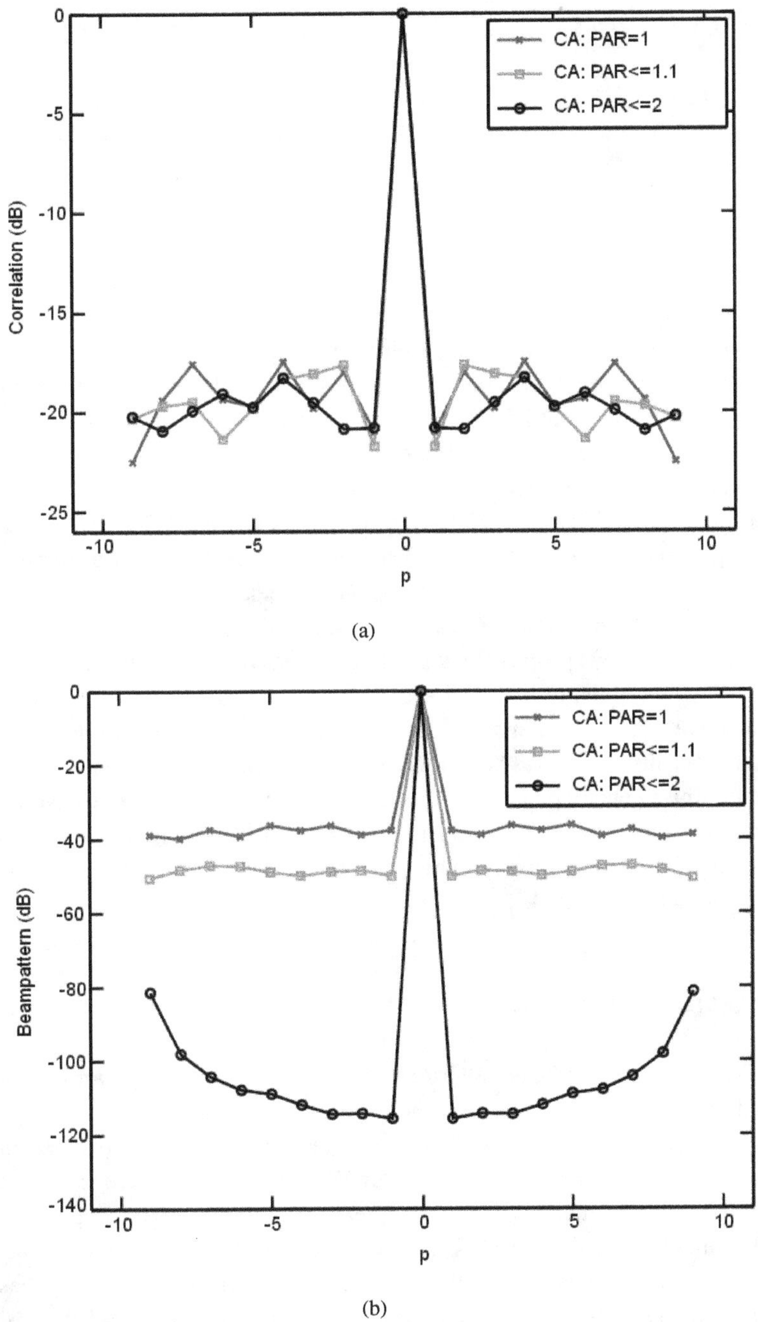

Figure 14.5 Correlation levels versus p for the CA synthesized waveform with $M = 10, N = 256$ ($\mathbf{R} \neq \mathbf{I}$). (a) $P = 1$ and (b) $P = 10$.

15 Wideband transmit beampattern synthesis

The *narrowband* transmit beampattern design problem has been discussed in Chapters 13 and 14; see also [Forsythe & Bliss 2005][Stoica et al. 2007][Fuhrmann & San Antonio 2008][Stoica, Li & Zhu 2008][Guo & Li 2008]. Most of the proposed methods first relate the desired beampattern to the covariance matrix of the transmit signals (see, e.g., Chapter 13), and then aim to design the signals that approximate the covariance matrix determined in the first stage (see, e.g., Chapter 14). In the *wideband* case, similar approaches have been proposed to design the power spectral density matrix [San Antonio & Fuhrmann 2005], but no signals have been synthesized due to the difficulty of imposing the unit-modulus or PAR constraints.

In this chapter we propose an algorithm named WB-CA (wideband beampattern CA) to design unimodular or low-PAR sequences for transmit beampattern synthesis in wideband active sensing systems. We do not formulate the problem in terms of the transmit spectral density matrix (as was done in [San Antonio & Fuhrmann 2005]), but instead directly link the beampattern to the signals through their Fourier transform. The design criterion is formulated in Section 15.1, which is followed by the algorithm description in Section 15.2. Simulation examples are shown in Section 15.3 and concluding remarks are given in Section 15.4.

15.1 Problem formulation

We focus on far-field beampattern synthesis for uniform linear arrays (ULA) as illustrated in Figure 15.1. (Note that the proposed method can be easily extended to the non-ULA case by using a more general steering vector than the one in (15.7); see, e.g., the steering vectors used in Chapter 17.) Suppose that there are M linearly spaced isotropic array elements and that the interelement spacing is d. The signal transmitted by the mth element is denoted as $s_m(t)$. Consider the beampattern in the far field at angle θ ($0° \leq \theta \leq 180°$) measured with respect to (w.r.t.) the array line. It is easy to see that the time delay between two neighboring elements is $d \cos\theta / c$, where c is the speed of wave propagation. We let $s_m(t) = x_m(t)e^{j2\pi f_c t}$ where f_c is the carrier frequency and $x_m(t)$ is the baseband signal whose spectral support is assumed to be included in the interval $[-B/2, B/2]$.

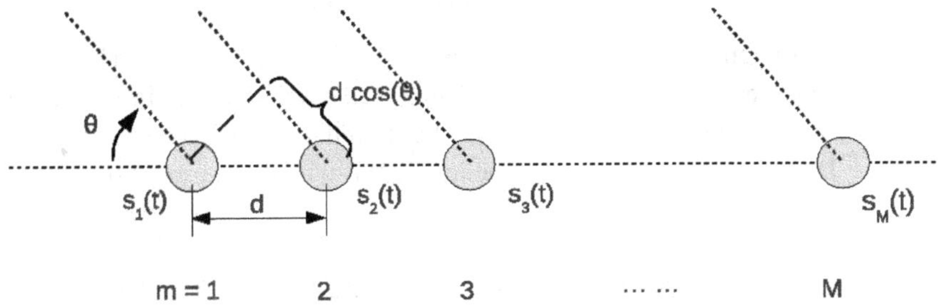

Figure 15.1 The ULA array configuration.

By using the above notation, the resulting far-field signal at angle θ can be written as

$$z_\theta(t) = \sum_{m=1}^{M} s_m\left(t - \frac{(m-1)d\cos\theta}{c}\right) \tag{15.1}$$
$$= \sum_{m=1}^{M} x_m\left(t - \frac{(m-1)d\cos\theta}{c}\right) e^{j2\pi f_c(t - \frac{(m-1)d\cos\theta}{c})}.$$

Suppose that the time support of $x_m(t)$ is $[0, \tau]$. Then the Fourier transform of $x_m(t)$ is given by

$$y_m(f) = \int_0^\tau x_m(t)e^{-j2\pi ft}\,dt, \quad f \in \left[-\frac{B}{2}, \frac{B}{2}\right] \tag{15.2}$$

and the inverse Fourier transform is accordingly

$$x_m(t) = \int_{-B/2}^{B/2} y_m(f)e^{j2\pi ft}\,df. \tag{15.3}$$

Substituting (15.3) into (15.1) yields

$$z_\theta(t) = \int_{-B/2}^{B/2} Y(\theta, f)e^{j2\pi(f+f_c)t}\,df \tag{15.4}$$

where

$$Y(\theta, f) = \sum_{m=1}^{M} y_m(f)e^{-j2\pi(f+f_c)\frac{(m-1)d\cos\theta}{c}}. \tag{15.5}$$

It follows from (15.4) that the beampattern at spatial angle θ and frequency $f + f_c$ can be defined as

$$P(\theta, f + f_c) = |Y(\theta, f)|^2 = |\mathbf{a}^H(\theta, f)\mathbf{y}(f)|^2, \quad f \in \left[-\frac{B}{2}, \frac{B}{2}\right], \tag{15.6}$$

where

$$\mathbf{a}(\theta, f) = \left[1 \;\; e^{j2\pi(f+f_c)\frac{d\cos\theta}{c}} \;\; \cdots \;\; e^{j2\pi(f+f_c)\frac{(M-1)d\cos\theta}{c}}\right]^T \tag{15.7}$$

and

$$\mathbf{y}(f) = \left[y_1(f)\ y_2(f) \cdots\ y_M(f)\right]^T. \tag{15.8}$$

Our problem is to design the signals $\{x_m(t)\}_{m=1}^M$ (band-limited to $[-B/2, B/2]$) so that the beampattern $P(\theta, f + f_c)$ in (15.6) matches a desired one. In the sequel, the baseband frequency range $[-B/2, B/2]$ is explicitly indicated when necessary.

Digital signal processing techniques deal with the sampled signal:

$$x_m(n) \triangleq x_m(t = nT_s), \quad n = 1, \ldots, N \tag{15.9}$$

where T_s is the symbol period that satisfies $T_s = 1/B$ and $N = \lfloor \tau / T_s \rfloor$. Then (15.2) becomes (by a slight abuse of notation)

$$y_m(fT_s) = T_s \sum_{n=1}^{N} x_m(nT_s) e^{-j2\pi n f T_s}, \quad f \in \left[-\frac{B}{2}, \frac{B}{2}\right]. \tag{15.10}$$

Since the interval for fT_s is $[-0.5, 0.5]$, it is enough to consider the DFT (discrete Fourier transform) of $\{x_m(n)\}$:

$$y_m(p) = \sum_{n=1}^{N} x_m(n) e^{-j2\pi \frac{(n-1)p}{N}}, \quad p = -\frac{N}{2}, \ldots, 0, \ldots, \frac{N}{2} - 1, \tag{15.11}$$

where N was assumed to be even (p will run from $-(N-1)/2$ to $(N-1)/2$ if N is odd). Note that in (15.11) we have dropped the multiplicative constant T_s from (15.10) since the scaling of $\{y_m(p)\}$ does not affect the proposed approach (see the discussions following (15.26)).

Similarly to the frequency grid considered above, we also use a grid with points denoted as $\{\theta_k\}_{k=1}^K$ for the spatial angle interval $[0°, 180°]$. For notational simplicity, let

$$\mathbf{a}_{kp} = \mathbf{a}\left(\theta_k, \frac{p}{NT_s}\right) \quad \text{(see (15.7))} \tag{15.12}$$

and

$$\mathbf{y}_p = \left[y_1(p)\ y_2(p) \cdots\ y_M(p)\right]^T \quad \text{(see (15.11))}. \tag{15.13}$$

Then it follows from (15.6) that the beampattern can be expressed on the discrete angle-frequency grid as

$$P_{kp} = |\mathbf{a}_{kp}^H \mathbf{y}_p|^2. \tag{15.14}$$

Letting d_{kp} denote the desired beampattern, we seek to solve the following beampattern matching problem:

$$\begin{aligned} &\min_{\{x_m(n)\}} \sum_{k=1}^{K} \sum_{p=-N/2}^{N/2-1} \left(d_{kp} - |\mathbf{a}_{kp}^H \mathbf{y}_p|\right)^2 \\ &\text{s.t. } \mathrm{PAR}(\mathbf{x}_m) \le \rho, \quad m = 1, \ldots, M \end{aligned} \tag{15.15}$$

where $\rho \geq 1$ is a predefined threshold and $\text{PAR}(\mathbf{x}_m)$ denotes the PAR (see (14.2)) of the mth sequence. As usual, we impose the energy constraint on the designed sequence: $\sum_{n=1}^{N} |x_m(n)|^2 = N$ for $m = 1, \ldots, M$.

The optimization problem in (15.15) is non-convex (and thus difficult in general) because of the PAR constraint. This non-convexity can be easily seen in the case of $\rho = 1$: then each $x_m(n)$ can only take values from the unit-circle, which is not a convex set. Global optimization algorithms, such as the simulated annealing method, are expected to be computationally too expensive for the problem in (15.15) due to its high dimensionality. In Section 15.2, an efficient cyclic algorithm will be proposed to search for the local minimum of (15.15).

Remark: Note that $\{x_m(n)\}$ is typically related to $\{x_m(t)\}$ through pulse shaping:

$$x_m(t) = \sum_{n=1}^{N} x_m(n)p(t-(n-1)T_s), \quad m = 1, \ldots, M \tag{15.16}$$

where $p(t)$ is the pulse (see also (1.1)). The spectrum of the baseband signal $x_m(t)$ would be confined to $[-B/2, B/2]$ only if $p(t)$ were a perfect Nyquist shaping pulse (i.e., a sinc function which is centered at 0 and has the first zero-crossing at T_s). The use of any practical shaping pulse such as a (truncated) raised cosine [Proakis 2001] will result in a leakage of the spectrum outside the desired range $[-B/2, B/2]$; these facts make (15.6) and (15.14) only approximately equivalent. We will examine the effect this approximation has on the design via an example in Section 15.3.

In addition, note that: i) the narrowband transmit beampattern design is just a special case of the wideband problem considered here and that ii) the receive beampattern design can be given a similar formulation which, however, differs in important ways from the transmit problem; see Appendix 15A and 15B for more details on these aspects. □

15.2 The proposed design methodology

Minimization of the criterion in (15.15) directly w.r.t. $\{x_m(n)\}$ appears to be a difficult task (unless the matrix $\begin{bmatrix}\mathbf{a}_{1p} & \cdots & \mathbf{a}_{Kp}\end{bmatrix}$ turns out to be semi-unitary for each value of p, which is hardly true in general). For this reason we adopt a two-stage design approach:

Stage 1 (beampattern to spectrum): First we solve (15.15) w.r.t. $\{\mathbf{y}_p\}$ considered to be general vectors in $\mathbb{C}^{M\times 1}$.

Stage 2 (spectrum to waveform): Then we fit the DFT of $\{\mathbf{x}(n)\}$ to the so-obtained $\{\mathbf{y}_p\}$, subject to the enforced PAR constraint on $\{\mathbf{x}(n)\}$.

Table 15.1. Stage 1 of WB-CA: Beampattern to Spectrum

Step 0: Initialize $\{\phi_{kp}\}$, such as $\{\phi_{kp}\} = 0$ or $\{\phi_{kp}\}$ are randomly generated variables uniformly distributed in $[0, 2\pi]$.

Step 1: For $\{\phi_{kp}\}$ set at their latest values (denoted as $\{\hat{\phi}_{kp}\}$), let

$$\mathbf{A}_p = \begin{bmatrix} \mathbf{a}_{1p}^H \\ \vdots \\ \mathbf{a}_{Kp}^H \end{bmatrix}, \quad \mathbf{b}_p = \begin{bmatrix} d_{1p}e^{j\hat{\phi}_{1p}} \\ \vdots \\ d_{Kp}e^{j\hat{\phi}_{Kp}} \end{bmatrix} \tag{15.19}$$

Then (15.18) can be written as $\sum_p \|\mathbf{b}_p - \mathbf{A}_p\mathbf{y}_p\|^2$. Thus the minimizer $\{\mathbf{y}_p\}$ is given by the least-squares estimate:

$$\hat{\mathbf{y}}_p = (\mathbf{A}_p^H\mathbf{A}_p)^{-1}\mathbf{A}_p^H\mathbf{b}_p, \quad p = -\frac{N}{2}, \ldots, 0, \ldots, \frac{N}{2} - 1. \tag{15.20}$$

Step 2: For $\{\mathbf{y}_p\}$ set at their latest values, the minimizer $\{\phi_{kp}\}$ is given by (see (15.17))

$$\hat{\phi}_{kp} = \arg(\mathbf{a}_{kp}^H\hat{\mathbf{y}}_p). \tag{15.21}$$

Iteration: Repeat steps 1 and 2 until convergence, e.g., until the change of $\{\phi_{kp}\}$ in two consecutive iterations is less than a predefined threshold.

15.2.1 Beampattern to spectrum

For a generic term $[d - |\mathbf{a}^H\mathbf{y}|]^2$ of (15.15) it holds that ($d \geq 0$)

$$\begin{aligned} &\min_{\phi} |de^{j\phi} - \mathbf{a}^H\mathbf{y}|^2 \\ &= \min_{\phi} \left\{d^2 + |\mathbf{a}^H\mathbf{y}|^2 - 2\mathrm{Re}\left[d|\mathbf{a}^H\mathbf{y}|\cos(\phi - \arg(\mathbf{a}^H\mathbf{y}))\right]\right\} \\ &= \left(d - |\mathbf{a}^H\mathbf{y}|\right)^2 \quad (\text{for } \phi = \arg(\mathbf{a}^H\mathbf{y})). \end{aligned} \tag{15.17}$$

Consequently we can get the minimizer of (15.15) from the $\{\mathbf{y}_p\}$ that minimize, along with the auxiliary variables $\{\phi_{kp}\}$, the following criterion:

$$\sum_k \sum_p \left|d_{kp}e^{j\phi_{kp}} - \mathbf{a}_{kp}^H\mathbf{y}_p\right|^2. \tag{15.18}$$

The above criterion can be conveniently minimized (w.r.t. $\{\mathbf{y}_p\}$ and $\{\phi_{kp}\}$) by the cyclic algorithm outlined in Table 15.1.

The algorithm in Table 15.1 monotonically decreases the criterion (15.18) at each iteration, and hence it monotonically decreases the original criterion in (15.15) as well. Thus it is bound to converge to at least a local minimum value of (15.15). The basic principle of the algorithm is related to the operation of the Gerchberg–Saxton algorithm as described in Appendix 2A (also see the references cited there).

Remark: It follows from the Parseval equality that the energy constraint on $\{x_m(n)\}$ imposes the following constraint on $\{\mathbf{y}_p\}$:

$$\sum_{p=-N/2}^{N/2-1} |y_m(p)|^2 = N \sum_{n=1}^{N} |x_m(n)|^2 = N^2, \quad m = 1, \ldots, M, \tag{15.22}$$

where $y_m(p)$ is the mth element of $\mathbf{y}_p$ (see (15.13)). Note that the steps in Table 15.1 omit the constraint in (15.22) for simplicity (observe that this constraint yields a coupling of $\{\mathbf{y}_p\}$, which therefore could no longer be determined separately as in (15.20)). Nonetheless, the proposed algorithm performs reasonably well likely because the energy constraint on $\{x_m(n)\}$ is considered in Stage 2 anyway (see below). □

15.2.2 Spectrum to waveform

In Stage 2 we aim to synthesize the waveform $\{\mathbf{x}(n)\}_{n=1}^{N}$, under the PAR constraint, so that its DFT approximates the $\{\hat{\mathbf{y}}_p\}_{p=-N/2}^{N/2-1}$ from Stage 1 as closely as possible.

We note first that the $\{\hat{\mathbf{y}}_p\}_{p=-N/2}^{N/2-1}$ obtained in Stage 1 have a phase ambiguity, which can be observed from the minimization criterion in (15.18): if $(\{\mathbf{y}_p\}, \{\phi_{kp}\})$ are minimizers of (15.18), then $(\{\mathbf{y}_p e^{j\psi_p}\}, \{\phi_{kp} + \psi_p\})$ are also minimizers of (15.18) for any ψ_p. This phase ambiguity results in fact from the original matching problem in (15.15): $\mathbf{y}_p$ and $\mathbf{y}_p e^{j\psi_p}$ lead to the same value of (15.15).

To exploit this phase flexibility associated with $\{\hat{\mathbf{y}}_p\}$, we introduce auxiliary variables $\{\psi_p\}_{p=-N/2}^{N/2-1}$ and minimize the following fitting criterion w.r.t. both $\{x_m(n)\}$ and $\{\psi_p\}$:

$$\sum_{p=-N/2}^{N/2-1} \left\| \hat{\mathbf{y}}_p^T e^{j\psi_p} - \left[1 \ e^{-j2\pi\frac{p}{N}} \ \cdots \ e^{-j2\pi\frac{(N-1)p}{N}}\right] \mathbf{X} \right\|^2, \tag{15.23}$$

where

$$\mathbf{X} = \left[\mathbf{x}_1 \ \mathbf{x}_2 \ \cdots \ \mathbf{x}_M\right] \tag{15.24}$$

$$= \begin{bmatrix} x_1(1) & x_2(1) & \cdots & x_M(1) \\ \vdots & \vdots & & \vdots \\ x_1(N) & x_2(N) & & x_M(N) \end{bmatrix}.$$

We further define

$$\mathbf{e}_p^H = \left[1 \ e^{-j2\pi\frac{p}{N}} \ \cdots \ e^{-j2\pi\frac{(N-1)p}{N}}\right], \quad p = -\frac{N}{2}, \ldots, \frac{N}{2} - 1, \tag{15.25}$$

$$\mathbf{F}^H = \begin{bmatrix} \mathbf{e}_{-N/2}^H \\ \vdots \\ \mathbf{e}_{N/2-1}^H \end{bmatrix}_{N\times N}, \quad \mathbf{S}^T = \begin{bmatrix} \hat{\mathbf{y}}_{-N/2}^T e^{j\psi_{-N/2}} \\ \vdots \\ \hat{\mathbf{y}}_{N/2-1}^T e^{j\psi_{N/2-1}} \end{bmatrix}_{N\times M}.$$

Then (15.23) can be written as

$$\|\mathbf{S}^T - \mathbf{F}^H\mathbf{X}\|^2 = N \left\| \frac{1}{N}\mathbf{F}\mathbf{S}^T - \mathbf{X} \right\|^2 \tag{15.26}$$

Table 15.2. Stage 2 of WB-CA: Spectrum to Waveform

Step 0: Initialize $\{\psi_p\}$, for instance as $\{\psi_p\} = 0$.
Step 1: For $\{\psi_p\}$ fixed at their latest values, the minimization of (15.26) w.r.t. $\{x_m(n)\}$ depends on the considered PAR constraint. Under the unit-modulus constraint (i.e., $|x_m(n)| = 1$), the minimization of (15.26) is immediate:

$$\hat{x}_m(n) = \exp[j \arg(\text{the } (n, m)\text{th element of } \mathbf{FS}^T)], \quad m = 1, \ldots, M, \quad n = 1, \ldots, N. \tag{15.27}$$

If instead the PAR $\leq \rho$ ($\rho > 1$) constraint is imposed, we need to solve M separate minimization problems:

$$\min_{\mathbf{x}_m} \|\mathbf{u}_m - \mathbf{x}_m\|^2 \quad \text{s.t. } \text{PAR}(\mathbf{x}_m) \leq \rho \tag{15.28}$$

(for $m = 1, \ldots, M$) where $\mathbf{u}_m$ is the mth column of $\frac{1}{N}\mathbf{FS}^T$. This problem can be solved efficiently by the "nearest-vector" method outlined in Section 4.2.
Step 2: For $\{x_m(n)\}$ fixed at their most recent values, the minimizer $\{\psi_p\}$ is given by (the derivation of (15.29) is similar to (15.17)):

$$\hat{\psi}_p = \arg(\hat{\mathbf{y}}_p^H \mathbf{v}_p), \quad p = -\frac{N}{2}, \ldots, \frac{N}{2} - 1, \tag{15.29}$$

where $\mathbf{v}_p^T$ is the $(p + N/2)$th row of $\mathbf{F}^H\mathbf{X}$.
Iteration: Repeat steps 1 and 2 until convergence.

where the equality comes from the fact that $1/\sqrt{N}\mathbf{F}$ is a unitary matrix.

Once again we use a cyclic algorithm to minimize (15.26) (w.r.t. $\{x_m(n)\}$ and $\{\psi_p\}$); see Table 15.2. Note that the required matrix calculations $\mathbf{F}^H\mathbf{X}$ and $\mathbf{FS}^T$ in Table 15.2 can be done by the FFT, which reduces the computation time.

It is easy to see that a possible scaling of $\mathbf{S}$ has no effect on (15.27). The same is true for (15.28) (which follows from the operation of the method in [Tropp et al. 2005] used to solve (15.28)). Therefore, we can choose the desired beampattern $\{d_{kp}\}$ in (15.18) without any concern for a possible normalization, as $\{\mathbf{y}_p\}$ will automatically scale (see (15.20)) to fit the chosen $\{d_{kp}\}$ and the scaling of $\{\mathbf{y}_p\}$ does not affect the synthesis of $\{x_m(n)\}$.

To summarize, the proposed two-stage design methodology, first determining $\{\mathbf{y}_p\}$ and then $\{x_m(n)\}$, basically reduces the problem in (15.18) to the design of N beamforming vectors $\{\mathbf{y}_p\}$, one for each frequency bin and then to matching them by the selection of $\{x_m(n)\}$. Note that there are $2MN$ real-valued elements in $\{\mathbf{y}_p\}$, and MN free variables in $\{x_m(n)\}$ under the unit-modulus constraint, and more than MN degrees of freedom if the PAR is allowed to be larger than 1. In addition, the $\{\psi_p\}$ provide N degrees of freedom. Hence we can expect a reasonable performance for the matching step of the proposed approach.

Although WB-CA relies on an iterative process, the updating formulas are relatively simple and the iteration turns out to converge very fast. For the numerical examples presented in the next section, the execution of the WB-CA algorithm coded in MATLAB takes only a few seconds on an ordinary PC.

15.3 Numerical examples

Unless stated otherwise, the following setting is used in this section: a ULA with $M = 10$ elements, the carrier frequency of the transmitted signal is $f_c = 1$ GHz, the bandwidth is $B = 200$ MHz and the number of symbols is $N = 64$. The symbol period is $T_s = 1/B$. The interelement spacing is given by $d = \frac{c}{2(f_c+B/2)}$, that is, half wavelength of the highest in-band frequency to avoid grating lobes. The spatial angle is divided into $K = 180$ grid points (i.e., one degree per grid step).

Remark: In practical applications the antenna elements of an array are typically mutually coupled. The interelement spacing d chosen above results in over-sampling (i.e., sampling interval less than half wavelength) for lower in-band frequencies and may render the mutual coupling effects non-negligible, which could lead to energy being coupled into transmitters. However, this issue lies outside the scope of this chapter (as it depends on the specific hardware implementation such as the system tolerance and antenna types), and we refer the interested reader to [Hui 2007][Frazer et al. 2007][Svantesson 1999] for discussions on decoupling. □

15.3.1 The idealized time-delayed case

It follows from (15.6) that we can steer the transmit beam towards the angle θ_0 by choosing the following signal spectrum:

$$\mathbf{y}(f) = \sqrt{N}\mathbf{a}(\theta_0, f), \quad f \in \left[-\frac{B}{2}, \frac{B}{2}\right] \tag{15.30}$$

where $\sqrt{N}$ is due to the energy constraint. Equation (15.30) leads to (see (15.14))

$$P_{kp} = N\left|\sum_{m=1}^{M} e^{j2\pi\left(\frac{p}{NT_s}+f_c\right)\frac{(m-1)d(\cos\theta_0-\cos\theta_k)}{c}}\right|^2, \tag{15.31}$$

where for a fixed value of f (i.e., p), the beam is steered in the direction of θ_0 as in the case of a narrowband phased array. The underlying signals, i.e., the inverse Fourier transform of (15.30), are given (up to a multiplicative constant) by

$$x_m(t) = \text{sinc}\left[\frac{\pi}{T_s}\left(t - \frac{(m-1)d\cos\theta_0}{c}\right)\right], \quad m = 1, \ldots, M, \tag{15.32}$$

where $\text{sinc}(t) = \sin(x)/x$. Note that such an idealized $\{x_m(t)\}$ has a very high PAR which is undesirable. Moreover, because $d/c = 1/(2f_c + B) \ll T_s$, the required time delay of $\frac{md\cos\theta_0}{c}$ can be too small to be readily implemented in practice, especially when θ_0 is close but not equal to 90°.

We show the beampattern $10\log_{10}(P_{kp}/N)$ in (15.31) for $\theta_0 = 120°$ as a 2D plot in Figure 15.2(a), as well as a 3D plot in Figure 15.2(b). The beampattern exhibits a clean mainlobe at θ_0 across the entire frequency range.

Remark: In the narrowband case, for a given ULA aperture, the transmit beampattern generated by a phased array has the smallest mainlobe width. In the above example, we used the impractical time-delayed $\{x_m(t)\}$ in (15.32) to get the phased array-like beampattern in (15.31), which thus has the narrowest mainlobe for every fixed frequency. Therefore we call it the "idealized time-delayed" case.

It has been assumed that for each array element an energy equal to N is emitted in all directions; and the energy constraint $\sum_p |y_m(p)|^2 = N^2$ in (15.22) indicates that on average $|y_m(p)|^2$ equals N. Therefore, were there only one array element, P_{kp} would equal N at every grid point in the angle-frequency plane (see, e.g., (15.31)). This is the reason why the normalization $10\log_{10}(P_{kp}/N)$ is used in all plots. Now that there are M transmit waveforms, the coherent sum gives $\max P_{kp} = \max |\mathbf{a}_{kp}^H \mathbf{y}_p|^2 = |M\sqrt{N}|^2 = M^2N$, which leads to $10\log_{10}(M^2N/N) = 20$ dB in the plot. In fact, in the above idealized time-delayed case, all M waveforms add coherently at $\theta = \theta_0$ and the energy is evenly distributed at $\theta = \theta_0$ for all frequencies, which produces a constant 20 dB mainlobe height (see Figure 15.2 and (15.31)). In other examples, however, the mainlobe height is not necessarily 20 dB and the upper limit of the colorbar always corresponds to the largest value in the plot. □

15.3.2 A narrow mainbeam

We use the proposed WB-CA algorithm to synthesize the following desired transmit beampattern:

$$d(\theta, f + f_c) = \begin{cases} 1, \ \theta = 120°, \\ 0, \ \text{elsewhere}, \end{cases} \quad \text{for all } f \in \left[-\frac{B}{2}, \frac{B}{2}\right], \tag{15.33}$$

that is, a beampattern with the mainlobe (as narrow as possible) located at 120° across the frequency support.

Stage 1 of WB-CA generates the DFT vectors $\{\hat{\mathbf{y}}_p\}_{p=-N/2}^{N/2-1}$, which are further normalized to preserve the total energy (i.e., normalized such that $\sum_p \|\mathbf{y}_p\|^2 = MN^2$). Figure 15.3 shows the beampattern P_{kp} that is calculated directly from these $\{\hat{\mathbf{y}}_p\}$. The so-obtained beampattern is quite similar to the idealized one in Figure 15.2. However, the underlying waveforms corresponding to Figure 15.3, given by the inverse DFT (IDFT) of $\{\hat{\mathbf{y}}_p\}_{p=-N/2}^{N/2-1}$, do not satisfy the energy and PAR requirement. Indeed, the M sequences obtained from the IDFT of $\{\hat{\mathbf{y}}_p\}$ have energies varying from 55.4 to 71.2 and PARs varying from 1.3 to 1.8. Note that such transmit sequences need to be scaled in practice so that the maximum energy does not exceed the system specifications, which will inevitably result in an energy loss.

We then proceed to Stage 2 of WB-CA and synthesize the sequences $\{\hat{x}_m(n)\}$ under the unit-modulus constraint. After that, we compute the DFT of $\{\hat{x}_m(n)\}$ and obtain the beampattern using (15.14); see Figure 15.4. It is clear that the strict unit-modulus con-

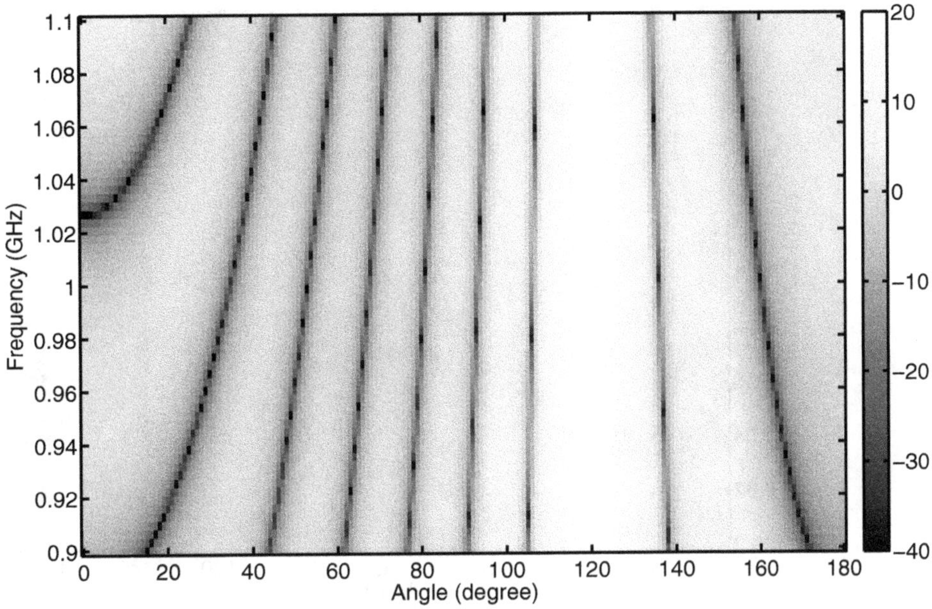

(a)

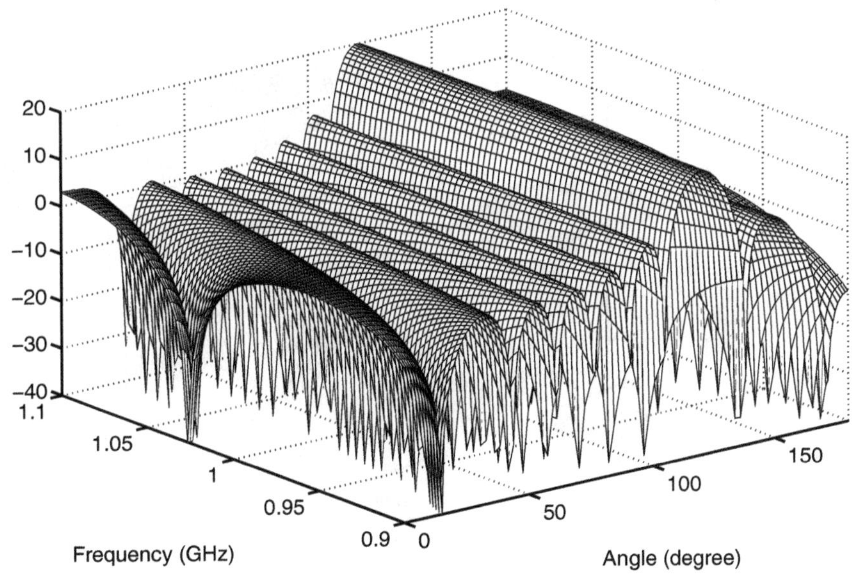

(b)

Figure 15.2 The idealized time-delayed beampattern in (15.31). (a) The 2D plot and (b) the 3D plot.

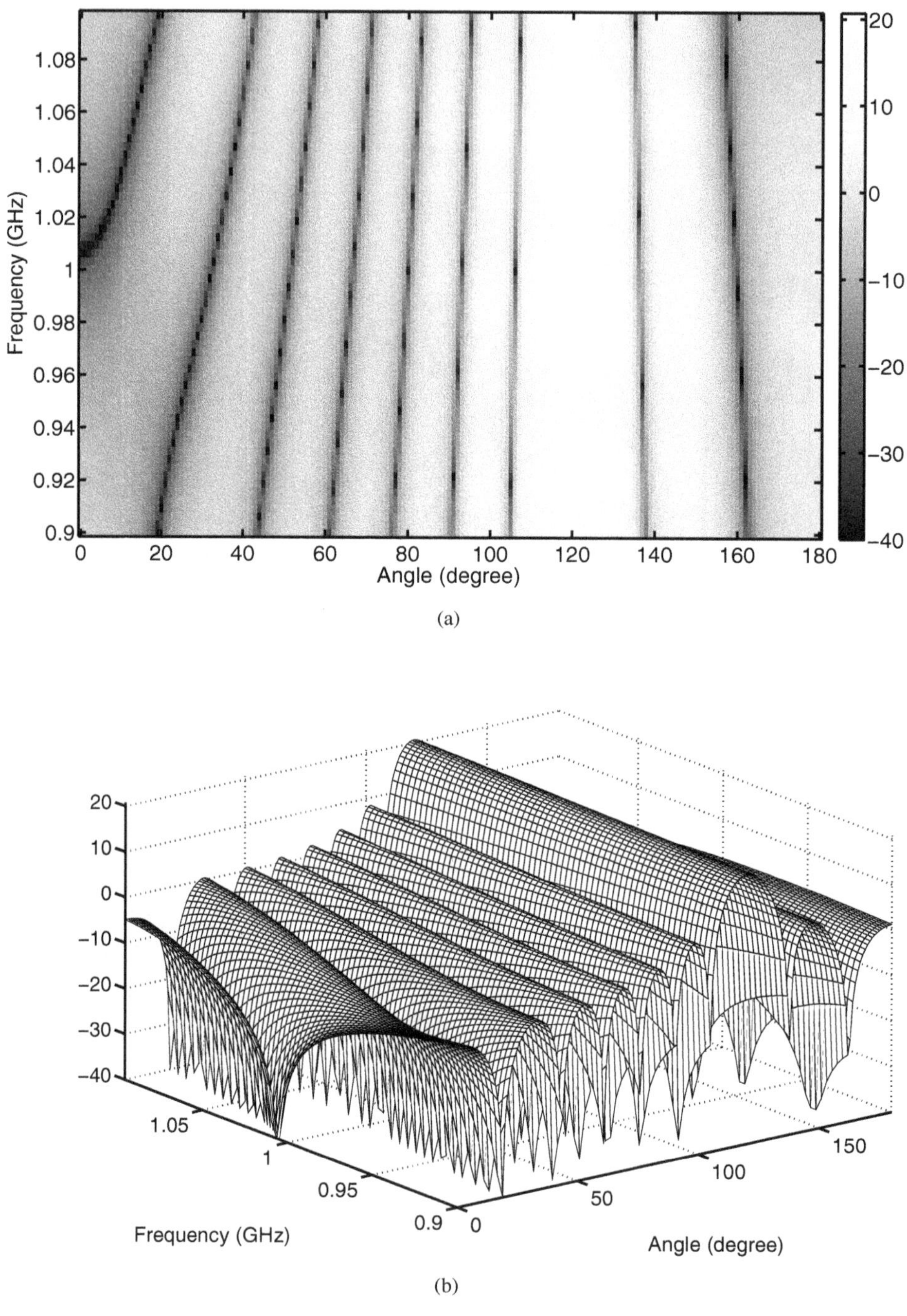

Figure 15.3 The WB-CA beampattern under only the total energy constraint. The desired beampattern is given in (15.33). (a) The 2D plot and (b) the 3D plot.

	Figure 15.3	Figure 15.4
Criterion in (15.15)	5987914	6048430

Table 15.3. Optimized criterion values for Figure 15.3 and 15.4

	Figure 15.7	Figure 15.8
Criterion in (15.15)	5698808	5619700

Table 15.4. Optimized criterion values for Figure 15.7 and 15.8

straint degrades the beampattern matching. Table 15.3 shows the minimum value of the fitting criterion (see (15.15)) associated with Figures 15.3 and 15.4, respectively.

Next we examine the beampattern of the continuous-time waveforms corresponding to the so-obtained $\{\hat{x}_m(n)\}$ (see the *Remark* at the end of Section 15.1). More specifically, we pass each $\{\hat{x}_m(n)\}_{n=1}^{N}$ $(m = 1, \ldots, M)$ through an FIR raised-cosine filter (with the roll-off factor equal to 0.5) to get the continuous-time signal $\hat{x}_m(t)$. The spectral density functions of $\{\hat{x}_m(t)\}$ are shown in Figure 15.5 in an overlapping manner, from which we observe that the spectrum is well contained within $[f_c - B/2, f_c + B/2]$ despite of a certain leakage outside the frequency range of interest. The beampattern of $\{\hat{x}_m(t)\}$, as defined in (15.6), is shown in Figure 15.6. Compared to Figure 15.4, the beampattern in Figure 15.6 is a poorer approximation of the desired one. As discussed in the Remark at the end of Section 15.1, a practical pulse shaping function renders (15.6) and (15.14) not exactly equivalent. Since the WB-CA algorithm aims to match (15.14) to the desired beampattern, the nonequivalence between (15.6) and (15.14) explains the degradation from Figure 15.4 to Figure 15.6.

15.3.3 Two mainbeams

In this example we consider the following desired beampattern:

$$d(\theta, f + f_c) = \begin{cases} 1, \ f_c - B/2 \leq f \leq f_c \text{ and } \theta = 120^\circ, \\ 1, \ f_c \leq f \leq f_c + B/2 \text{ and } \theta = 60^\circ, \\ 0, \ \text{elsewhere.} \end{cases} \tag{15.34}$$

The WB-CA beampattern under the unit-modulus constraint is shown in Figure 15.7 and that under the PAR ≤ 2 constraint is shown in Figure 15.8. While Figure 15.7 already provides a reasonably good beampattern matching, relaxing the PAR from 1 to 2 leads to the visibly better result in Figure 15.8 due to more degrees of freedom in the waveform design. Table 15.4 shows this performance improvement in terms of the corresponding fitting criterion values.

15.3.4 A wide mainbeam

In both examples above, we focused on achieving a beampattern where mainlobe(s) were as narrow as possible. Specifically the idealized phased array-like beampattern in Figure 15.2, which has the narrowest possible mainlobe, was well approximated by

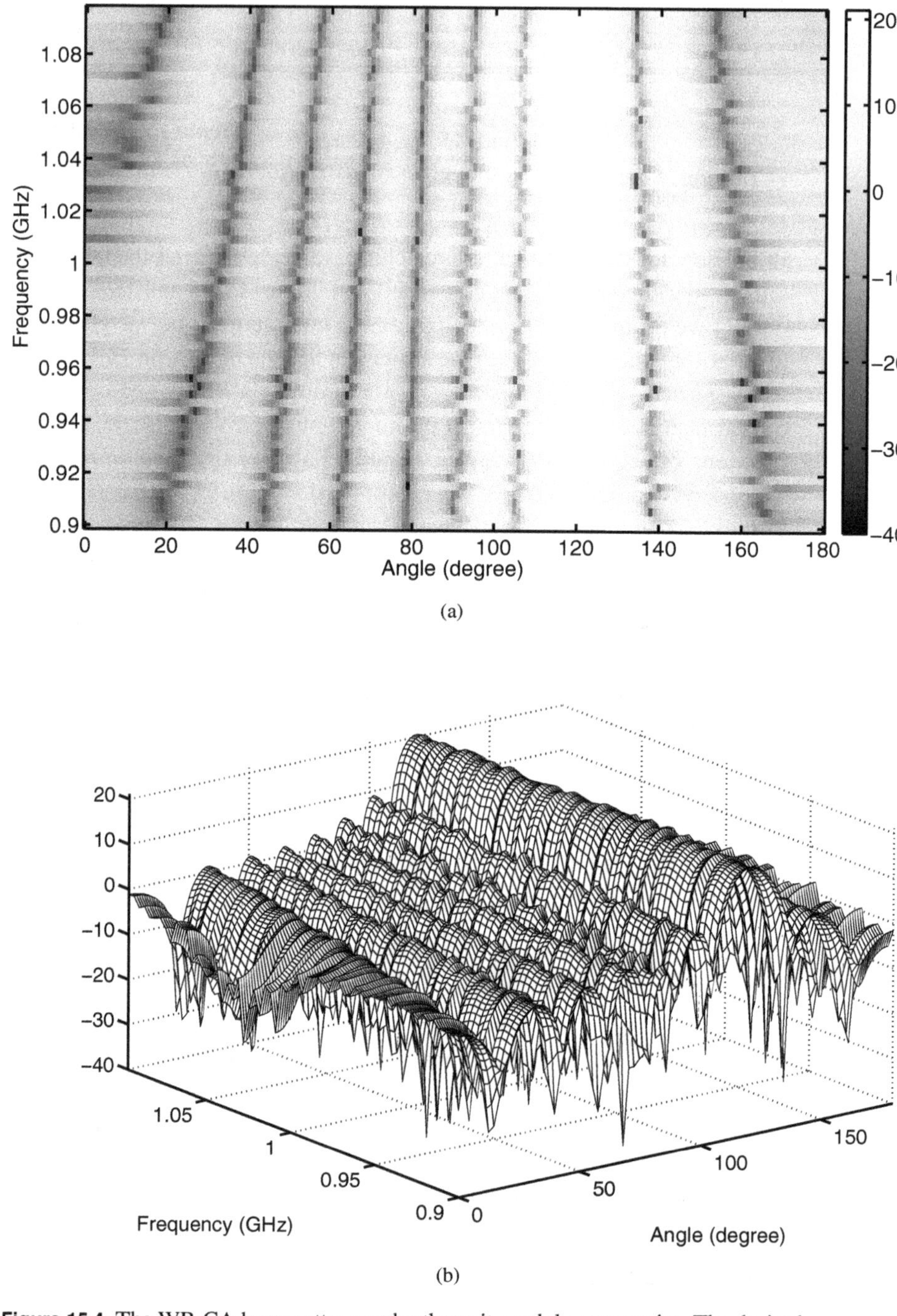

Figure 15.4 The WB-CA beampattern under the unit-modulus constraint. The desired beampattern is given in (15.33). (a) The 2D plot and (b) the 3D plot.

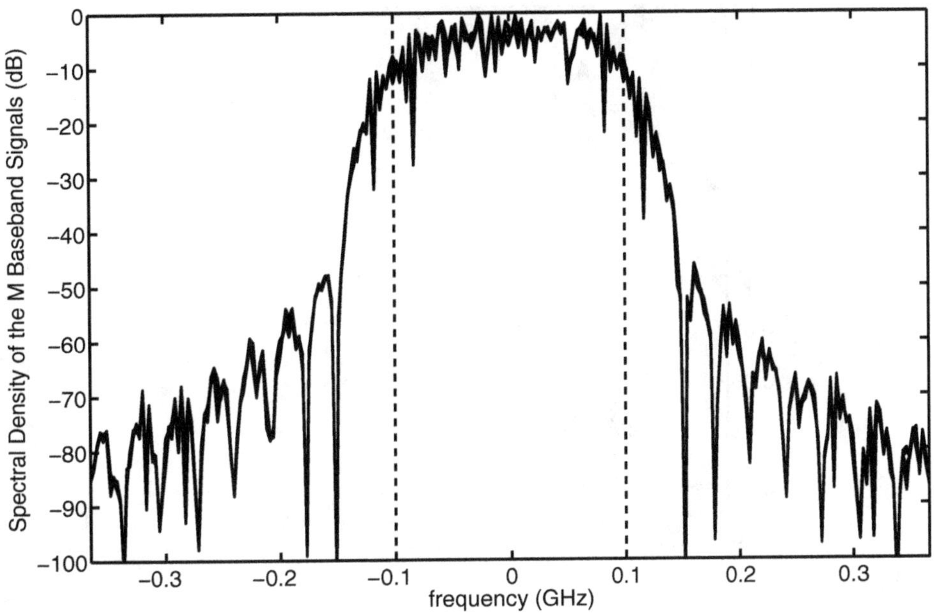

Figure 15.5 The overlaid spectral densities of the continuous waveforms corresponding to the sequences used in Figure 15.4. The two vertical dashed lines represent the boundaries of the frequency range of interest.

using practical waveforms in Section 15.3.2. If we want to obtain a narrower mainlobe, we have to use a larger value of M, i.e., more transmit antenna elements.

Here we consider instead the following beampattern with a wide mainlobe:

$$d(\theta, f + f_c) = \begin{cases} 1, & 100^\circ \leq \theta \leq 140^\circ, \\ 0, & \text{elsewhere}, \end{cases} \qquad \text{for all } f \in \left[-\frac{B}{2}, \frac{B}{2}\right]. \tag{15.35}$$

The corresponding WB-CA beampattern under the unit-modulus constraint is shown in Figure 15.9 and that under the PAR ≤ 2 constraint in Figure 15.10. We observe that the mainlobe in Figure 15.9 or 15.10 has an almost constant width for different frequencies, unlike the mainlobe in Figure 15.2 whose width tends to (slightly) shrink as the frequency increases (the well-known beam squint phenomenon). Also note the "mainlobe splitting" in Figure 15.9 or 15.10. Had we synthesized an even wider mainlobe than (15.35), the splitting would have been more severe (e.g., the mainlobe can split twice so that there are three local maxima in the mainlobe area).

In all of the above examples the bandwidth B was set to 200 MHz. A larger bandwidth means more constraints and thus the beampattern matching is expected to become more difficult. To illustrate this fact, we repeat the example corresponding to Figure 15.4 except that the bandwidth is now equal to 350 MHz. The result is shown in Figure 15.11, where the beampattern is more irregular than in Figure 15.4.

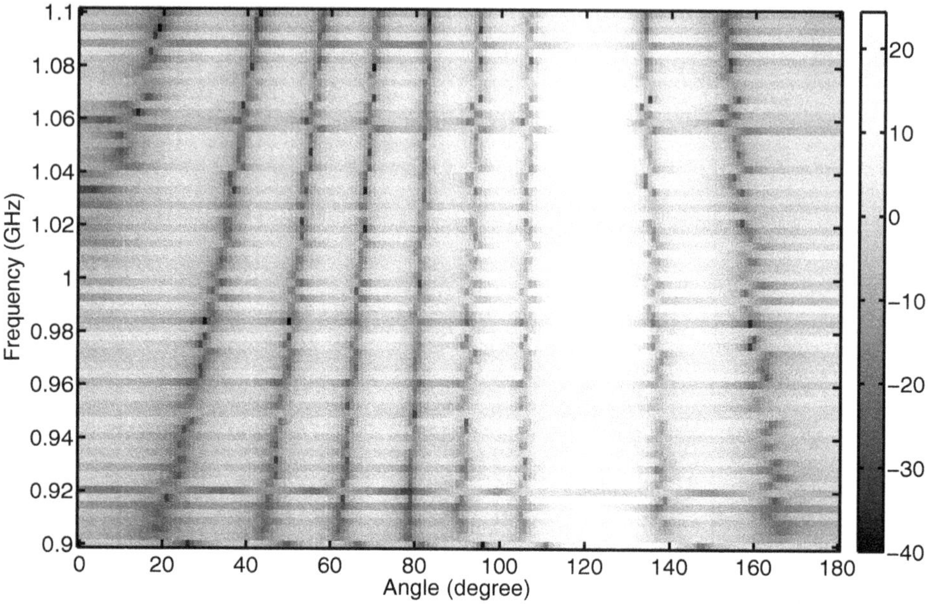

(a)

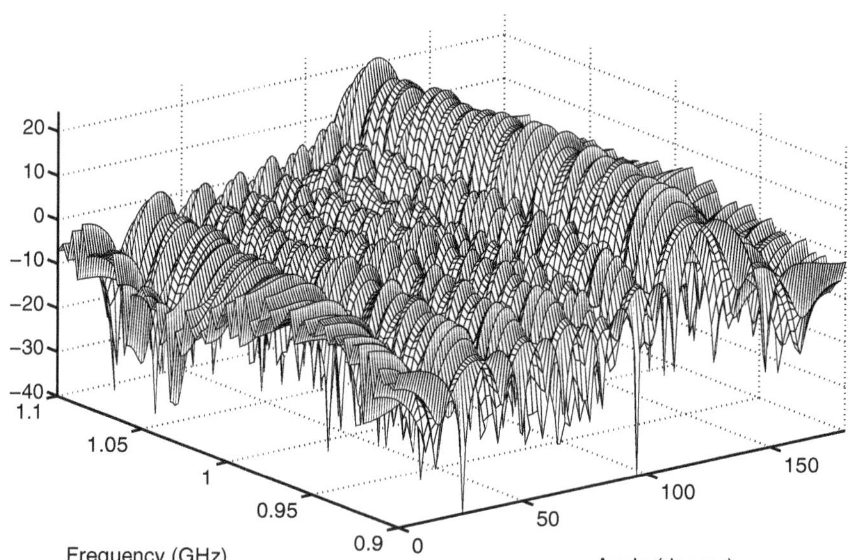

(b)

Figure 15.6 The WB-CA beampattern of the continuous waveforms corresponding to the sequences used in Figure 15.4. The desired beampattern is given in (15.33). (a) The 2D plot and (b) the 3D plot.

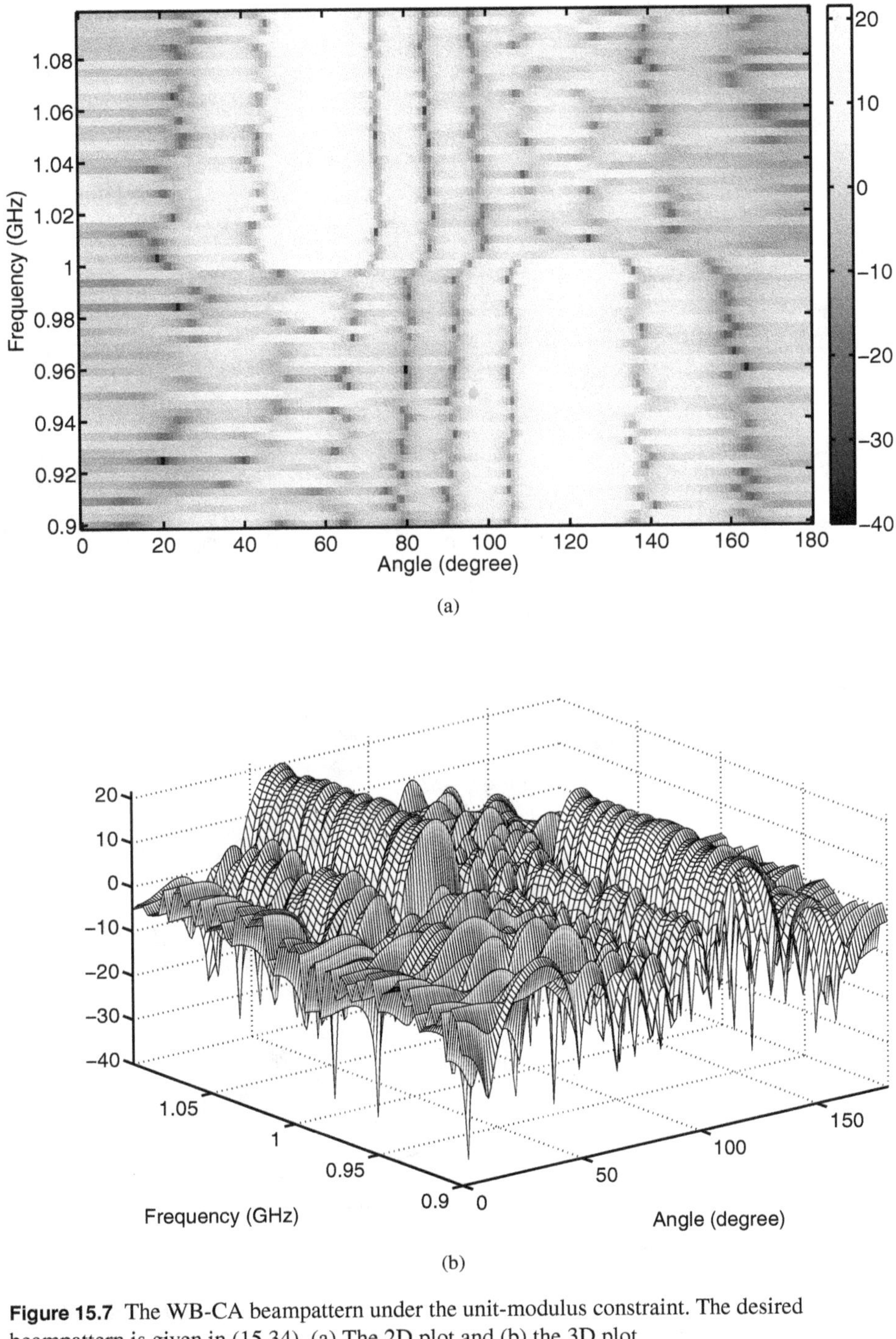

Figure 15.7 The WB-CA beampattern under the unit-modulus constraint. The desired beampattern is given in (15.34). (a) The 2D plot and (b) the 3D plot.

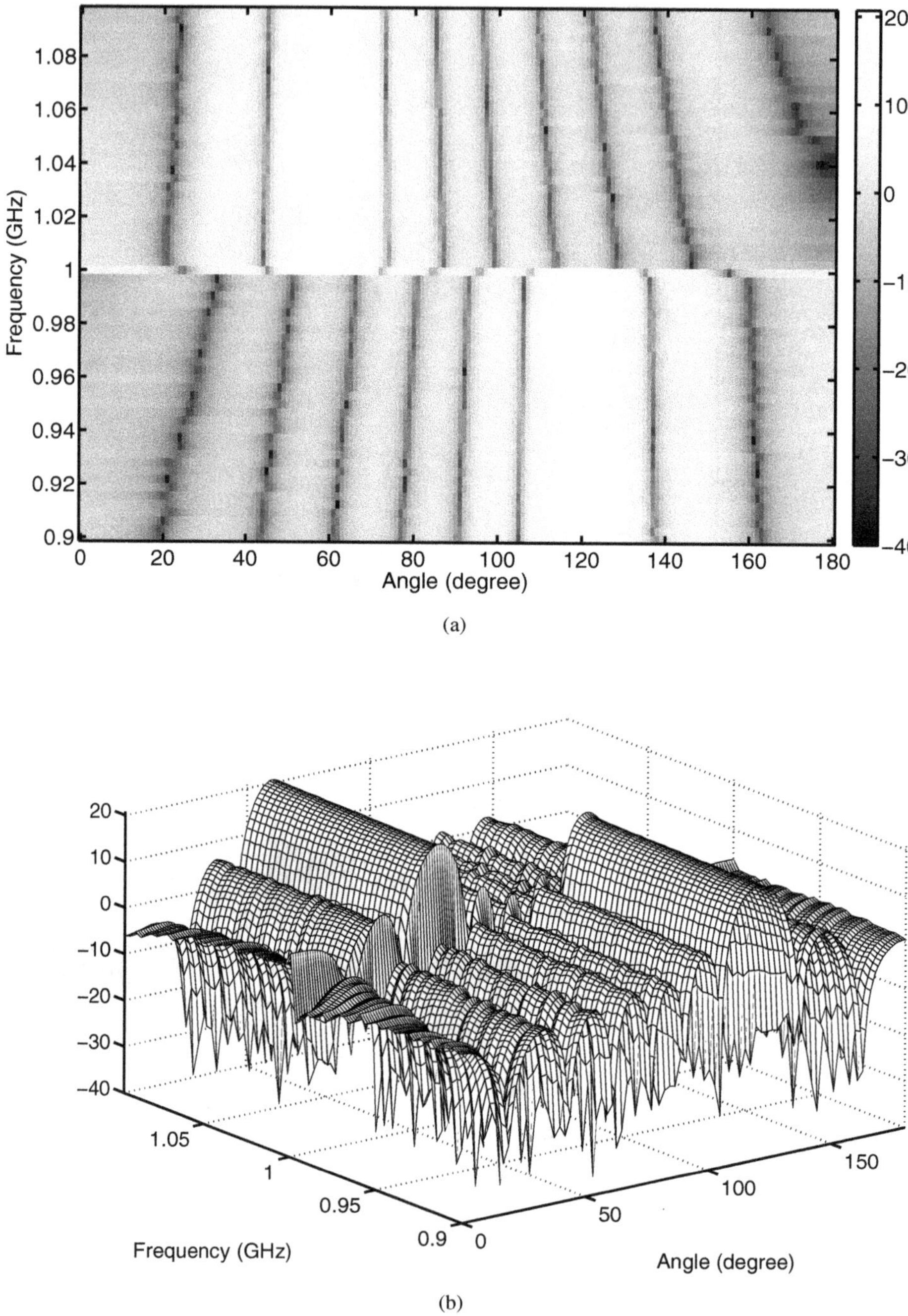

Figure 15.8 The WB-CA beampattern under the PAR ≤ 2 constraint. The desired beampattern is given in (15.34). (a) The 2D plot and (b) the 3D plot.

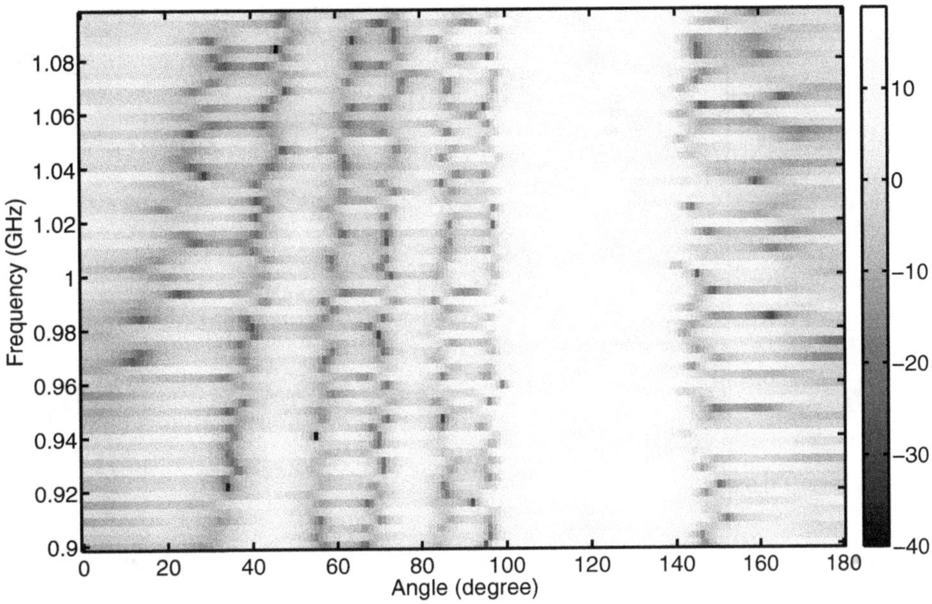

(a)

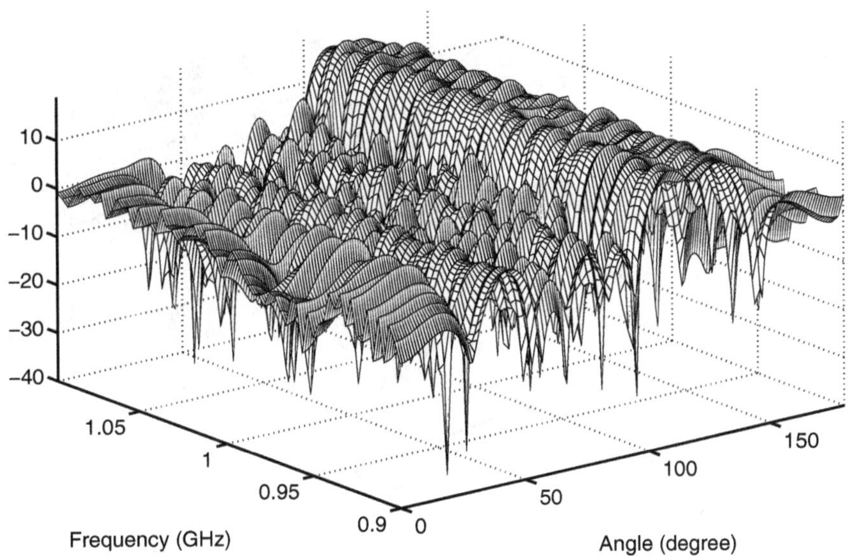

(b)

Figure 15.9 The WB-CA beampattern under the unit-modulus constraint. The desired beampattern is given in (15.35). (a) The 2D plot and (b) the 3D plot.

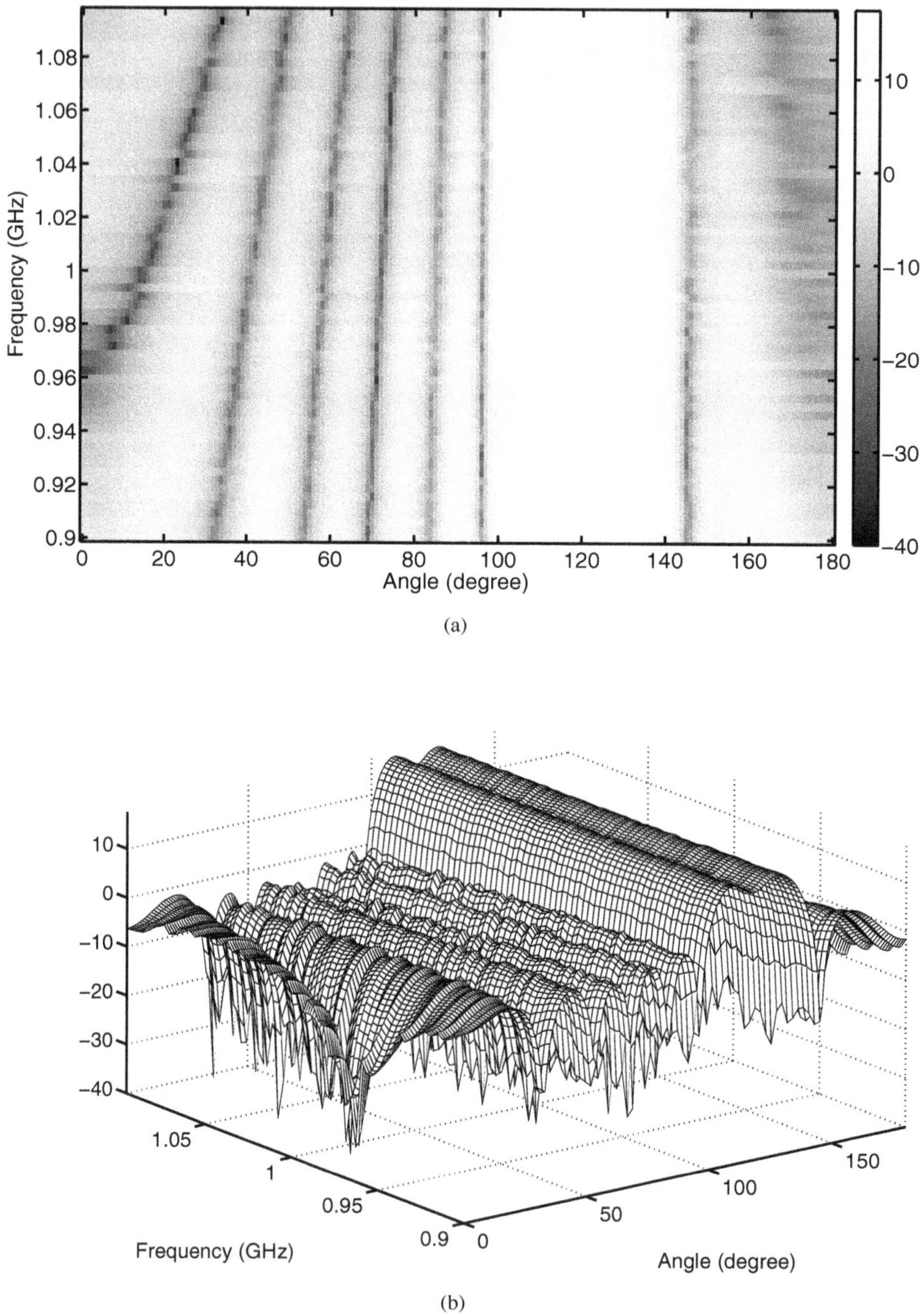

Figure 15.10 The WB-CA beampattern under the PAR ≤ 2 constraint. The desired beampattern is given in (15.35). (a) The 2D plot and (b) the 3D plot.

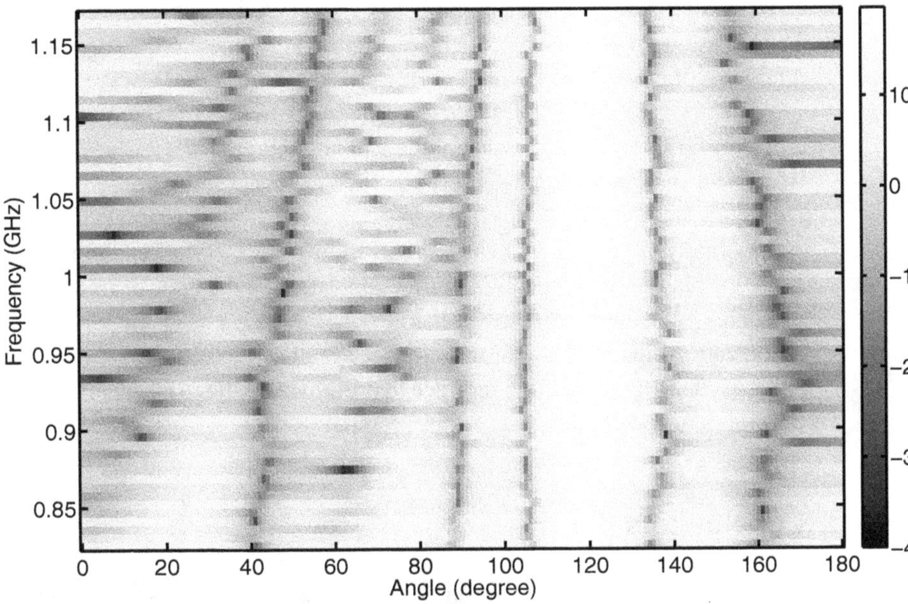

(a)

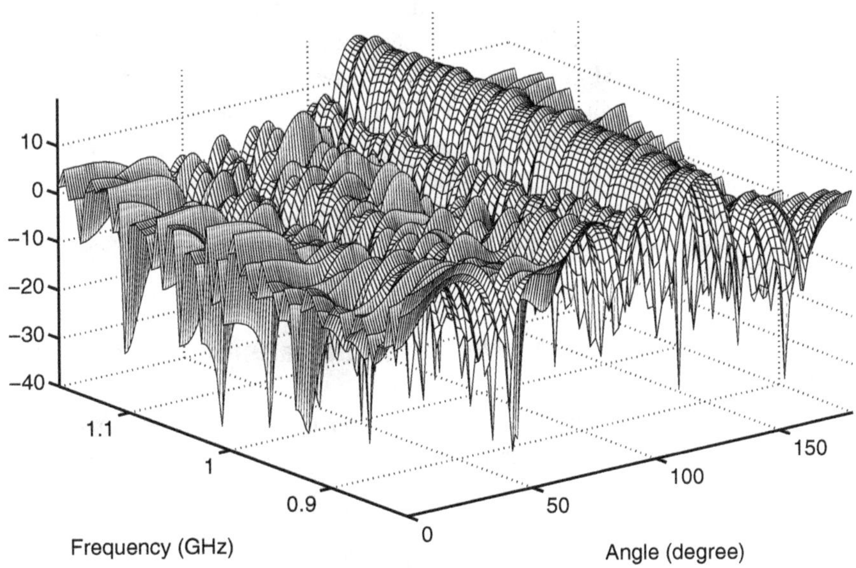

(b)

Figure 15.11 The WB-CA beampattern for the same settings as in Figure 15.4, excepting that the bandwidth B is changed from 200 to 350 MHz. (a) The 2D plot and (b) the 3D plot.

Regarding choosing N (the number of transmitted symbols), we note that increasing N does not necessarily lead to a better beampattern matching. The reason is that while a larger N increases the number of degrees of freedom of the waveform $\{x_m(n)\}$, it also increases proportionally the number of elements in $\{\mathbf{y}_p\}$ that are to be matched in Stage 2 of WB-CA (see the discussion at the end of Section 15.2). At the same time, N cannot be chosen too small because the frequency grid should be dense enough to cover the frequency support finely.

We finally point out that the initialization of WB-CA (i.e., step 0 in both Stages) does not play an important role in the algorithm performance. In all of the numerical examples presented, randomly generated phases were used for initialization; different initializations led to different waveforms but all of these waveforms had similar beampatterns. This also signifies the fact that the beampattern matching problem is highly multi-modal and challenging.

15.4 Conclusions

In this chapter we have proposed an algorithm named WB-CA to synthesize transmit beampatterns for wideband array systems. The waveform diversity in such systems can be exploited to synthesize various types of wideband beampatterns. At the same time, practical waveform constraints such as unit-modulus or low PAR have been taken into account. The WB-CA algorithm uses a cyclic approach to minimize a beampattern matching criterion, is computationally efficient and its local convergence is guaranteed. Several numerical examples have been included to demonstrate the effectiveness of WB-CA in generating practical waveforms for matching a desired beampattern.

WB-CA was derived under the setting of 1D ULA (see Figure 15.1) but it can be easily adapted for use in the 1D non-ULA case by employing a more general steering vector than the one in (15.7); see, e.g., Chapter 17. However for 2D arrays, the WB-CA optimization problem becomes harder due to an increased spatial dimension.

Appendix 15A Narrowband transmit beampattern

In the narrowband case, $B \ll f_c$ and therefore the distribution of energy versus frequency f is of less interest. Instead, the total energy over f is the quality of interest, which is given by (see (15.14))

$$P(\theta_k) = \sum_{p=-N/2}^{N/2-1} P_{kp} = \sum_{p=-N/2}^{N/2-1} |\mathbf{a}_{kp}^H \mathbf{y}_p|^2. \tag{15.36}$$

Since the interelement spacing d is on the order of the carrier wavelength, the narrowband assumption $B \ll f_c$ implies that $\frac{fd\cos\theta}{c} \sim 0$ $(-B/2 \le f \le B/2)$, which means that the steering vector $\mathbf{a}_{kp}$ is independent of frequency. Thus its subscript p can be dropped and (15.36) becomes

$$P(\theta_k) = \mathbf{a}_k^H \left(\sum_{p=-N/2}^{N/2-1} \mathbf{y}_p \mathbf{y}_p^H \right) \mathbf{a}_k = \mathbf{a}_k^H \left\{ \sum_{u=1}^{N} \sum_{v=1}^{N} \mathbf{x}(u)\mathbf{x}^H(v) \left[\sum_{p=-N/2}^{N/2-1} e^{-j2\pi \frac{(u-v)p}{N}} \right] \right\} \mathbf{a}_k$$

$$= \mathbf{a}_k^H \left(N \sum_{n=1}^{N} \mathbf{x}(n)\mathbf{x}^H(n) \right) \mathbf{a}_k, \qquad (15.37)$$

where $\mathbf{x}(n)$ is defined as

$$\mathbf{x}(n) = \left[x_1(n)\ x_2(n) \cdots\ x_M(n) \right]^T, \quad n = 1, \ldots, N. \qquad (15.38)$$

The result in (15.37), up to a multiplicative constant, coincides with the narrowband beampattern expression used in Chapter 13. Also note that although the WB-CA algorithm holds for both wideband and narrowband cases, the algorithms discussed in Chapters 13 and 14 are still recommended for the narrowband beampattern design owing to the straightforward SDP formulation used in Chapter 13 that nicely permits a global solution of the covariance matrix.

Appendix 15B Receive beampattern

Parallelling the discussion in Section 15.1, we briefly formulate here the receive beampattern synthesis problem for wideband signals.

Suppose that a wideband signal $g(t)e^{j2\pi f_c t}$ with frequency band $[f_c - B/2, f_c + B/2]$ is impinging from angle θ ($0° \le \theta \le 180°$) on a ULA. Let $G(f)$ denote the Fourier transform of $g(t)$. The signal received at the mth array element can be written as

$$r_m(t) = g\left(t - \frac{(m-1)d\cos\theta}{c}\right) e^{j2\pi f_c \left(t - \frac{(m-1)d\cos\theta}{c}\right)} \qquad (15.39)$$

$$= \int_{-B/2}^{B/2} G(f) e^{-j2\pi(f+f_c)\frac{(m-1)d\cos\theta}{c}} e^{j2\pi(f+f_c)t}\, df.$$

Let $H_m(f)$ denote the frequency response of the FIR filter used to process the demodulated signal $r_m(t)e^{-j2\pi f_c t}$. Then the receive beampattern can be expressed in the frequency domain as

$$A(\theta, f + f_c) = \left| \sum_{m=1}^{M} H_m(f) e^{-j2\pi(f+f_c)\frac{(m-1)d\cos\theta}{c}} \right|^2, \quad f \in \left[-\frac{B}{2}, \frac{B}{2} \right], \qquad (15.40)$$

where $G(f)$ is omitted because it is the same for all array elements. The receive beampattern synthesis problem can be stated as designing a set of M filters $\{h_m(t)\}_{m=1}^M$ (the Fourier transform of $h_m(t)$ is $H_m(f)$) such that $A(\theta, f + f_c)$ matches a desired pattern.

As pointed out in the Introduction, there is no essential constraint on $\{h_m(t)\}_{m=1}^M$, the design of which can therefore be done by a host of approaches, such as classic filter design methods [Ward et al. 1996] or convex optimization [Lebret & Boyd 1997]. On the other hand, the transmit beampattern design, which has been the topic of this chapter (as well the previous two chapters), is much harder because of the PAR constraint and elemental antenna power constraint, despite the fact that (15.6) and (15.40) have the same form.

Part IV

Diverse application examples

16 Radar range and range–Doppler imaging

We have shown a radar SAR imaging example in Section 3.4.5 where we demonstrated the performance improvement of using the newly designed Multi-WeCAN sequences compared with existing sequences. In this chapter we will present a few more radar imaging examples. In particular, we will show that for some cases such as those with low SNR or closely-spaced targets, even careful construction of the transmit waveform, when coupled with a matched filter at the receiver, may not provide sufficient clutter or noise reduction. To address these situations, we turn our attention in this chapter to the receiver side of a radar system. Besides the matched filter, we will review the instrumental variable filter (see also Chapter 8) and an iterative adaptive approach [Yardibi et al. 2010], both of which bring improvements in interference suppression at the expense of a loss in SNR and/or an increased computational complexity.

16.1 Problem formulation

Let $s(t)$ denote the transmitted waveform (see (1.1)) that comprises N subpulses. We represent $s(t)$ by the vector $\mathbf{x}$, whose components correspond to the phase-coded amplitude of each subpulse

$$\mathbf{x} = \left[x(1)\ x(2)\ \cdots\ x(N)\right]^T. \tag{16.1}$$

Due to hardware constraints such as limitations of power amplifiers, the components of $\mathbf{x}$ are commonly restricted to being constant modulus; see (1.4).

Targets in the scene are represented by their RCSs (radar cross sections) $\{\kappa_{r,l}\}$ with $r = 1, \ldots, R$ denoting the range bin and $l = 1, \ldots, L$ denoting the Doppler bin. Then the received signal $\mathbf{y}_{r'}$ (aligned with the transmitted waveform's reflection from a range bin of interest r') can be modeled as

$$\mathbf{y}_{r'} = \kappa_{r',l'}\tilde{\mathbf{x}}_{l'} + \sum_{\substack{n=-N+1 \\ (r'+n,l)\neq(r',l')}}^{N-1} \sum_{l=1}^{L} \kappa_{r'+n,l}\mathbf{J}_n\tilde{\mathbf{x}}_l + \boldsymbol{\epsilon} \tag{16.2}$$

where $\kappa_{r',l'}$ refers to the reflection coefficient of the target of interest and $\boldsymbol{\epsilon}$ refers to the noise component of the received signal. Further, $\mathbf{J}_n$ is a shift matrix designed to temporally align the reflected signal from a target that lies n range bins away from the

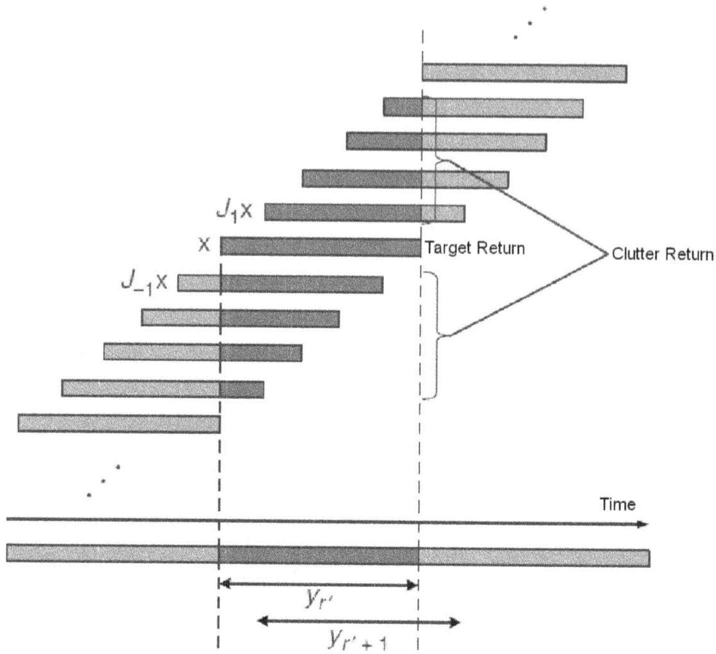

Figure 16.1 Received signal aligned with the return from a target in range bin r'.

bin of interest:

$$\mathbf{J}_n = \begin{bmatrix} \mathbf{0} & & \mathbf{0} \\ \vdots & & \\ 1 & & \\ & \ddots & \\ \mathbf{0} & & \underbrace{1\,0\,\cdots\,0}_{n+1} \end{bmatrix}_{N\times N} = \mathbf{J}_{-n}^T, \quad n = 0, 1, \ldots, N-1. \tag{16.3}$$

We assume that, in general, $\kappa_{r,l} = 0$ for any r such that $r \notin \{1, \ldots, R\}$. We let $\tilde{\mathbf{x}}_l = \mathbf{x} \odot \mathbf{a}_l$ denote the Doppler shifted waveform where

$$\mathbf{a}_l = \begin{bmatrix} 1 & e^{j\omega_l} & \cdots & e^{j\omega_l(N-1)} \end{bmatrix}^T, \quad l = 1, \ldots, L \tag{16.4}$$

and ω_l represents the Doppler frequency for the lth Doppler bin (we assume L bins divide the Doppler interval of interest). We illustrate the model for $\mathbf{y}_{r'}$ in Figure 16.1 (the Doppler shifting is not shown). The problem of interest is the successful estimation of the unknown target coefficients given by $\{\kappa_{r',l'}\}$.

16.2 Receiver design

In the previous chapters we have described many different waveforms such as CAN and WeCAN that are designed to provide low correlations and thus allow for good clutter suppression at the receiver. For some cases, however, even a careful construction of the radar's transmit waveform, when coupled with a matched filter at the receiver, might not provide sufficient sidelobe reduction. To address these situations, we turn our attention in this section to the receiver stage of an active sensing system. We begin our discussion by reviewing the matched filter and by motivating the need for more advanced receiver designs.

16.2.1 Matched filter

As discussed in Chapter 1, a matched filter works by amplifying the signal of interest in the received signal and by reducing the noise component, which is usually assumed to be uncorrelated with the transmitted sequence. In the presence of stochastic additive white noise, in fact, a matched filter provides the highest SNR (signal-to-noise ratio) performance. If the transmitted waveform, or waveform set, has good correlation properties, a matched filter will also weaken the reflected signals from targets in neighboring range cells to the one of interest.

After the matched filter is applied to $\mathbf{y}_{r'}$, the least-squares estimate for the reflection coefficient $\kappa_{r',l'}$ is given by

$$\hat{\kappa}_{r',l'} = \frac{\sum_{n=1}^{N} \tilde{x}_{l'}^{*}(n) y_{r'}(n)}{\sum_{n=1}^{N} |\tilde{x}_{l'}(n)|^2} = \frac{\tilde{\mathbf{x}}_{l'}^{H} \mathbf{y}_{r'}}{||\tilde{\mathbf{x}}_{l'}||^2}. \tag{16.5}$$

Similar estimates can be generated for the other targets in the scene by reformulating the model for the received signal (16.2) (so that $\mathbf{y}_{r'}$ is aligned with the return from a range bin of interest r' for $r' = 1, \ldots, R$).

If there were no interference terms in (16.2) (i.e., if $\kappa_{r,l} = 0$ for any $\{r, l\} \neq \{r', l'\}$), then the matched filter would provide a highly accurate estimate of $\kappa_{r',l'}$. When interference terms (clutter) are present in the received signal, which is commonly the case in practice, then the performance of the matched filter for estimation will depend directly on the correlation properties of the transmitted sequence(s). In Chapters 2 and 3, we described several cyclic approaches that can be used to design sequences or sequence sets with low correlations.

The auto-correlation of a waveform $s(t)$ represents the matched filter's temporal response to a target with negligible Doppler shift (a stationary target relative to the radar). If a target is moving, we have instead to consider the ambiguity function of the signal, as discussed in Chapter 6. In Figure 6.5, we showed the ambiguity function of a CAN sequence (initialized with a random sequence) of length $N = 50$. As we can see, the ambiguity function of the CAN waveform resembles a "thumbtack" in shape. Although a thumbtack form is desirable, since this shape can lead to improved Doppler resolution, the total volume underneath the function remains fixed (see (6.2)). Since we are unable to design a sequence that has zero sidelobes for all time delays and Doppler shifts, we instead seek to replace the matched filter with more advanced receiver designs.

16.2.2 Instrumental variable (IV) receive filter

The instrumental variable (IV) method (also called a mismatched filter), a more general approach for estimating $\kappa_{r',l'}$, can be used to significantly lower sidelobes at the cost of reduced SNR; see [Ackroyd & Ghani 1973][Zoraster 1980][Stoica, Li & Xue 2008] and Chapter 8. Temporarily neglecting Doppler effects (so that $L = 1$, $\omega_l = 0$ and $\tilde{\mathbf{x}}_l = \mathbf{x}$), the IV estimate of $\kappa_{r'}$ is given by

$$\hat{\kappa}_{r'} = \frac{\mathbf{z}^H \mathbf{y}_{r'}}{\mathbf{z}^H \mathbf{x}}, \tag{16.6}$$

where $\mathbf{z}$ denotes the IV receive filter. In the case of $\mathbf{z} = \mathbf{x}$, (16.6) reduces to the matched filter estimate of $\kappa_{r'}$. In general, and unlike the matched filter, the elements of $\mathbf{z}$ are not restricted to be unimodular, since this vector is only designed for the purpose of estimation. Also, we note that IV filters can be precomputed offline. From a computational standpoint, therefore, IV certainly adds a minimal burden to the receiver, as the complexity of its application is comparable to that of the matched filter. We assume herein that $\mathbf{z}$ is a vector of length N, although by padding the transmit waveform with zeros, a longer IV vector could be designed to improve sidelobe reduction even more (at a cost of further reduced SNR).

We consider the IV formulation given in [Stoica, Li & Xue 2008]. The goal of the IV approach is to find a vector $\mathbf{z}$ that minimizes the ISL metric (see (2.1) for the original definition of ISL) in the case of using IV filters:

$$\mathrm{ISL}_{\mathrm{IV}} = \frac{\sum_{k=-(N-1),k\neq 0}^{N-1} \left|\mathbf{z}^H \mathbf{J}_k \mathbf{x}\right|^2}{\left|\mathbf{z}^H \mathbf{x}\right|^2}. \tag{16.7}$$

By applying the Cauchy–Schwartz inequality, the minimum value of $\mathrm{ISL}_{\mathrm{IV}}$ can be shown to be achieved when $\mathbf{z} = \mathbf{R}_{\mathrm{IV}}^{-1}\mathbf{x}$, where

$$\mathbf{R}_{\mathrm{IV}} = \sum_{\substack{k=-N+1\\ k\neq 0}}^{N-1} \mathbf{J}_k \mathbf{x}\mathbf{x}^H \mathbf{J}_k^T. \tag{16.8}$$

When motion is present in the scene, we assume that the Doppler shifts of the targets in the scene $\{\omega_l\}_{l=1}^L$ are assumed to lie within an uncertainty interval denoted by $\Omega = [\omega_a, \omega_b]$ (where $\omega_b > \omega_a$ and where we choose L such that $\{\omega_l\}_{l=1}^L$ covers Ω). Since no knowledge is assumed of the targets' Doppler shifts, other than that they belong to Ω, the ISL criterion (16.7) can be rewritten as [Stoica, Li & Xue 2008]

$$\mathrm{ISL}_{\mathrm{IV,D}} = \sum_{\substack{k=-(N-1)\\ k\neq 0}}^{N-1} \left(\frac{1}{\omega_b - \omega_a}\right) \frac{\int_\Omega \left|\mathbf{z}_{l'}^H \mathbf{J}_k \tilde{\mathbf{x}}(\omega)\right|^2 d\omega}{\left|\mathbf{z}_{l'}^H \tilde{\mathbf{x}}_{l'}\right|^2}, \tag{16.9}$$

where $\mathbf{z}_{l'}$ refers to the receive filter for Doppler bin l' and $\tilde{\mathbf{x}}(\omega)$ denotes the Doppler shifted waveform (for the Doppler frequency ω). When the Doppler uncertainty interval Ω becomes larger, the minimum achievable value of $\mathrm{ISL}_{\mathrm{IV,D}}$ could become significantly greater than that of $\mathrm{ISL}_{\mathrm{IV}}$. Intuitively, this is due to the fact that the designs based on $\mathrm{ISL}_{\mathrm{IV,D}}$ are more conservative, as they try to optimize the ISL metric averaged over the entire set Ω. For this reason, the IV approach does not perform well when Doppler effects are nonnegligible.

16.3 Iterative adaptive approach (IAA)

To provide higher resolution in the nonnegligible Doppler case, at the cost of increased computational complexity at the receiver, we now explore a more advanced estimation technique. The iterative adaptive approach (IAA), first presented in [Yardibi et al. 2010], was shown to offer improved resolution and interference rejection performance. IAA is a nonparametric and user parameter-free weighted least-squares algorithm. In [Yardibi et al. 2010], IAA was shown to perform well for applications in channel estimation for communications, radar and sonar range–Doppler imaging, and passive array sensing. Whereas some data-adaptive algorithms require a significant number of snapshots to obtain accurate target estimates, IAA was shown to achieve good performance even with a single data vector. We briefly summarize the algorithm here.

Consider the model for $\mathbf{y}_{r'}$ in (16.2). The goal of IAA is to minimize the following weighted least-squares cost function with respect to a target of interest $\kappa_{r',l'}$:

$$||\mathbf{y}_{r'} - \kappa_{r',l'}\tilde{\mathbf{x}}_{l'}||^2_{\mathbf{Q}^{-1}_{r',l'}}, \tag{16.10}$$

where $||\mathbf{u}||^2_{\mathbf{Q}^{-1}} \triangleq \mathbf{u}^H\mathbf{Q}^{-1}\mathbf{u}$. The interference covariance matrix for a target of interest $\kappa_{r',l'}$ is denoted by $\mathbf{Q}_{r',l'}$, and is defined as

$$\mathbf{Q}_{r',l'} = \mathbf{R}_{\mathrm{IAA}}(r') - |\kappa_{r',l'}|^2\tilde{\mathbf{x}}_{l'}\tilde{\mathbf{x}}_{l'}^H, \tag{16.11}$$

where

$$\mathbf{R}_{\mathrm{IAA}}(r') = \sum_{r=-N+1}^{N-1}\sum_{l=1}^{L} |\kappa_{r'+r,l}|^2\mathbf{J}_r\tilde{\mathbf{x}}_l\tilde{\mathbf{x}}_l^H\mathbf{J}_r^H. \tag{16.12}$$

The weighted least-squares estimate for a target of interest $\kappa_{r',l'}$, after some simplification, is given by

$$\hat{\kappa}_{r',l'} = \frac{\tilde{\mathbf{x}}_{l'}^H\mathbf{R}^{-1}_{\mathrm{IAA}}(r')\mathbf{y}_{r'}}{\tilde{\mathbf{x}}_{l'}^H\mathbf{R}^{-1}_{\mathrm{IAA}}(r')\tilde{\mathbf{x}}_{l'}}, \quad l' = 1,\ldots,L, \quad r' = 1,\ldots,R. \tag{16.13}$$

Since the estimate (16.13) depends on the covariance matrix $\mathbf{R}_{\mathrm{IAA}}(r')$, which in turn depends on the target amplitudes, the algorithm uses an iterative approach, which is summarized in Table 16.1. The target coefficients are initialized using the matched filter approach outlined in Section 16.2.1. To estimate targets in other range bins, we simply redefine $\mathbf{y}_{r'}$, which represents the length-N signal vector aligned with the received

Table 16.1. The IAA algorithm for range–Doppler imaging

Step 0: Initialization

$$\hat{\kappa}_{r',l'} = \frac{1}{N}\tilde{\mathbf{x}}_{l'}^H \mathbf{y}_{r'}, \quad l' = 1, \ldots, L, \quad r' = 1, \ldots, R. \tag{16.14}$$

Step 1: Compute the covariance matrix for $r' = 1, \ldots, R$:

$$\mathbf{R}_{\text{IAA}}(r') = \sum_{r=-N+1}^{N-1} \sum_{l=1}^{L} |\hat{\kappa}_{r'+r,l}|^2 \mathbf{J}_r \tilde{\mathbf{x}}_l \tilde{\mathbf{x}}_l^H \mathbf{J}_r^H. \tag{16.15}$$

Step 2: Update the estimates for all range bins $r' = 1, \ldots, R$ and Doppler bins $l' = 1, \ldots, L$:

$$\hat{\kappa}_{r',l'} = \frac{\tilde{\mathbf{x}}_{l'}^H \mathbf{R}_{\text{IAA}}^{-1}(r')\mathbf{y}_{r'}}{\tilde{\mathbf{x}}_{l'}^H \mathbf{R}_{\text{IAA}}^{-1}(r')\tilde{\mathbf{x}}_{l'}}. \tag{16.16}$$

Iteration: Repeat steps 1 and 2 until convergence, e.g., repeat them T_{IAA} times.

reflection from a range bin of interest r'. IAA typically converges after about ten iterations (which corresponds to $T_{\text{IAA}} = 10$ in Table 16.1); a local convergence proof of IAA was presented in [Roberts et al. 2010].

16.4 Numerical examples

We present two radar imaging examples, one with stationary targets and one with mobile targets. As CAN sequences were shown to contain lower correlation sidelobes than the other existing transmit sequences, we consider only Frank sequences (see (1.23)), for comparison purposes, and CAN sequences (initialized by Frank sequences) in the following.

For each example, we will assume circularly symmetric independent and identically distributed additive complex Gaussian noise with zero mean and variance σ^2. The SNR is defined as $10\log_{10}(1/\sigma^2)$.

16.4.1 Negligible Doppler example

In the first example, we consider a stationary scene with $R = 512$ equally spaced range bins. We place three targets in the scene: one target at range bin 200 with amplitude −7 dB, one target at range bin 308 with amplitude −17 dB, and one target at range bin 320 with amplitude 0 dB. The transmit waveforms are designed for $N = 256$. The SNR is set at 20 dB. The true target locations are indicated in each figure using circles.

We show the result using a Frank sequence and a matched filter at the receiver in Figure 16.2(a). As evidenced, the two stronger targets are successfully identified using this scheme. The third, weaker target, however, appears within the sidelobes of the strongest

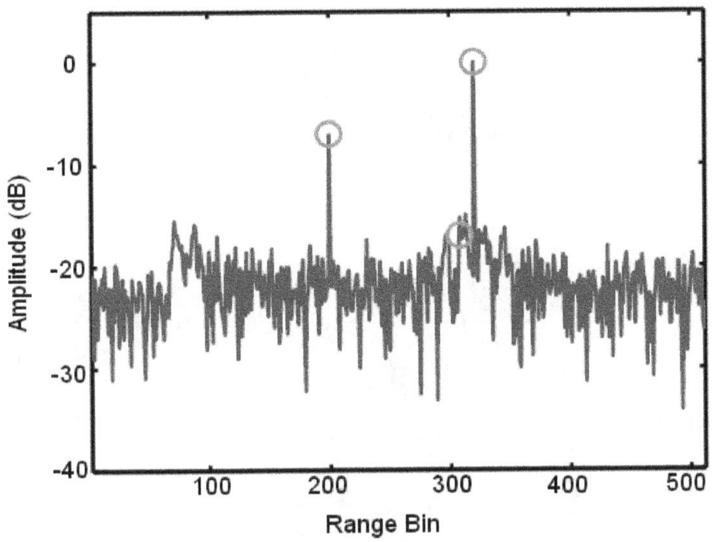

(a)

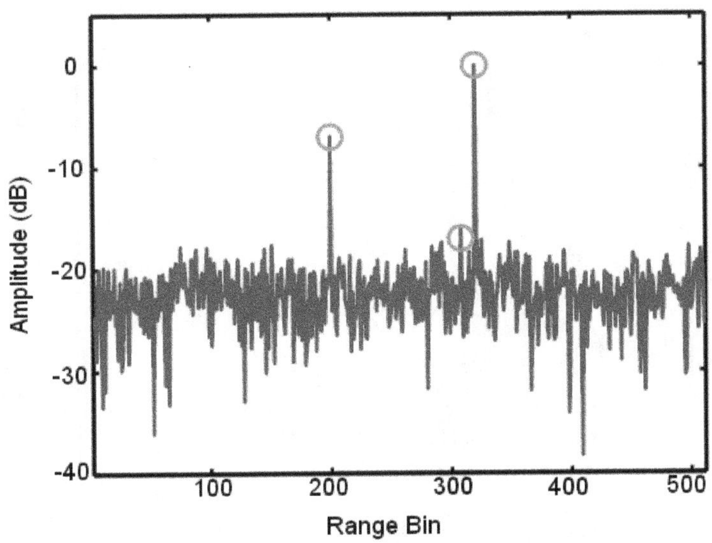

(b)

Figure 16.2 Range profiles for $N = 256$ and SNR = 20 dB. (a) Frank sequence with matched filter at the receiver. (b) CAN sequence with matched filter at the receiver. Each true target location is indicated using a circle.

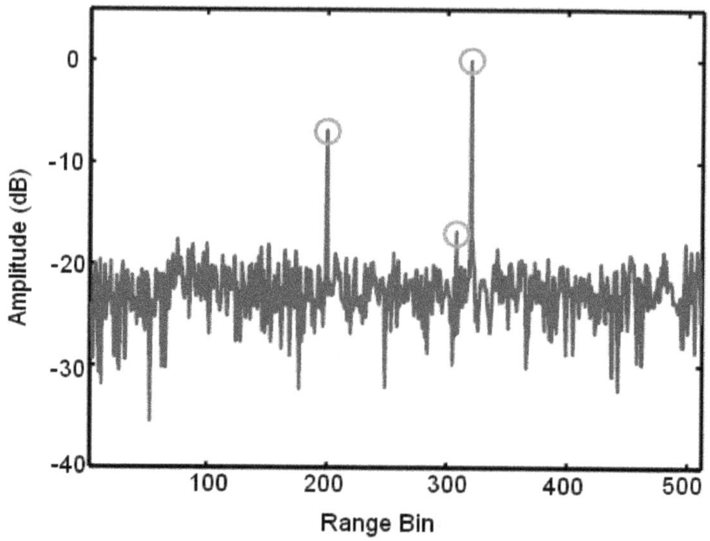

(a)

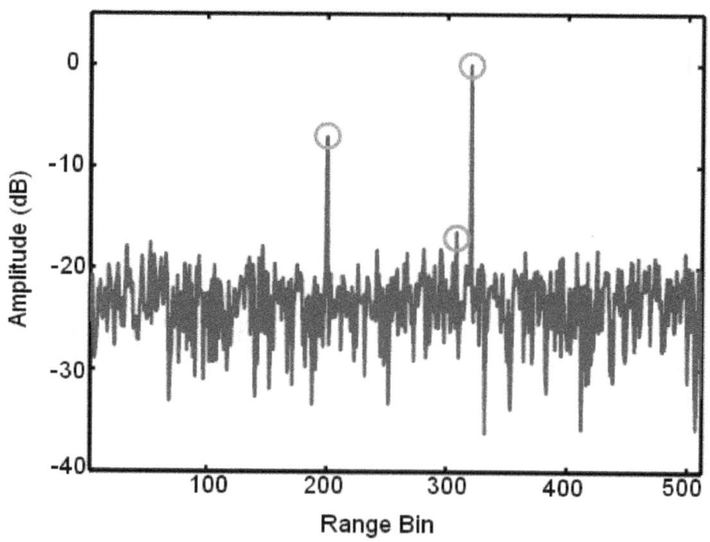

(b)

Figure 16.3 Range profiles for $N = 256$ and SNR = 20 dB. (a) CAN sequence with IV receive filter. (b) CAN sequence with IAA at the receiver. Each true target location is indicated using a circle.

target, and the matched filter does not produce a peak at the true target location. In Figure 16.2(b), we again use a matched filter, but now transmit a CAN sequence. For this case, the sidelobes are reduced and a peak is now discernible at the location of the weakest target. We use the CAN sequence for Figures 16.3(a) and 16.3(b), too.

We adopt an IV receive filter (with length N) in Figure 16.3(a). When compared with the matched filter result in Figure 16.2(b), IV produces lower leakage and furthermore shows well-separated peaks at the true target locations. As mentioned, IV filters can be precomputed offline, so this technique does not have a higher computational burden than the matched filter. The IAA result is shown in Figure 16.3(b). In this case, IAA achieves a similar performance to the IV filter, but at the cost of increased computational efforts at the receiver.

16.4.2 Non-negligible Doppler example

In this example, we evaluate the different design strategies for the non-negligible Doppler case. The Doppler shift of a target is represented by $\Phi_l = \omega_l N(180°/\pi)$ where $l = 1, \ldots, L$. The scene contains $R = 100$ equally spaced range bins and $L = 37$ Doppler bins with 5° separation between bins (we define Ω by setting $\Phi_1 = -90°$ and $\Phi_L = 90°$). We consider three targets in the scene. The first target is located at range bin 60 with Doppler shift $-10°$ and amplitude 10 dB. The second and third targets have Doppler shift 10° and amplitude 30 dB, and are located at range bin 40 and range bin 65, respectively. The SNR is set at 10 dB (again, assuming circularly symmetric i.i.d. noise), and the length of the transmit sequence is set at $N = 36$. The true target locations are again indicated in each figure using circles, which is colored according to its corresponding amplitude.

We transmit a Frank sequence and a CAN sequence for Figures 16.4(a) and 16.4(b), respectively, with a matched filter at the receiver. As evidenced, the matched filter performs poorly for this non-negligible Doppler case, as the target estimates have high sidelobes, and the weaker target (10 dB) is not identified. The result for the CAN sequence appears to have slightly lower sidelobes. In the remaining two figures, then, we use only CAN sequences for transmission.

The matched filter is replaced with an IV filter at the receiver in Figure 16.5(a). As expected, the IV filter does not perform well for mobile targets. The sidelobes are slightly reduced compared to the matched filter result, but large sidelobes continue to dominate the image. The IAA result is shown in Figure 16.5(b). In comparison to the matched filter and IV filter, IAA significantly reduces the sidelobes and produces a peak at each of the true target locations (again, at the cost of increased computation).

16.5 Conclusions

In this chapter we demonstrated how CAN sequences, which are designed to have low correlation properties, can benefit radar range and range–Doppler imaging. We showed that when further improvements in resolution and interference suppression are needed

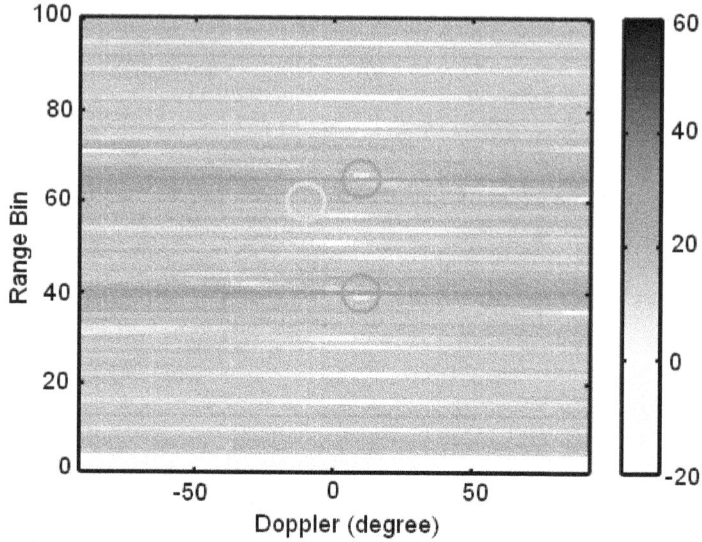

(a)

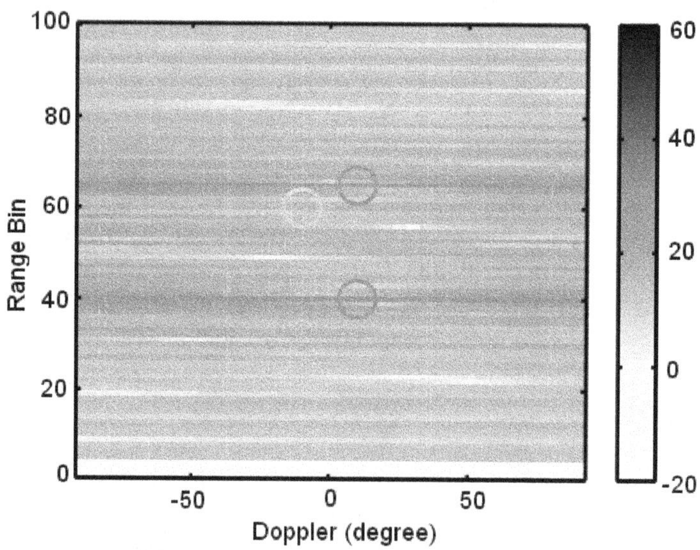

(b)

Figure 16.4 Range-Doppler images for $N = 36$ and SNR = 10 dB. (a) Frank sequence with matched filter at the receiver. (b) CAN sequence with matched filter at the receiver. Each true target location is indicated using a circle.

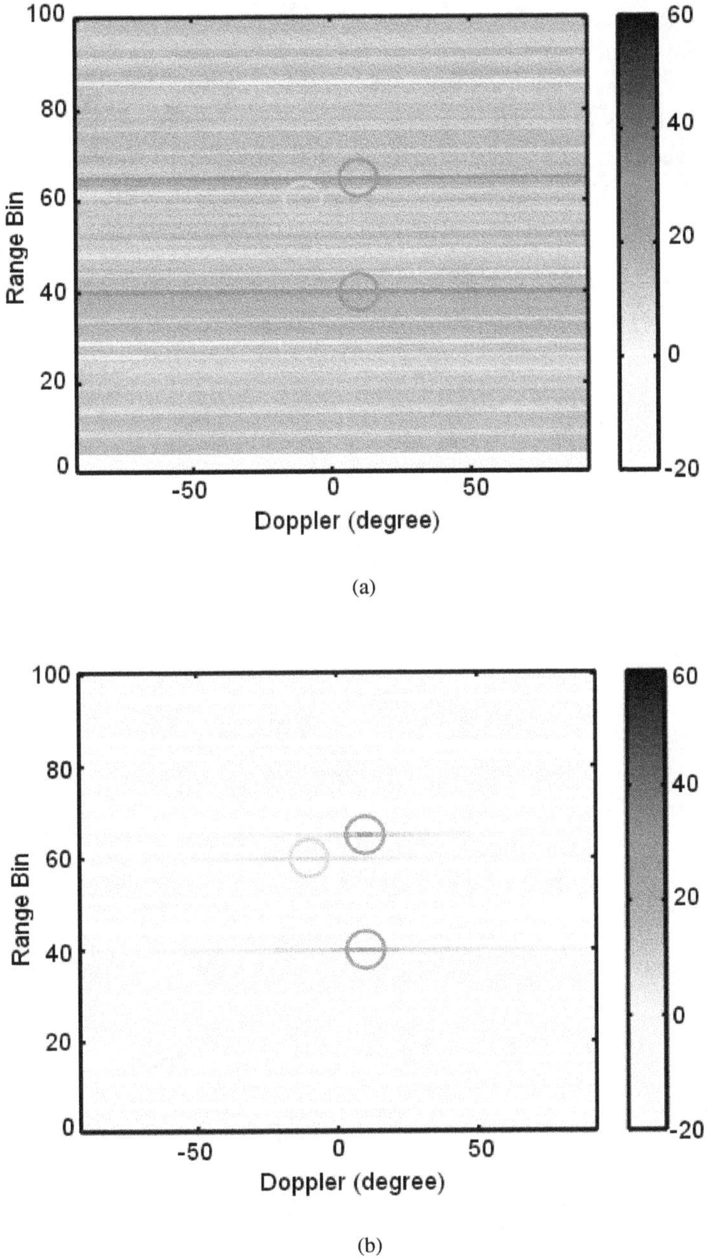

Figure 16.5 Range-Doppler images for $N = 36$ and SNR = 10 dB. (a) CAN sequence with IV receive filter. (b) CAN sequence with IAA at the receiver. Each true target location is indicated using a circle.

and cannot be achieved at the signal design stage, better signal processing (compared to the vanilla matched filter) is required at the receiver. The IV filter, which does not induce any additional computational burden, was shown to provide improved performance in the negligible Doppler (stationary target) case compared to a matched filter. When motion is present in the scene, IAA, at the cost of an increased computational burden, was shown to produce higher resolution and more accurate target estimates.

17 Ultrasound system for hyperthermia treatment of breast cancer

The development of breast cancer imaging techniques such as microwave imaging [Meaney et al. 2000][Guo et al. 2006], ultrasound imaging [Szabo 2004][Kremkau 1993], thermal acoustic imaging [Kruger et al. 1999] and MRI has improved the ability to visualize and accurately locate a breast tumor without the need for surgery [Gianfelice et al. 2003]. This has led to the possibility of noninvasive local hyperthermia treatment of breast cancer. Many studies have been performed to demonstrate the effectiveness of local hyperthermia on the treatment of breast cancer [Falk & Issels 2001][Vernon et al. 1996]. A challenge in the local hyperthermia treatment is heating the malignant tumors to a temperature above 43°C for about thirty to sixty minutes, while at the same time maintaining a low temperature in the surrounding healthy breast tissue region.

There are two major classes of local hyperthermia techniques: microwave hyperthermia [Fenn et al. 1996] and ultrasound hyperthermia [Diederich & Hynynen 1999]. The penetration of microwave in biological tissues is poor. Moreover, the focal spot generated by microwave is unsatisfactory at the normal/cancerous tissue interface because of the long wavelength of the microwave. Ultrasound can achieve much better penetration depths than microwave. However, because its acoustic wavelength is very short, the focal spot generated by ultrasound is very small (millimeter or submillimeter in diameter) compared with a large tumor (centimeter in diameter on the average). Thus, many focal spots are required for complete tumor coverage, which results in a long treatment time and missed cancer cells.

In this chapter we present a waveform-diversity-based ultrasound hyperthermia technique for the treatment of breast cancer. As discussed in Chapter 13, transmitting multiple different waveforms via a transducer array offers more flexibility for transmit beampattern design than the standard phased array technique. By designing the transmitted signal (spatial) covariance matrix under the uniform elemental power constraint, the waveform diversity can be exploited to maximize the power deposition at the entire tumor region while minimizing the impact on the surrounding healthy tissue region.

We validate the proposed algorithm on a 2D breast model with an embedded tumor. The model includes breast tissue, skin, and chest wall. The finite-difference time-domain (FDTD) method is used to simulate the acoustic field and the temperature distribution within the breast. We show via numerical simulations that the proposed method can provide the necessary temperature gradients required for effective hyperthermia treatment of the tumor while maintaining a low temperature level at the surrounding healthy tissue.

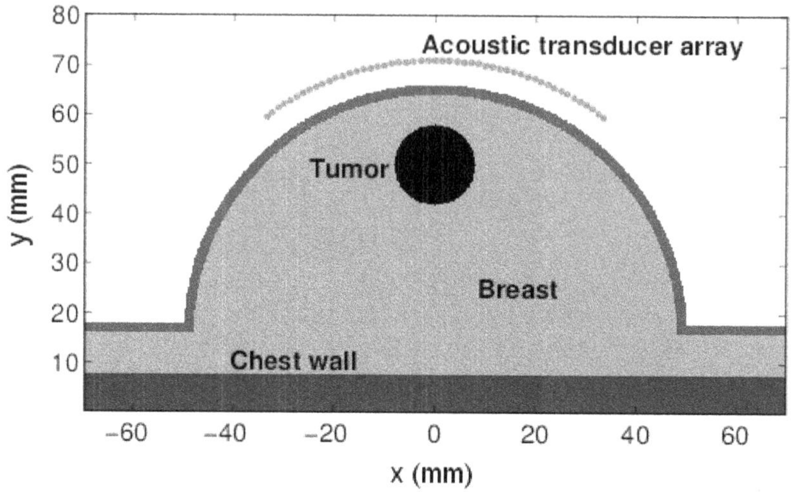

Figure 17.1 Breast model and transducer array

17.1 Waveform diversity based ultrasound hyperthermia

We consider the ultrasound hyperthermia system shown in Figure 17.1. Let $\mathbf{r}_0$ denote the center location of the tumor, which is assumed to be estimated accurately a priori using breast cancer imaging techniques. There are M acoustic transducers deployed around the breast at locations $\mathbf{r}_m$ $(m = 1, 2, \ldots, M)$. Let $x_m(n)$ $(n = 1, 2, \ldots, N)$ denote the discrete-time baseband signal transmitted by the mth acoustic transducer, where N denotes the number of samples of each transmitted signal pulse.

We assume that the transmitted acoustic signals are narrowband and each acoustic transducer is omnidirectional. The baseband signal at a location $\mathbf{r}$ inside the breast can be described by

$$y(\mathbf{r}, n) = \sum_{m=1}^{M} \frac{e^{-j2\pi f_c \tau_m(\mathbf{r})}}{||\mathbf{r}_m - \mathbf{r}||^{1/2}} x_m(n), \qquad n = 1, 2, \ldots, N, \tag{17.1}$$

where f_c is the carrier frequency,

$$\tau_m(\mathbf{r}) = \frac{||\mathbf{r}_m - \mathbf{r}||}{c} \tag{17.2}$$

is the time needed by the signal emitted via the mth transducer to arrive at the location $\mathbf{r}$, with c being the sound speed inside the breast tissues, and $1/||\mathbf{r}_m - \mathbf{r}||^{1/2}$ is the propagation attenuation of the acoustic wave. Let

$$\mathbf{a}(\mathbf{r}) = \left[\frac{e^{j2\pi f_c \tau_1(\mathbf{r})}}{||\mathbf{r}_1 - \mathbf{r}||^{1/2}} \quad \frac{e^{j2\pi f_c \tau_2(\mathbf{r})}}{||\mathbf{r}_2 - \mathbf{r}||^{1/2}} \quad \cdots \quad \frac{e^{j2\pi f_c \tau_M(\mathbf{r})}}{||\mathbf{r}_M - \mathbf{r}||^{1/2}} \right]^T \tag{17.3}$$

be the steering vector and let

$$\mathbf{x}(n) = [x_1(n) \quad x_2(n) \quad \cdots \quad x_M(n)]^T . \tag{17.4}$$

Then (17.1) can be rewritten as

$$y(\mathbf{r}, n) = \mathbf{a}^H(\mathbf{r})\mathbf{x}(n), \quad n = 1, 2, \ldots, N. \tag{17.5}$$

The power of the transmitted signals at location $\mathbf{r}$, which is also called the transmit beampattern (see Chapter 13), is given by

$$P(\mathbf{r}) = \mathrm{E}\left\{y(\mathbf{r}, n)y^H(\mathbf{r}, n)\right\} = \mathbf{a}^H(\mathbf{r})\mathbf{R}\mathbf{a}(\mathbf{r}), \tag{17.6}$$

where $\mathbf{R}$ is the covariance matrix of $\mathbf{x}(n)$, i.e.,

$$\mathbf{R} = \mathrm{E}\left\{\mathbf{x}(n)\mathbf{x}^H(n)\right\}. \tag{17.7}$$

The transmit beampattern is a function of the location $\mathbf{r}$.

The purpose of the waveform diversity technique is to focus the acoustic power onto the entire tumor region while minimizing the peak power level in the surrounding healthy breast tissue region. The corresponding beampattern design problem is to choose the covariance matrix $\mathbf{R}$, under the uniform elemental power constraint

$$R_{mm} = \frac{c}{M}, \quad m = 1, 2, \ldots, M, \tag{17.8}$$

where R_{mm} denotes the (m, m)th element of $\mathbf{R}$ and c is the total transmitted power, such that it

(a) achieves a predetermined main-beam width matched to the entire tumor region (to within 10% of the power deposited at the tumor center);
(b) minimizes the peak sidelobe level in a prescribed region (e.g., the surrounding healthy breast tissue region).

This problem can be formulated as (see Section 13.2.4 of Chapter 13 for a similar formulation):

$$\begin{aligned}
&\min_{t,\mathbf{R}} \ (-t) \\
&\text{s.t. } \mathbf{a}^H(\mathbf{r}_0)\mathbf{R}\mathbf{a}(\mathbf{r}_0) - \mathbf{a}^H(\mu)\mathbf{R}\mathbf{a}(\mu) \geq t, \quad \forall\, \mu \in \Omega_B, \\
&\quad \mathbf{a}^H(\nu)\mathbf{R}\mathbf{a}(\nu) \geq 0.9\mathbf{a}^H(\mathbf{r}_0)\mathbf{R}\mathbf{a}(\mathbf{r}_0), \quad \forall\, \nu \in \Omega_T, \\
&\quad \mathbf{a}^H(\nu)\mathbf{R}\mathbf{a}(\nu) \leq 1.1\mathbf{a}^H(\mathbf{r}_0)\mathbf{R}\mathbf{a}(\mathbf{r}_0), \quad \forall\, \nu \in \Omega_T, \\
&\quad \mathbf{R} \geq 0, \\
&\quad R_{mm} = \frac{c}{M}, \quad m = 1, 2, \ldots, M,
\end{aligned} \tag{17.9}$$

where Ω_T and Ω_B denote the tumor region and the surrounding healthy breast tissue (sidelobe region), respectively.

As shown in Chapter 13, this beampattern design problem is a semidefinite program (SDP) that can be efficiently solved in polynomial time using public domain software. Once $\mathbf{R}$ is determined, a signal sequence $\{\mathbf{x}(n)\}$ that has $\mathbf{R}$ as its covariance matrix can be synthesized using the algorithm discussed in Chapter 14 or simply as (whenever the PAR of the signal is not much of a concern):

$$\mathbf{x}(n) = \mathbf{R}^{1/2} w(n), \quad n = 1, 2, \ldots, N, \tag{17.10}$$

where $\{w(n)\}$ is a sequence of i.i.d. random vectors with mean zero and covariance matrix $\mathbf{I}$, and $\mathbf{R}^{1/2}$ denotes a square root of $\mathbf{R}$. By transmitting the above $\mathbf{x}(n)$ using the acoustic transducer array, we can get a desired high acoustic power deposition over the entire tumor region while minimizing the power deposition in the surrounding healthy breast tissue region. If the PAR of the transmitted signal is of concern, the technique introduced in Chapter 14 can be used to synthesize the sequence set $\{\mathbf{x}(n)\}$.

17.2 Numerical results

The 2D breast model we consider is a 10cm-diameter semicircle, which includes breast tissues, skin, and chest wall (see Figure 17.1). The acoustic properties of the breast tissues are assumed random with a variation of $\pm 5\%$ around the nominal values. A tumor 16 mm in diameter is embedded below the skin with the tumor center at $x = 0$ mm, $y = 50$ mm. There are 51 acoustic transducers deployed uniformly around the breast model. The distance between neighboring transducers is 1.5 mm (half the wavelength of the carrier frequency). An acoustic wave with frequency 500 KHz is used as the carrier frequency.

The FDTD method [Yuan et al. 1999][Katsibas & Antonopoulos 2004] was employed to calculate the acoustic power deposition in the breast, as well as the temperature distribution in the tissues; see [Guo & Li 2008] for details.

We demonstrate the performance of the proposed waveform diversity based method via several numerical examples. The conventional delay-and-sum (DAS) beamforming method (also called phased array) is also applied to the same scenario, and its results are compared with those of the waveform diversity based method. The DAS beamformer transmits the same waveform through all transducers using the weight vector

$$\mathbf{w} = \frac{\mathbf{a}(\mathbf{r}_O)}{\|\mathbf{a}(\mathbf{r}_O)\|^2}. \tag{17.11}$$

The associated beampattern is

$$P(\mathbf{r}) = \left|\mathbf{a}^H(\mathbf{r})\,\mathbf{w}\right|^2. \tag{17.12}$$

which is a special case of (17.6) corresponding to $\mathbf{R} = \mathbf{w}\mathbf{w}^H$.

Figure 17.2 shows the calculated transmit beampatterns for Ω_B and Ω_T defined by the breast model under consideration. Figure 17.2(a) is the beampattern corresponding to

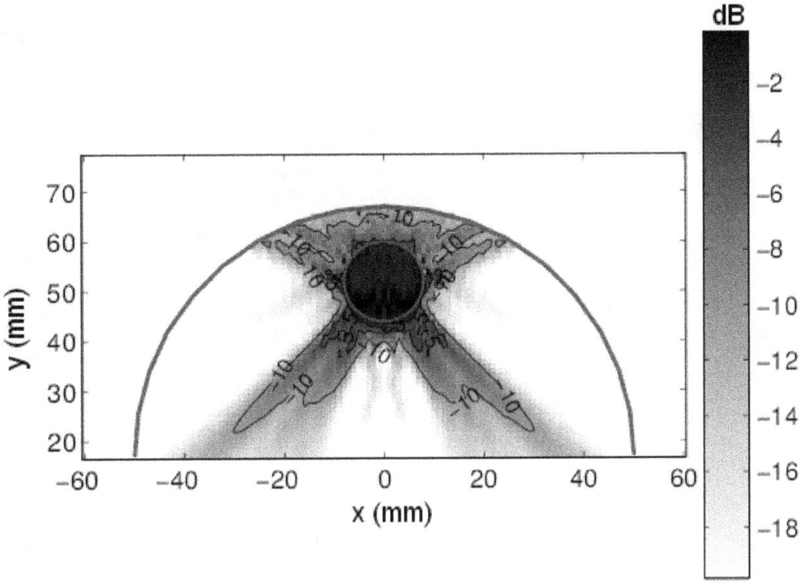

(a)

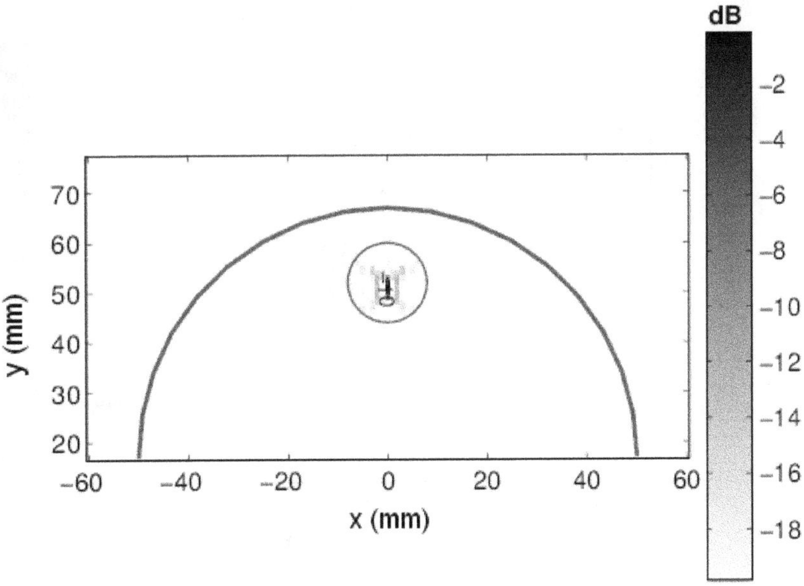

(b)

Figure 17.2 The beampattern obtained using (a) waveform diversity and (b) DAS.

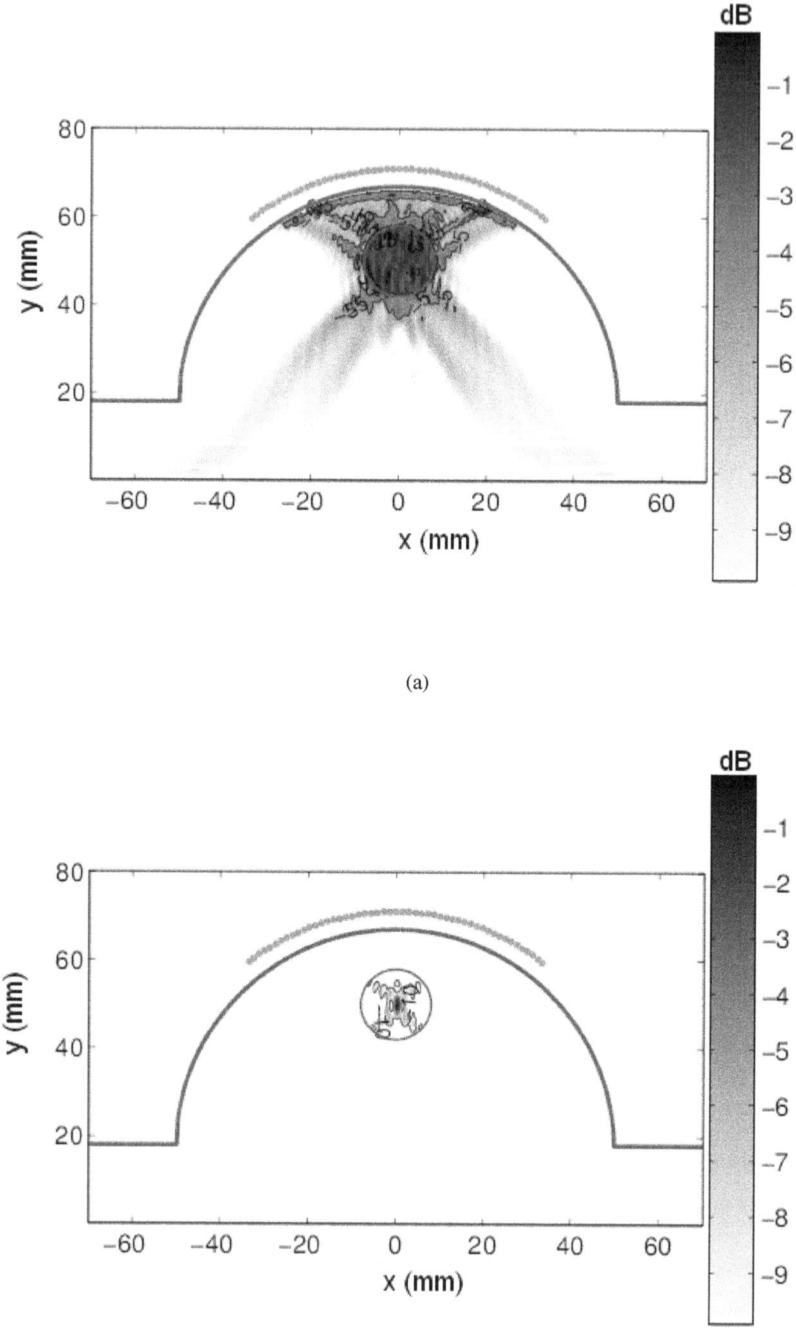

(a)

(b)

Figure 17.3 Power densities with (a) waveform diversity and (b) DAS.

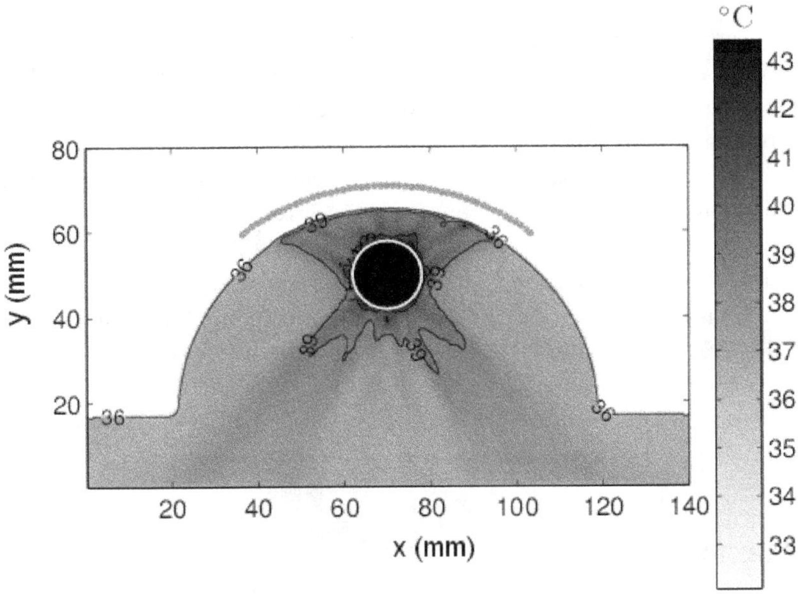

(a)

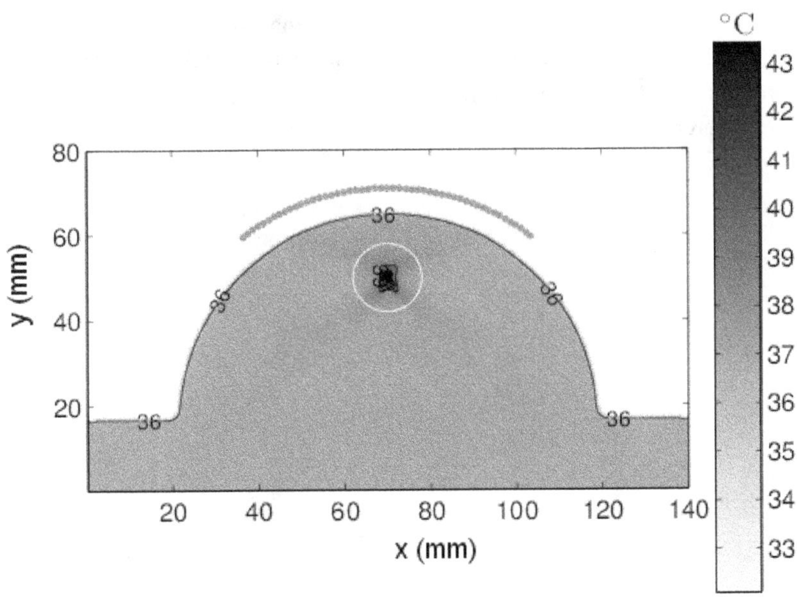

(b)

Figure 17.4 Temperature distribution for (a) waveform diversity and (b) DAS.

the waveform diversity technique which is calculated by using (17.6) with the optimal covariance matrix **R** determined by using (17.9). The figure shows that the 3dB main beam matches the tumor region quite well, and that the sidelobe level is low. Figure 17.2(b) is the DAS beampattern which is calculated using (17.12). As can be seen, this beampattern is rather narrow, and only focuses on the central region of the tumor.

Figures 17.3(a) and 17.3(b) show the acoustic power densities within the breast for the waveform diversity technique and DAS, respectively. The acoustic power densities in Figure 17.3 agree well with the beampatterns in Figure 17.2, with the waveform diversity technique giving a focal spot matched to the entire tumor region.

Figure 17.4 shows the temperature distributions within the breast model. Figure 17.4(a), which is the result of waveform diversity, shows that the entire tumor region is heated to a temperature greater than 43°C while maintaining the surrounding healthy tissues at a low temperature level (below 40°C). In contrast to this, DAS only heats a small region of the tumor to a temperature greater than 43°C.

17.3 Conclusions

We have presented a waveform-diversity-based ultrasound hyperthermia technique for the treatment of breast cancer. By choosing the covariance matrix of the transmitted waveforms properly, this method can provide a focal spot that covers the entire tumor region while minimizing the impact on the surrounding healthy breast region. Simulation examples have been included to show that this method has better acoustic power deposition than its conventional DAS counterpart, and can provide the necessary temperature gradients required for effective hyperthermia treatment of breast cancer. More recent results on this topic can be found in [Zeng et al. 2010].

18 Covert underwater acoustic communications – coherent scheme

Achieving reliable communication over underwater acoustic (UWA) channels is a challenging problem owing to the scarce bandwidth available and the double spreading phenomenon, i.e., spreading in both time (multipath delay spread) and frequency domains (Doppler spread) [Kilfoyle & Baggeroer 2000]. Delay and Doppler spreading is inherent to many practical communication channels, but are considerably amplified in UWA environments [Stojanovic et al. 1994]. Double spreading complicates the receiver structure and makes it difficult to extract the desired symbols from the received signals.

Telemetry systems adopting direct-sequence spread-spectrum (DSSS) based modulation techniques are conventionally referred to as operating at low data rates. Existing literature regarding low data rate UWA communications include [Palmese et al. 2007][Stojanovic et al. 1998][Hursky et al. 2006][Yang & Yang 2008][Stojanovic & Freitag 2000][Blackmon et al. 2002][Ritcey & Griep 1995][Stojanovic & Freitag 2004][Sozer et al. 1999][Tsimenidis et al. 2001][Iltis & Fuxjaeger 1991]. By sacrificing the data rate, DSSS techniques exploit frequency diversity in the frequency-selective UWA channel and benefit from spreading gain to allow many co-channel users. At the receiver side, decentralized reception schemes encompass nonlinear equalization including hypothesis-feedback equalization [Stojanovic & Freitag 2000], and linear equalization including RAKE receivers [Tse & Viswanath 2005]. Performance comparisons of hypothesis-feedback equalization and RAKE reception are presented in [Blackmon et al. 2002].

In this chapter, we consider a single user scenario with a *coherent* RAKE reception scheme (the *noncoherent* schemes, which do not require a channel estimation, will be discussed in Chapter 19). Although the coherent RAKE receiver cannot completely combat severe inter- or intrasymbol interferences in a frequency-selective UWA channel, the adverse effects of the interferences can be alleviated by carefully designing the spreading waveforms [Ling, Yardibi, Su, He & Li 2009]. *Good waveform design*, which accounts for practical concerns such as the modulation scheme, the channel characteristics, etc., allows for a simple and efficient reception scheme (RAKE, for example). An ideal spreading waveform, whose aperiodic correlations are zero at certain time lags, can effectively suppress inter- or intrasymbol interferences. The m-sequence, for example, is a popular spreading waveform employed in UWA channels due to its good correlation properties [Yang & Yang 2008]. M-sequences, as well as most other existing practical spreading waveforms, are constructed in a deterministic and systematic manner with strict constraints on the chip length. These features, as will be discussed in the sequel, make such waveforms unattractive for covert UWA communications, which requires either low probability of detection (LPD) or at least low probability of interception (LPI).

Since the processing gain cannot be very large due to channel variations, the incoming chip SNR (signal-to-noise ratio) must be increased to maintain satisfactory detection performance with coherent RAKE. A direct consequence of boosting the incoming chip SNR is degraded LPD performance (i.e., the presence of the transmitted signal can be detected more easily by an eavesdropper). To protect the privacy of the transmitted signal, another metric, referred to as low probability of interception (LPI) [Yang & Yang 2008], can be considered. LPI can be ensured in a variety of ways, for example, via the use of an off-the-shelf encryption technique during the source coding or channel coding stage. In this chapter, we investigate the LPI property solely from a spreading waveform design aspect. A spreading waveform that is constructed in a deterministic and systematic manner, such as an m-sequence, is not a viable candidate waveform since an eavesdropper can exhaustively search through the set of all possible waveforms to find the waveform in question. A more favorable spreading waveform would possess unrestricted phase values (not from a finite alphabet) and flexible length. Note that, solely from an LPI point of view, a random phase spreading waveform (i.e., the phase of each involved chip is independently and uniformly distributed between 0 and 2π) is an attractive candidate waveform. As we will show in numerical examples, however, the detection performance of different realizations of the random phase waveform exhibits considerable variations due to the unoptimized correlations. Fortunately, starting with a random phase initialization, the CAN and WeCAN algorithms presented in Chapter 3 can be adopted to refine the waveform correlation properties. In this way, aside from the flexible length and the arbitrary phase values originally possessed by the random phase waveform, the so-obtained waveform is further entailed with good correlation properties.

18.1 Problem formulation

Suppose that a transmitter maps source bits to payload symbols using Gray coded quadrature phase-shift keying (QPSK), and each QPSK payload symbol in $\{s_n\}$ is spread by a common unimodular spreading waveform $\mathbf{x}$ consisting of P chips, i.e., $\mathbf{x} = [\,x_1 \;\cdots\; x_P\,]^T$. The resulting phase-modulated waveforms $\{s_n\mathbf{x}\}$ are then up-converted to the carrier frequency and transmitted via UWA channels in the presence of a strong noise background.

We assume a block-fading channel, in which the channel impulse response (CIR) remains stationary over at least one symbol period. We let $\mathbf{h}_n = [\,h(n,1) \;\cdots\; h(n,R)\,]^T$ characterize the CIR vector over the nth symbol period (i.e., during the transmission of $s_n\mathbf{x}$) with R resolved channel taps ($P > R$ in general). We further assume that sampling and synchronization procedures have already been employed, and that the sampled complex baseband signals are available at the receiver. Note that although our emphasis is on QPSK modulation schemes, the derivations provided in the following sections can be easily extended to a general M-ary phase-shift keying (PSK) case.

By confining our attention to the detection of the nth QPSK payload symbol s_n, the problem can be formulated as (the same analysis is repeated for all QPSK payload symbols of interest):

$$\mathbf{y}_n = \mathbf{X}_n \mathbf{h}_n + \mathbf{e}_n, \tag{18.1}$$

where

$$\mathbf{y}_n = [\, y_1 \; \cdots \; y_{P+R-1} \,]^T \tag{18.2}$$

contains the $P + R - 1$ synchronized measured data samples (i.e., y_1, the first element of $\mathbf{y}_n$, maps to $s_n x_1$, and so on). Further,

$$\mathbf{e}_n = [\, e_1 \; \cdots \; e_{P+R-1} \,]^T \tag{18.3}$$

represents additive noise (thermal or hardware related noise, interferences or jamming, as well as the overwhelming ambient noise of the sea). The elements of $\mathbf{e}_n$ are assumed to be circularly symmetric independent and identically distributed (i.i.d.) complex-valued Gaussian random variables with zero mean and variance σ^2, denoted as $\mathbf{e}_n \sim \mathcal{CN}(0, \sigma^2 \mathbf{I})$. The matrix $\mathbf{X}_n \in \mathcal{C}^{(P+R-1)\times R}$ in (18.1) contains multiple shifted replicas of the phase-modulated spreading waveforms and is given by

$$\mathbf{X}_n = \begin{bmatrix} s_n x_1 & s_{n-1} x_P & \cdots & s_{n-1} x_{P-R+2} \\ \vdots & s_n x_1 & & s_{n-1} x_{P-R+3} \\ s_n x_P & \vdots & \ddots & \vdots \\ s_{n+1} x_1 & s_n x_P & & s_{n-1} x_P \\ \vdots & s_{n+1} x_1 & \ddots & s_n x_1 \\ s_{n+1} x_{R-2} & \vdots & \ddots & \vdots \\ s_{n+1} x_{R-1} & s_{n+1} x_{R-2} & & s_n x_P \end{bmatrix}, \tag{18.4}$$

where s_{n-1} and s_{n+1} denote, respectively, the symbols transmitted before and after the symbol of current interest.

The problem is then to estimate the QPSK symbol s_n given the measurement vector $\mathbf{y}_n$ and the known spreading waveform $\mathbf{x}$. As mentioned previously, coherent RAKE reception is employed herein. We are particularly interested in designing a waveform $\mathbf{x}$ that not only facilitates the reception scheme considered, but also ensures LPI communications.

18.2 Spreading waveform synthesis

In this section, we first explore the characteristics of common spreading waveforms that facilitate coherent RAKE reception. Specifically, we assess the impact of the correlation (the aperiodic auto-correlation) properties of the spreading waveform on the outputs of

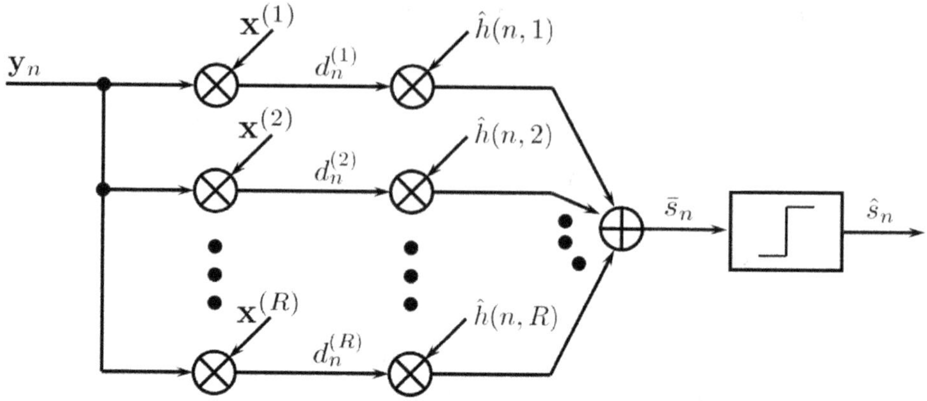

Figure 18.1 Structure of a coherent RAKE detector.

each RAKE finger. Then, we consider two of the algorithms presented in Chapter 3 to generate a spreading waveform with desirable characteristics.

The matrix $\mathbf{X}_n$ in (18.4) can be decomposed as follows to isolate the contribution of s_n from that of its adjacent symbols s_{n-1} and s_{n+1}:

$$\mathbf{X}_n = s_n\mathbf{C} + s_{n-1}\mathbf{B} + s_{n+1}\mathbf{A}, \tag{18.5}$$

where the dimensions of $\mathbf{A}$, $\mathbf{B}$ and $\mathbf{C}$ conform with that of $\mathbf{X}_n$. The matrix $\mathbf{C}$ contains only the shifted replicas of $\mathbf{x}$ that are relevant to the symbol of current interest s_n:

$$\mathbf{C} = \begin{bmatrix} x_1 & & \mathbf{0} \\ \vdots & \ddots & \\ x_P & & x_1 \\ & \ddots & \vdots \\ \mathbf{0} & & x_P \end{bmatrix}. \tag{18.6}$$

The matrices $\mathbf{B}$ and $\mathbf{A}$ are composed of the residual chips associated with s_{n-1} and s_{n+1}, respectively:

$$\mathbf{B} = \begin{bmatrix} 0 & x_P & \cdots & x_{P-R+2} \\ 0 & 0 & & x_{P-R+3} \\ \vdots & \vdots & \ddots & \vdots \\ 0 & 0 & \cdots & x_P \\ & & \mathbf{0} & \end{bmatrix}, \quad \mathbf{A} = \begin{bmatrix} & & \mathbf{0} & \\ x_1 & \cdots & 0 & 0 \\ \vdots & \ddots & \vdots & \vdots \\ x_{R-2} & & 0 & 0 \\ x_{R-1} & \cdots & x_1 & 0 \end{bmatrix}. \tag{18.7}$$

Note that $\mathbf{A}$, $\mathbf{B}$ and $\mathbf{C}$ are all independent of the symbol index n.

Figure 18.1 shows the structure of a conventional coherent RAKE detector. The received measurement vector $\mathbf{y}_n$ is first multiplied with the vector $\mathbf{x}^{(l)} \in C^{(P+R-1)\times 1}$, which is the shifted version of the spreading waveform $\mathbf{x}$ associated with the lth channel tap (i.e., the tap represented by $h(n,l)$ over the nth symbol period). More specifically, the vector $\mathbf{x}^{(l)}$ is the lth column of $\mathbf{C}$ in (18.6):

$$\mathbf{x}^{(l)} = \Big[\underbrace{0 \cdots 0}_{l-1}\ x_1\ x_2 \cdots x_P\ \underbrace{0 \cdots 0}_{R-l}\Big]^T, \tag{18.8}$$

where $r = 1, \cdots, R$. The correlation function of the spreading waveform $\mathbf{x}$ is defined in the usual way (see (1.15)):

$$r_k = \sum_{n=k+1}^{P} x_n x_{n-k}^* = r_{-k}^*, \quad k = 0, \cdots, P-1, \tag{18.9}$$

where $r_0 = P$ due to the unimodular nature of $\{x_p\}_{p=1}^{P}$.

It can be easily verified that

$$\mathbf{x}^{(l)H}\mathbf{C} = \Big[r_{l-1} \cdots r_1\ r_0\ r_1^* \cdots r_{R-l}^*\Big], \tag{18.10}$$

$$\mathbf{x}^{(l)H}\mathbf{B} = \Big[\underbrace{0 \cdots 0}_{l}\ r_{P-1} \cdots r_{P-R+l}\Big] \tag{18.11}$$

and

$$\mathbf{x}^{(l)H}\mathbf{A} = \Big[r_{P-l+1}^* \cdots r_{P-1}^*\ \underbrace{0 \cdots 0}_{R-l+1}\Big]. \tag{18.12}$$

On the basis of (18.1) and (18.10)–(18.12), the output of a RAKE finger $d_n^{(l)}$ (see Figure 18.1) follows:

$$\begin{aligned} d_n^{(l)} =& \mathbf{x}^{(l)H}\mathbf{y}_n \\ =& \mathbf{x}^{(l)H}\left(s_n\mathbf{C} + s_{n-1}\mathbf{B} + s_{n+1}\mathbf{A}\right)\mathbf{h}_n + \mathbf{x}^{(l)H}\mathbf{e}_n \\ =& \sum_{q=1}^{l-1}\left(s_n r_{l-q} + s_{n+1} r_{P-l+q}^*\right) h(n,q) + \\ & \sum_{q=l+1}^{R}\left(s_n r_{q-l}^* + s_{n-1} r_{P+l-q}\right) h(n,q) + s_n r_0 h(n,l) + e_n^{(l)}, \end{aligned} \tag{18.13}$$

where $r = 1, 2, \ldots, R$ and $e_n^{(l)} = \mathbf{x}^{(l)H}\mathbf{e}_n$ is distributed as $\mathcal{CN}(0, r_0\sigma^2)$. Note that correlated vectors $\{\mathbf{x}^{(l)}\}$ will translate into correlated noise $\{e_n^{(l)}\}$.

Then $\{d_n^{(l)}\}_{r=1}^{R}$ over the nth symbol period are weighted by appropriate channel taps and summed to form the symbol estimate $\bar{s}_n$ (see Figure 18.1):

$$\bar{s}_n = \frac{\sum_{r=1}^{R} d_n^{(l)}\hat{h}^*(n,l)}{r_0\sum_{r=1}^{R}|\hat{h}(n,l)|^2} = \frac{\sum_{r=1}^{R} d_n^{(l)}\hat{h}^*(n,l)}{r_0\|\hat{\mathbf{h}}_n\|^2}. \tag{18.14}$$

The hard decision $\hat{s}_n$ is obtained by slicing $\bar{s}_n$; see Figure 18.1.

For a general frequency-selective channel with $R > 1$ ($R = 1$ leads to a flat fading channel), the correlation properties of $\mathbf{x}$ are quite important (see (18.13)). Therefore, a spreading sequence with good correlation properties is desirable. In the absence of CIR information at the transmitter, which is the usual case in UWA communications, the ideal correlation function is

$$r_k = 0, \quad k \in [1, R-1] \cup [P-R+1, P-1]. \tag{18.15}$$

We assume that $P > 2R - 2$. The correlation function $\{r_k\}$ over $k \in [R, P-R]$ has no impact on the RAKE performance.

The use of such an ideal spreading waveform leads to uncorrelated $\{e_n^{(l)}\}$ and simplifies (18.13) to

$$d_n^{(l)} = s_n r_0 h(n,l) + e_n^{(l)}, \quad r = 1, \ldots, R. \tag{18.16}$$

Therefore an ideal spreading waveform effectively decomposes an R-tap frequency-selective channel into R parallel and independent flat-fading channels that do not interfere with each other. As a consequence, there is no interference across the RAKE fingers and the symbol estimate is given by (assuming a perfect channel estimate, i.e., $\mathbf{h}_n = \hat{\mathbf{h}}_n$):

$$\bar{s}_n = s_n + \frac{\sum_{r=1}^{R} e_n^{(l)} h^*(n,l)}{r_0 \|\mathbf{h}_n\|^2}. \tag{18.17}$$

Using the fact that $r_0 = P$ and $|s_n| = 1$, and letting $\text{SNR} = \frac{\|\mathbf{h}_n\|^2}{\sigma^2}$ denote the incoming chip SNR before RAKE processing, we can readily check from (18.17) that the SNR is increased by a factor of P at the output of a coherent RAKE. The chip length P, therefore, is also referred to as the processing gain in the DSSS literature [Tse & Viswanath 2005]. Note that in (18.17) $\sum_{r=1}^{R} e_n^{(l)} h^*(n,l) \sim \mathcal{CN}(0, r_0 \|\mathbf{h}_n\|^2 \sigma^2)$. This fact is a direct consequence of $\{e_n^{(l)}\}$ being uncorrelated, which, as previously mentioned, is true when the spreading waveform satisfies (18.15).

By assuming that the spreading waveform satisfies (18.15), and therefore that (18.17) and the above mentioned properties of the noise in this equation hold true, the bit error rate (BER) performance corresponding to the QPSK modulation scheme is given by [Proakis 2001]

$$P_{\text{BER}} = \frac{1}{2}\text{erfc}\left(\sqrt{\frac{P \times \text{SNR}}{2}}\right), \tag{18.18}$$

where $\text{erfc}(\cdot)$ represents the complementary error function.

As previously mentioned, UWA environments (especially the time-varying nature of underwater medium) constrain the values of P that can be used, as the block fading assumption can be easily violated if a long waveform is adopted [Yang & Yang 2008]. For this reason, a spreading waveform with a relatively short chip length is more suitable for UWA environments.

The CAN and WeCAN algorithms introduced in Chapter 2 can be used to approximately achieve the goal presented in (18.15). Both algorithms make use of a cyclic approach to efficiently minimize correlation-related criteria. Moreover, through different random phase initializations, different waveforms can be obtained. As will be shown in the next section, flexible length and random phase values ensure LPI, whereas the optimized correlation properties facilitate the coherent RAKE reception in the sense of suppressing the inter- and intrasymbol interferences. Both features make these waveforms desirable especially for covert UWA applications.

Of the two algorithms considered, WeCAN aims to suppress the correlations over the lag interval of interest (i.e., $k \in [1, R-1] \cup [P-R+1, P-1]$), and is used under the assumption that $P > 2R - 2$, see (18.15). This implicitly requires a priori information on the channel tap number R. For practical UWA communications, R may not be available prior to the experiment or the relationship $P > 2R - 2$ might not hold (but we still assume that $P > R$), which would be the case when a short spreading waveform is used in a severe time-dispersive channel. In such cases we should instead aim to suppress the correlation levels over the *entire* time lag interval (i.e., $[1, P-1]$). For this purpose, the CAN algorithm can be applied.

18.3 Numerical examples

We aim to compare the BER performance when four different spreading waveforms are employed. Note that the chip length P for CAN, WeCAN and random phase waveforms can be arbitrarily chosen. However, we choose $P = 63$ to meet the length constraint imposed by the m-sequence. The correlation levels of the four waveforms are plotted in Figures 18.2 and 18.3; the correlation level is defined as (see also (2.43)):

$$\text{correlation level} = 20 \log_{10} \frac{|r_p|}{P} \text{ dB}, \quad p = 0, 1, \ldots, P-1, \tag{18.19}$$

where r_p is given by (18.9). Note that the CAN and WeCAN waveforms in Figures 18.3(a) and 18.3(b), respectively, are generated using the random phase waveform associated with Figure 18.2(b) to initialize the algorithms. By considering the simulated time-invariant frequency-selective CIR shown in Figure 18.4 with $R = 20$ resolved taps, we are particularly interested in suppressing the correlation levels over the lags $[-62, -44] \cup [-19, -1] \cup [1, 19] \cup [44, 62]$ (indicated via the vertical dash-dot lines in Figures 18.2 and 18.3). Overall, the WeCAN waveform gives the lowest correlation levels over the lag ranges of interest, while the random phase waveform exhibits the highest.

Next, we proceed with the evaluation of the BER performance. The selected information sequence consists of 1000 QPSK payload symbols and each symbol is spread by a common spreading waveform. The transmitted signal propagates through the frequency-selective channel shown in Figure 18.4, followed by the coherent RAKE receiver outlined in Figure 18.1. The incoming measurements are constructed according to (18.1). Fifty different random phase waveforms are used in this example and they are obtained as follows. We first generate 1000 independent random phase waveforms, calculate the peak sidelobe level (PSL) of their correlation functions and then keep the waveforms corresponding to the 50 lowest PSL values among the 1000 candidates. (The lowest PSL of the 1000 candidates is -19.02 dB, the corresponding correlation function being shown in Figure 18.2(b).) These 50 selected random phase waveforms are used to initial-

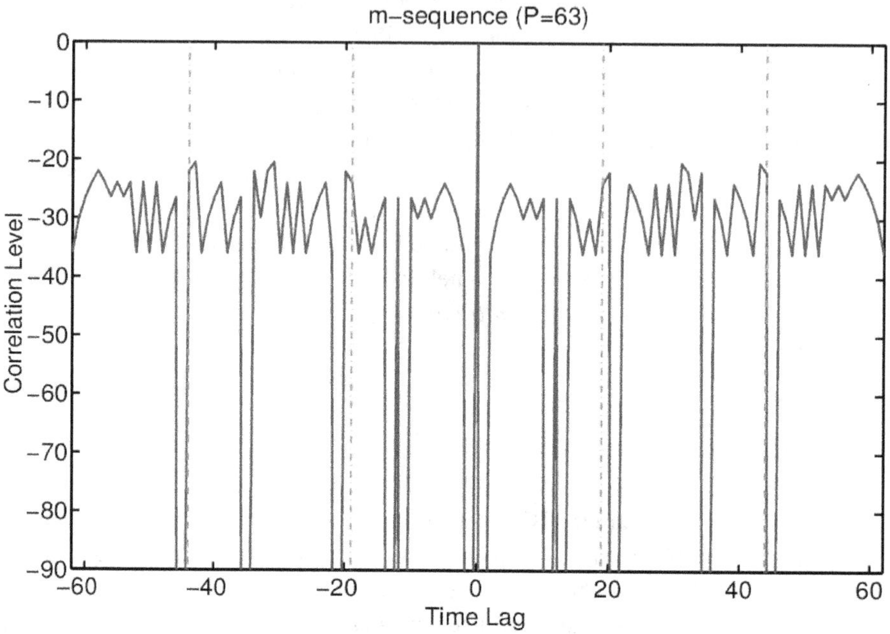

(a)

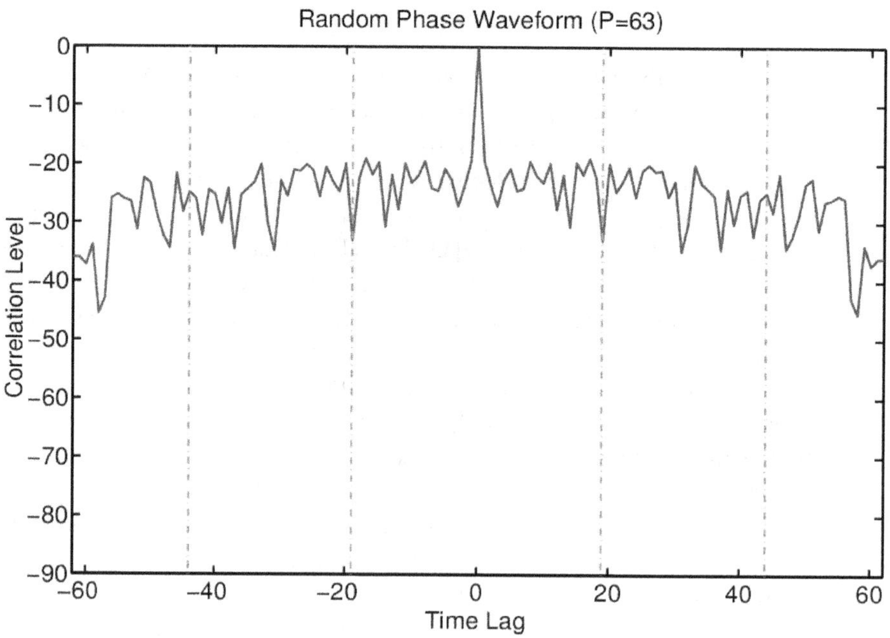

(b)

Figure 18.2 Correlation levels of the spreading waveforms with $P = 63$. The vertical dash-dot lines indicate the lag intervals $[-62, -44] \cup [-19, -1] \cup [1, 19] \cup [44, 62]$, over which we want to suppress the correlation levels. (a) M-sequence. (b) Random phase sequence.

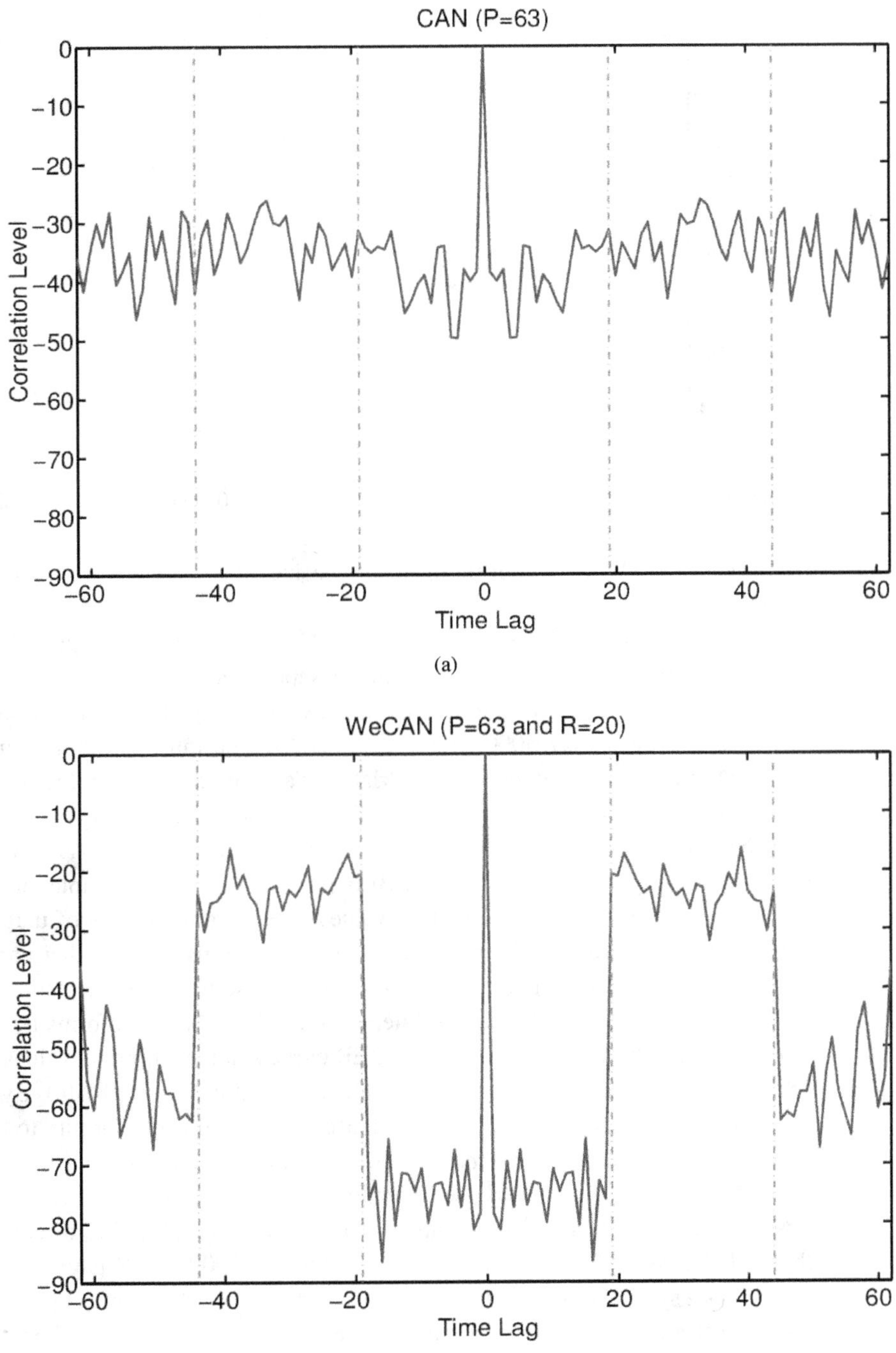

(a)

(b)

Figure 18.3 Correlation levels of the spreading waveforms with $P = 63$. The vertical dash-dot lines indicate the lag intervals $[-62, -44] \cup [-19, -1] \cup [1, 19] \cup [44, 62]$, over which we want to suppress the correlation levels. (a) CAN sequence. (b) WeCAN sequence. Note that the random phase sequence in Figure 18.2(b) is used to initialize the CAN and WeCAN algorithms that yield the sequences leading to figures here.

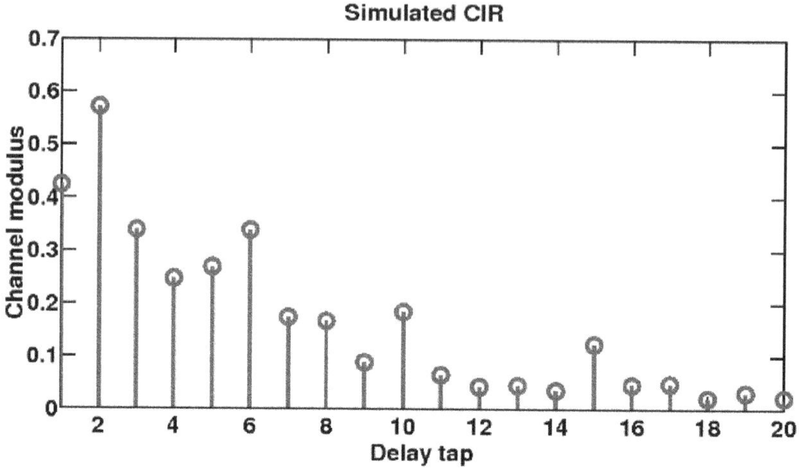

Figure 18.4 The modulus of the simulated CIR where $R = 20$ channel taps are considered.

ize the CAN and WeCAN algorithms to synthesize 50 CAN waveforms and 50 WeCAN waveforms.

We first assume that the receiver knows the CIR. The resulting empirical BER curves for the four types of waveforms are shown superimposed in the subplots of Figures 18.5(a)-18.5(c), along with the theoretical BER given by (18.18). Each point in these figures is averaged over 5000 Monte-Carlo trials. The information sequence and the noise pattern vary independently from trial to trial. We can see from Figure 18.5(a) that the theoretical BER curve can be reasonably regarded as an average detection performance of the 50 selected random phase waveforms. The random phase waveform, however, exhibits significant variations in BER performance. For example, at SNR = -12 dB, the span of the 50 empirical BER values exceeds three orders of magnitude. The performance variations of the 50 related CAN waveforms, on the other hand, are considerably reduced (see Figure 18.5(b)) owing to the suppressed correlation levels. The low correlation levels at the lags of interest of the WeCAN waveforms translate into a close similarity between the theoretical BER curve and the empirical values; see Figure 18.5(c). The BER curve for the m-sequence is plotted in Figure 18.5(c). By comparing Figures 18.2–18.3 and Figures 18.5(a)-18.5(c), we note that conformity to the theoretical values depends on the goodness of the correlation properties.

Next, we proceed to assess the detection performance when the receiver does not possess perfect CIR information and has to estimate it in a training-directed mode. To this end, 5 QPSK pilot symbols are added before the 1000 QPSK payload symbols and these 5 pilot symbols (or 500 chips after spreading) are used to conduct training-directed channel estimation. The channel estimation algorithm used is the so-called sparse learning via iterative minimization (SLIM) [Ling, Tan, Yardibi, Li, He & Nordenvaad 2009]. Performing RAKE detection using the estimated CIR, results in the empirical BER performance shown in Figures 18.6(a)-18.6(c) for the four waveforms, along with the theoretical BER corresponding to perfect spreading waveform and perfect CIR infor-

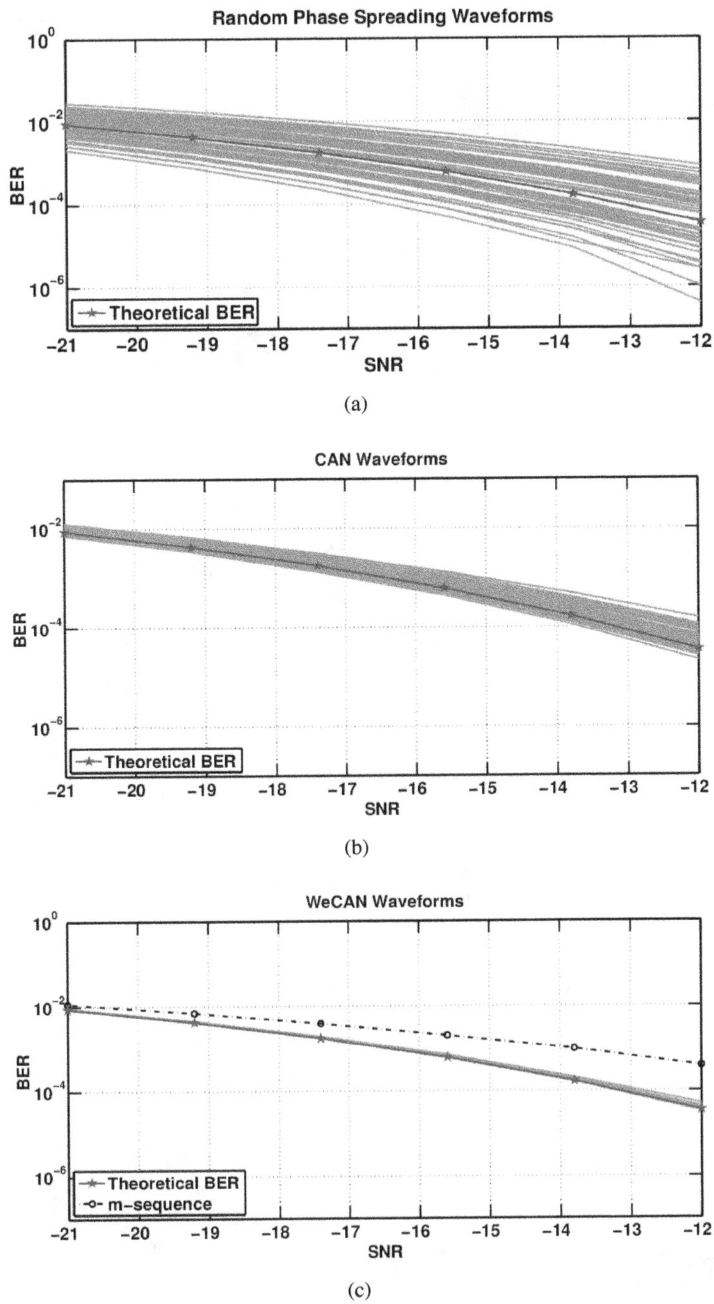

Figure 18.5 The 50 empirical BER curves (along with the theoretical BER values) for four types of spreading waveforms with $P = 63$ in the case that the receiver knows the CIR. Each point is averaged over 5000 Monte-Carlo runs. (a) Random phase waveforms. (b) CAN waveforms. (c) WeCAN waveforms and m-sequence. Note that the 50 CAN and 50 WeCAN waveforms considered are generated by using the 50 random phase waveforms in (a) to initialize the algorithms.

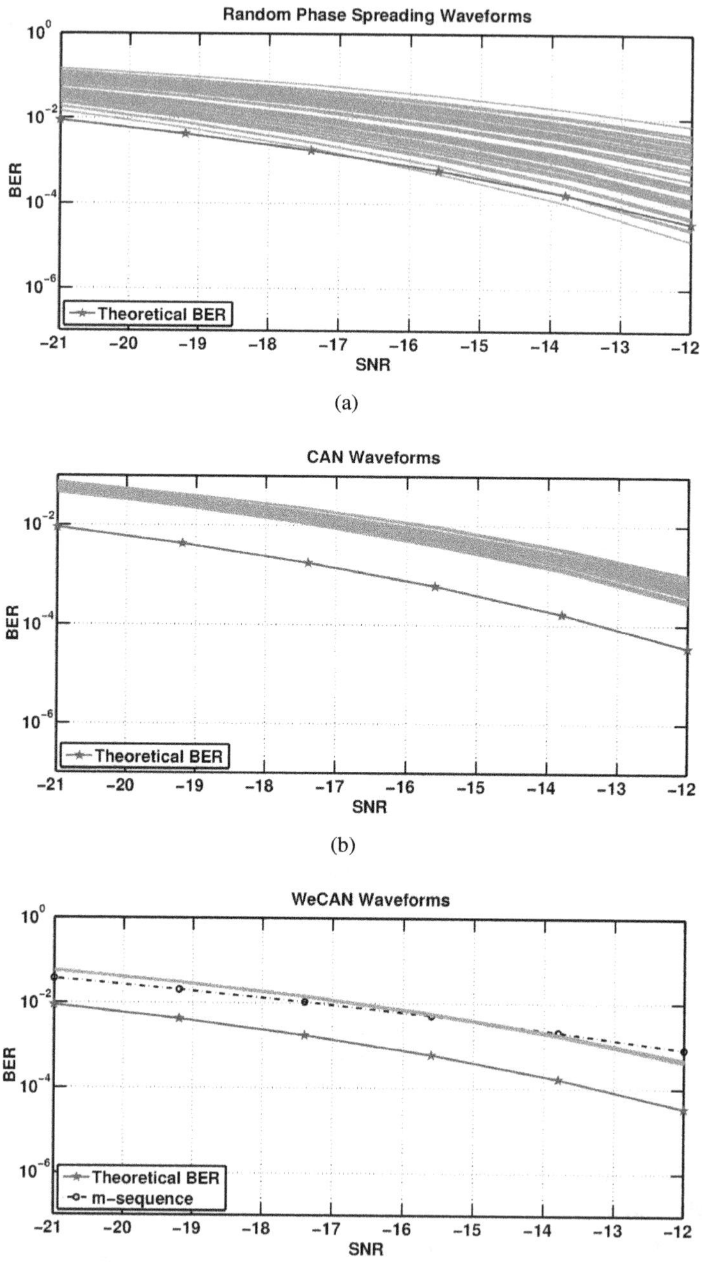

Figure 18.6 The 50 empirical BER curves (along with the theoretical BER values) for four types of spreading waveforms with $P = 63$ in the case that the receiver estimates the CIR using 5 QPSK pilot symbols. Each point is averaged over 5000 Monte-Carlo runs. (a) Random phase waveforms. (b) CAN waveforms. (c) WeCAN waveforms and m-sequence. Note that the 50 CAN and 50 WeCAN waveforms considered are generated by using the 50 random phase waveforms in (a) to initialize the algorithms.

mation. Each point is averaged over 5000 Monte-Carlo trials, in which the information sequence (including the 5 QPSK pilot symbols) and the noise vary independently from one trial to another. By comparing Figures 18.5(a)–18.5(c) and 18.6(a)-18.6(c), one observes that the presence of the CIR estimation error shifts the empirical BER curves upward by approximately one order of magnitude. For a time-invariant channel, the gap between the theoretical and empirical BER curves for CAN and WeCAN waveforms would diminish if more pilot symbols were used to conduct training-directed channel estimation.

Although random phase spreading waveforms help ensure LPI communications, their large variations in detection performance make them rather unappealing (since it is hard to predict the resulting performance of a specific realization of a random phase waveform, as evidenced in Figures 18.5(a) and 18.6(a)). On the other hand, the optimized correlation levels of WeCAN and CAN waveforms lead to consistent BER performance, making them preferable over their random phase counterpart. (Note that the m-sequence also gives good BER performance but is not suitable for covert communication due to its deterministic nature.)

18.4 Conclusions

We have considered covert UWA communication schemes that adopt a DSSS-based modulation technique and a coherent RAKE reception. We have shown that the WeCAN and CAN algorithms can be used to synthesize the spreading waveforms. The so-obtained waveforms not only possess good correlation properties that lead to a satisfactory BER performance despite the frequency-selective nature of the UWA channel, but also have randomness properties that help protect the privacy of the transmitted information.

19 Covert underwater acoustic communications – noncoherent scheme

In Chapter 18 we discussed covert UWA communications with a *coherent* RAKE receiver, in which case the CAN or WeCAN sequences were shown to be the probing waveforms of choice. As pointed out before, low chip SNR communication over a time-varying UWA channel precludes accurate channel estimation. This could render coherent detection schemes ineffective [Smadi & Prabhu 2004][Yang & Yang 2008] and thus make noncoherent schemes more favorable. In this chapter we examine covert UWA communications with *noncoherent* schemes and the waveform design issues thereof. More specifically, two types of noncoherent transceiver designs are addressed: orthogonal modulation and differential phase-shift keying (DPSK), both coupled with a DSSS technique and a noncoherent RAKE reception [Brennan 1959]. Although only binary information sequences are considered here, the derivations can be easily extended to a general M-ary case.

For orthogonal modulation, spreading waveform sets with low correlation properties (both auto- and cross-correlation) such as the Gold sequence set [Gold 1968] are desired for combating the multipath nature of UWA channels. For DPSK modulation, a common design is to adopt the cyclic prefix so that the receiver can eliminate the intersymbol interference by ignoring the prefix chips before proceeding with symbol detection [Pursley 1977][Tse & Viswanath 2005]. Such a design requires waveforms with low or even zero *periodic* correlations over certain lags, an example of which is the Frank sequence (see (1.23)). However, the aforementioned sequences (the Gold and Frank sequences) belong to the class of unimodular polyphase sequences that are constructed from fixed formulas with restricted lengths and/or fixed phase constellations. Such polyphase (especially binary) sequences have a significant probability of interception or detection by a brute force search and thus are not favored from the LPI/LPD point of view. On the contrary, the Multi-WeCAN sequence sets (see Chapter 3) and the PeCAN sequences (see Chapter 9) possess LPI or LPD properties due to arbitrary lengths and unconstrained phases lying between 0 and 2π, at the same time having desired correlation properties.

19.1 RAKE energy-based detection of orthogonal signals

Suppose that a transmitter maps each symbol (bit) in a binary information sequence to one of two orthogonal spreading waveforms $\{\mathbf{x}_i\}_{i=1}^{2}$ ($\mathbf{x}_1^H\mathbf{x}_2 = 0$). Each spreading waveform consists of P unimodular chips (i.e., $\mathbf{x}_i = \left[x_i(1) \cdots x_i(P)\right]^T$ where $i \in \{1,2\}$) and the information is conveyed by choosing one out of the two candidate spreading waveforms. Let us assume that the spreading waveform $\mathbf{x}_1$ is transmitted over one symbol (bit) interval. We confine our focus on detecting this waveform (the discussion for $\mathbf{x}_2$ is similar). The problem can then be formulated as

$$\mathbf{y} = \mathbf{X}\mathbf{h} + \mathbf{e}, \tag{19.1}$$

where $\mathbf{y} = \left[y(1) \cdots y(P+R-1)\right]^T$ and $\mathbf{e} = \left[e(1) \cdots e(P+R-1)\right]^T$ represent, respectively, the vectors of the incoming measurements and the noise terms. We assume the noise to be a complex-valued white Gaussian random process with zero mean and variance σ^2, denoted as $\mathbf{e} \sim \mathcal{CN}(0, \sigma^2\mathbf{I})$. The channel impulse response (CIR) vector is $\mathbf{h} = \left[h(1) \cdots h(R)\right]$. The matrix $\mathbf{X}$ contains multiple shifted replicas of the transmitted chips and is given by

$$\mathbf{X} = \begin{bmatrix} x_1(1) & x_i(P) & \cdots & x_i(P-R+3) & x_i(P-R+2) \\ \vdots & x_1(1) & & \vdots & x_i(P-R+3) \\ x_1(P) & \vdots & \ddots & x_i(P) & \vdots \\ x_j(1) & x_1(P) & & x_1(1) & x_i(P) \\ \vdots & x_j(1) & \ddots & \vdots & x_1(1) \\ x_j(R-2) & \vdots & \ddots & x_1(P) & \vdots \\ x_j(R-1) & x_j(R-2) & & x_j(1) & x_1(P) \end{bmatrix}_{(P+R-1)\times R}, \tag{19.2}$$

where $\mathbf{x}_i$ and $\mathbf{x}_j$ denote the waveforms transmitted before and after the current waveform, respectively. Subscripts $i, j \in \{1,2\}$, but their values cannot be determined due to the randomness nature of the original bit sequence.

Expressing the received measurements as in (19.1), the problem of interest reduces to determining which waveform out of the two candidates is transmitted ($\mathbf{x}_1$ in this example) given the measurements $\mathbf{y}$. Since we focus on noncoherent detection in this chapter, a knowledge of $\mathbf{h}$ is not assumed by the detection procedure.

The structure of an energy-based RAKE detector for binary orthogonal waveforms is shown in Figure 19.1. This structure can be roughly divided into two stages. The first stage projects the incoming measurements $\mathbf{y}$ onto vector $\mathbf{x}_m^{(r)} \in \mathcal{C}^{(P+R-1)\times 1}$, which is a shifted version of the waveform $\mathbf{x}_m$ associated with the rth channel tap (i.e., the path represented by $h(r)$):

$$\mathbf{x}_m^{(r)} = [\underbrace{0 \cdots 0}_{r-1}\; x_m(1)\, x_m(2) \cdots x_m(P)\; \underbrace{0 \cdots 0}_{R-r}]^T, \quad m = 1,2, \quad r = 1,\ldots,R. \tag{19.3}$$

Similarly to what was done in Chapter 18, we decompose the matrix $\mathbf{X}$ in (19.2) in order to assess the impact of the spreading waveforms on the output of each finger in the RAKE receiver. We rewrite the matrix $\mathbf{X}$ as $\mathbf{X} = \mathbf{X}_1 + \mathbf{A}_{ij}$, where the dimensions of $\mathbf{X}_1$ and $\mathbf{A}_{ij}$ conform with those of $\mathbf{X}$. The matrix $\mathbf{X}_1$ contains only the shifted replicas

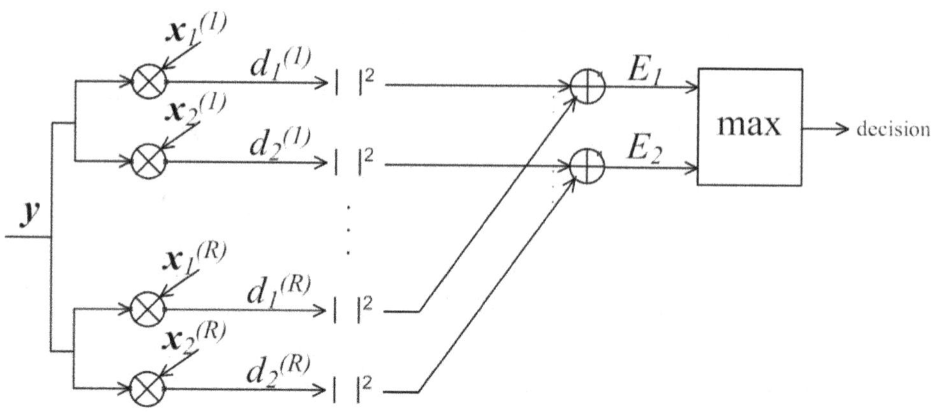

Figure 19.1 Structure of an energy-based RAKE detector for binary orthogonal waveforms.

of $\mathbf{x}_1$:

$$\mathbf{X}_1^T = \begin{bmatrix} x_1(1) \cdots x_1(P) & & \mathbf{0} \\ & \ddots & \ddots & \\ \mathbf{0} & & x_1(1) \cdots x_1(P) \end{bmatrix}, \tag{19.4}$$

while $\mathbf{A}_{ij} = \mathbf{X} - \mathbf{X}_1$ is composed of the chips of the adjacent waveforms $\mathbf{x}_i$ and $\mathbf{x}_j$. Then it can be easily verified that

$$\mathbf{x}_m^{(r)H}\mathbf{X}_1 = [r_{1m}(r-1) \cdots r_{1m}(1)\, r_{1m}(0)\, r_{m1}^*(1) \cdots r_{m1}^*(R-r)], \tag{19.5}$$

$$\mathbf{x}_m^{(r)H}\mathbf{A}_{ij} = [r_{mj}^*(P-r+1) \cdots r_{mj}^*(P-1)\, 0\, r_{im}(P-1) \cdots r_{im}(P-R+r)], \tag{19.6}$$

where

$$r_{\tilde{i}\tilde{j}}(k) = \sum_{n=k+1}^{P} x_{\tilde{i}}(n)x_{\tilde{j}}^*(n-k), \quad \tilde{i},\tilde{j} = 1,2, \quad k = 0, \cdots, P-1, \tag{19.7}$$

denotes the correlation of the waveform set $\{\mathbf{x}_1, \mathbf{x}_2\}$ (see also (3.1)). When $\tilde{i}$ equals $\tilde{j}$, the quantity $r_{\tilde{i}\tilde{i}}(k)$ is the auto-correlation of $\mathbf{x}_{\tilde{i}}$; otherwise, $r_{\tilde{i}\tilde{j}}(k)$ shows the cross-correlation between $\mathbf{x}_{\tilde{i}}$ and $\mathbf{x}_{\tilde{j}}$.

Using (19.5) and (19.6) we can express the output of a RAKE finger $d_m^{(r)}$, i.e., the projection of $\mathbf{y}$ onto $\mathbf{x}_m^{(r)}$, as

$$\begin{aligned} d_m^{(r)} &= \frac{\mathbf{x}_m^{(r)H}\mathbf{y}}{\|\mathbf{x}_m\|} = \frac{\mathbf{x}_m^{(r)H}\left(\mathbf{X}_1 + \mathbf{A}_{ij}\right)\mathbf{h}}{\|\mathbf{x}_m\|} + \frac{\mathbf{x}_m^{(r)H}\mathbf{e}}{\|\mathbf{x}_m\|} \\ &= \left\{ \sum_{q=1}^{r-1} [r_{1m}(r-q) + r_{mj}^*(P-r+q)]h(q) + r_{1m}(0)h(r) + \right. \\ &\quad \left. \sum_{q=r+1}^{R} [r_{m1}^*(q-r) + r_{im}(P-q+r)]h(q) \right\} \frac{1}{\|\mathbf{x}_m\|} + e_m^{(r)}, \end{aligned} \tag{19.8}$$

where $i, j, m = 1, 2,\ r = 1, 2, \cdots, R$, and $e_m^{(r)} = \mathbf{x}_m^{(r)H}\mathbf{e}/\|\mathbf{x}_m\|$. The normalization factor $1/\|\mathbf{x}_m\|$ is such that $e_m^{(r)} \sim \mathcal{CN}(0, \sigma^2)$ given that $\mathbf{e} \sim \mathcal{CN}(0, \sigma^2\mathbf{I})$.

A frequency-selective channel scatters the signal power over R resolved channel paths, and the second stage of the RAKE receiver combines the scattered energy along each candidate vector by performing equal gain combining (EGC) [Tse & Viswanath 2005][Brennan 1959]:

$$E_m = \sum_{r=1}^{R} \left|d_m^{(r)}\right|^2, \quad m = 1, 2. \tag{19.9}$$

By expressing the finger outputs in terms of the auto- and cross-correlation functions, (19.8) offers insights into formulating the design criteria for the spreading waveforms. The contributions of the auto- and cross-correlation functions are carried over into the finger output $d_m^{(r)}$ through a weighted sum, with the channel taps being the weights. In the absence of prior information on $\mathbf{h}$, the best we can do is to design two spreading waveforms such that, excepting the quantities $\{r_{ii}(0)\}_{i=1}^{2}$ (which equal P by definition), all the other correlation functions involved in (19.8) are 0. Mathematically, we want

$$\begin{cases} r_{11}(k) = r_{22}(k) = 0, & k \in \{1, \ldots, R-1\} \cup \{P-R+1, \ldots, P-1\}, \\ r_{12}(k) = r_{21}(k) = 0, & k \in \{0, \ldots, R-1\} \cup \{P-R+1, \ldots, P-1\}, \end{cases} \tag{19.10}$$

where it is assumed that $P > 2R - 2$. The Multi-WeCAN algorithm discussed in Chapter 3 can be used to design sequence sets with low correlations in the region of interest indicated by (19.10).

Inserting (19.10) into (19.8) and recalling that $r_{11}(0) = r_{22}(0) = P$, we have:

$$d_m^{(r)} = \begin{cases} \sqrt{P}h(r) + e_1^{(r)}, & m = 1, \\ e_2^{(r)}, & m = 2. \end{cases} \tag{19.11}$$

Thus two waveforms that perfectly satisfy (19.10) decompose the original multipath problem into R independent flat-fading channels that do not interfere with each other. In this ideal case, the BER performance can be expressed as [Simon & Alouini 1998]:

$$P_{\text{BER}} = \frac{e^{-\lambda}}{2^{2R-1}} \sum_{k=0}^{R-1} \left[\frac{1}{k!} \sum_{n=1}^{R-1-k} \binom{2R-1}{n} \right] \lambda^k, \tag{19.12}$$

where $\lambda = \frac{P}{2}\text{SNR}$ and

$$\text{SNR} = \frac{\|\mathbf{h}\|^2}{\sigma^2} \tag{19.13}$$

is the received chip SNR before RAKE processing.

19.2 RAKE demodulator for DPSK signals

DPSK is a widely-used encoding scheme, in which information is conveyed in the phase difference between two successive encoded symbols. Depending on the channel conditions, a DPSK detection scheme can be conducted either coherently (predetection maxi-

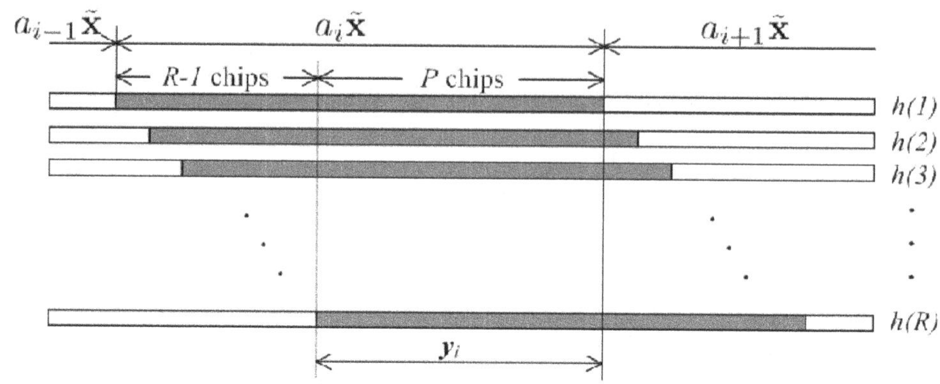

Figure 19.2 Phase-modulated spreading waveforms propagating over a frequency-selective channel with R resolved paths.

mal ratio combiner) or noncoherently (postdetection EGC) [Wang & Moeneclaey 1992]. In general, a noncoherent approach is preferred for covert communication applications, in which the reliability of the communication heavily depends on the processing gain achieved by adopting a DSSS technique and the assumption that the channel taps remain stable over at least two successive symbol periods [Proakis 2001][Yang & Yang 2008].

Let $\{b_i\}$ and $\{a_i\}$ be the original information symbols and the DPSK encoded symbols, respectively. For expositional simplicity, we consider binary symbols with the entries of $\{b_i\}$ and $\{a_i\}$ being either 1 or -1. The results developed in this section, however, can be extended to a general M-ary symbol scenario [Proakis 2001].

Given a binary information sequence $\{b_i\}$, $\{a_i\}$ is constructed recursively as follows

$$a_i = b_i a_{i-1}, \quad i = 1, 2, \ldots, \tag{19.14}$$

with symbol a_0 being either -1 or 1. In typical DSSS applications, the encoded DPSK symbols (bits) $\{a_i\}$ phase-modulate (or, multiply) a common waveform, say $\tilde{\mathbf{x}}$, before transmission over a UWA channel.

It is worth pointing out the two major differences between the DPSK scheme considered here and the orthogonal modulation approach presented in the previous section. Firstly, the former approach only needs one waveform, in contrast to the two distinct orthogonal waveforms required in the latter case. Secondly, in the DPSK scenario, the transmitted information is determined by comparing the phase difference between two successive encoded symbols, whereas for orthogonal modulation, it is determined by choosing a waveform from a set of two candidates.

Figure 19.2 shows a phase-modulated spreading waveform $a_i\tilde{\mathbf{x}}$ propagating over R resolved paths. The multipath effects are reflected by the shifted replicas of $a_i\tilde{\mathbf{x}}$, where $\tilde{\mathbf{x}}$ has chip length $R + P - 1$ and is phase-modulated by the encoded symbol a_i. Due to the delay spreading, for a frequency-selective channel with R resolved paths, the intersymbol interference from the waveform $a_{i-1}\tilde{\mathbf{x}}$ in the previous symbol period extends over the first $R - 1$ chips of $a_i\tilde{\mathbf{x}}$. The receiver can simply ignore this intersymbol interference by considering only the output $\mathbf{y}_i$ over the remaining P chips, where the effects of

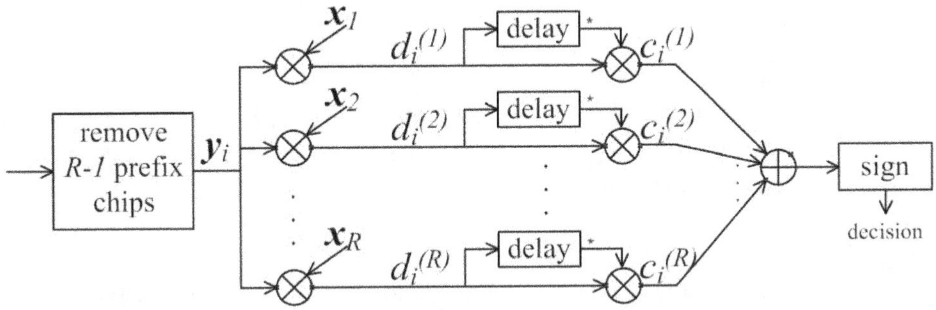

Figure 19.3 Noncoherent RAKE demodulator for binary DPSK signals.

$a_{i-1}\tilde{\mathbf{x}}$ have died out. By considering $\mathbf{y}_i$, the intrasymbol interference becomes the only possible source of interference. Also note that such a cyclic scheme reduces the data rate by a factor of $(R-1)/(P+R-1)$ due to the disposal of $R-1$ prefix chips [Tse & Viswanath 2005].

Reminiscent of the cyclic prefix adopted in OFDM (orthogonal frequency-division multiplexing) [Tse & Viswanath 2005], the common waveform $\tilde{\mathbf{x}}$ can be mathematically expressed as:

$$\tilde{\mathbf{x}} = [\ \underbrace{x(P-R+2)\ \cdots\ x(P)}_{R-1 \text{ prefix chips}}\ x(1)\ x(2)\ \cdots\ x(P)\ ]^T, \tag{19.15}$$

where the first $R-1$ elements are the prefix chips, and they are a copy of the last $R-1$ chips of $\tilde{\mathbf{x}}$. This way, based on Figure 19.2, the incoming measurements $\mathbf{y}_i \in C^{P\times 1}$ can be represented as:

$$\mathbf{y}_i = a_i\mathbf{X}\mathbf{h} + \mathbf{e}_i, \quad i = 1, 2, \ldots, \tag{19.16}$$

where

$$\mathbf{X} = \begin{bmatrix} x(1) & x(P) & \cdots & x(P-R+2) \\ x(2) & x(1) & \cdots & x(P-R+3) \\ \vdots & \vdots & & \vdots \\ x(P) & x(P-1) & \cdots & x(P-R+1) \end{bmatrix}_{P\times R}, \tag{19.17}$$

the CIR vector $\mathbf{h}$ has the same definition as in (19.1), and $\mathbf{e}_i \in C^{P\times 1}$ is the noise vector in the ith symbol period, which follows the distribution $\mathcal{CN}(0, \sigma^2\mathbf{I})$. Note that $\mathbf{X}$ and $\mathbf{h}$ are independent of the symbol period i. By investigating (19.17), we can see that the cyclic prefix scheme entrusts $\mathbf{X}$ with a cyclic shift property (see also (9.1)). That is, the rth column of $\mathbf{X}$, say $\mathbf{x}_r$, is derived by cyclically rotating the first column $\mathbf{x}_1$ by $r-1$ chips, where $r = 2, \cdots, R$.

A noncoherent RAKE receiver structure corresponding to the considered modulation scheme is illustrated in Figure 19.3. The receiver first block removes the $R-1$ prefix chips, followed by an array of R RAKE fingers designed to combine multipath arrivals. During the ith symbol period, the rth finger projects $\mathbf{y}_i$ onto the vector $\mathbf{x}_r$ (i.e., the rth column of $\mathbf{X}$, see (19.17)), generating

$$d_i^{(r)} = \mathbf{x}_r^H \mathbf{y}_i, \quad r = 1, \ldots, R. \tag{19.18}$$

Each RAKE finger is followed by a differential phase decoder, which correlates $d_i^{(r)}$ with $d_{i-1}^{(r)}$. The quantity $d_{i-1}^{(r)}$ is similarly derived to (19.18), but in the preceding signaling interval. Denote the output of this correlation as $c_i^{(r)}$, where

$$c_i^{(r)} = d_{i-1}^{(r)*} d_i^{(r)}, \quad r = 1, \ldots, R. \tag{19.19}$$

The EGC sums the quantities $\{c_i^{(r)}\}_{r=1}^R$ to generate a sufficient statistic for estimating the information bit b_i [Proakis 2001].

We can see that if the columns of $\mathbf{X}$ are orthogonal to each other, (19.18) can be written as

$$d_i^{(r)} = a_i P h(r) + \mathbf{x}_r^H \mathbf{e}_i, \quad r = 1, \ldots, R, \tag{19.20}$$

which means that the finger outputs are not contaminated by any intrasymbol inference. Consequently, we are particularly interested in designing a waveform with P chips such that its cyclically rotated versions (which appear in the $\mathbf{X}$ matrix) are orthogonal to each other. This requirement is equivalent to designing a waveform with zero periodic auto-correlation sidelobes:

$$\tilde{r}_k = 0, \quad k = 1, \ldots, R - 1, \tag{19.21}$$

where $\{\tilde{r}_k\}$ denotes the periodic (auto) correlation function (see also (1.16) in Chapter 1)

$$\tilde{r}_k = \sum_{n=1}^{P} x(n) x^*((n-k) \bmod P), \quad k = -(P-1), \cdots, 0, \cdots, (P-1). \tag{19.22}$$

The PeCAN algorithm in Chapter 9 can be used to design unimodular sequences with periodic auto-correlation sidelobes that are practically zero.

Under the assumption that (19.21) is exactly satisfied, the RAKE finger output is given by (19.20), and correspondingly, $c_i^{(r)}$ can be expressed as:

$$c_i^{(r)} = a_{i-1}^* a_i |h(r)|^2 P^2 + \Delta e, \tag{19.23}$$

where $\Delta e = Ph(r)^* a_{i-1}^* \mathbf{x}_r^* \mathbf{e}_i + Ph(r) a_i \mathbf{e}_{i-1}^* \mathbf{x}_r + \mathbf{e}_{i-1}^* \mathbf{x}_r \mathbf{x}_r^* \mathbf{e}_i$.

In a flat-fading channel with $R = 1$, the detection is based on the sign of the sufficient statistic $\mathrm{Re}\left(c_i^{(1)}\right)$ [Proakis 2001]

$$\hat{x}_i = \mathrm{sign}\left[\mathrm{Re}\left(c_i^{(1)}\right)\right]. \tag{19.24}$$

For a frequency-selective channel with $R > 1$, on the other hand, the sufficient statistic becomes $\mathrm{Re}\left(\sum_{r=1}^R c_i^{(r)}\right)$. Similarly to (19.24), the detection is based on (see Figure 19.3) [Proakis 2001]:

$$\hat{x}_i = \mathrm{sign}\left[\mathrm{Re}\left(\sum_{r=1}^{R} c_i^{(r)}\right)\right]. \tag{19.25}$$

The BER performance of this detection scheme follows from [Proakis 2001][Simon & Alouini 1998]:

$$P_{\text{BER}} = \frac{e^{-\tilde{\lambda}}}{2^{2R-1}} \sum_{k=0}^{R-1} \left[\frac{1}{k!} \sum_{n=1}^{R-1-k} \binom{2R-1}{n} \right] \tilde{\lambda}^k, \tag{19.26}$$

where $\tilde{\lambda} = P\text{SNR}$ and SNR is as defined in (19.13).

19.3 The impact of *P* and *R* on performance and an enhanced RAKE scheme

In the previous two sections, BER expressions were derived for both binary orthogonal modulation and binary DPSK modulation schemes coupled with a DSSS technique and RAKE reception. The impact of *P* and *R* on the BER performance is analyzed in this section. The numerical analysis leads to the discouraging conclusion that the reliability of the communication schemes is questionable, since the UWA frequency-selective channel requires a large *R* value, which leads to severe degradation in the BER performance. This degradation can be compensated for, to some extent, by using a longer spreading waveform(s) (i.e., a larger *P*). The time-varying nature of the UWA channel, however, constrains the largest *P* value that we can use. This dilemma leads us to the idea of exploiting the sparse nature of the UWA channel: we identify the path corresponding to the principal arrival that enjoys the highest SNR, and use only the output from this RAKE finger to make decisions.

19.3.1 Impact of *P* and *R* on the BER performance

The SNR defined above in (19.13) is an important metric for evaluating the covertness of communications. For example, a received signal transmitted at a level of SNR < -8 dB within the signal band is hard to detect by an unalerted eavesdropper [Yang & Yang 2008]. By comparing (19.12) and (19.26), we can see that for binary orthogonal modulation, SNR should be doubled to achieve the same BER as DPSK. Figure 19.4 therefore gives several BER curves with different values of *P* and *R* only for DPSK modulation. Note once again that these curves are derived by assuming perfect spreading waveform(s).

We observe from Figure 19.4 that for a fixed processing gain *P*, a flat-fading channel with $R = 1$ gives the best BER performance, and that the larger the value of *R*, the more severe degradation the BER performance suffers (see also [Proakis 2001]). In UWA applications, the *R* value is determined by several factors, such as the acoustic conditions, the chip rate, the sampling scheme at the receiver side and the communication range. For a typical frequency-selective UWA channel, a value of 50 or more for *R* is quite normal. For $R = 50$, we can see that for DPSK, using a spreading waveform with $P = 100$, the BER equals 0.5422% at an SNR of −8 dB.

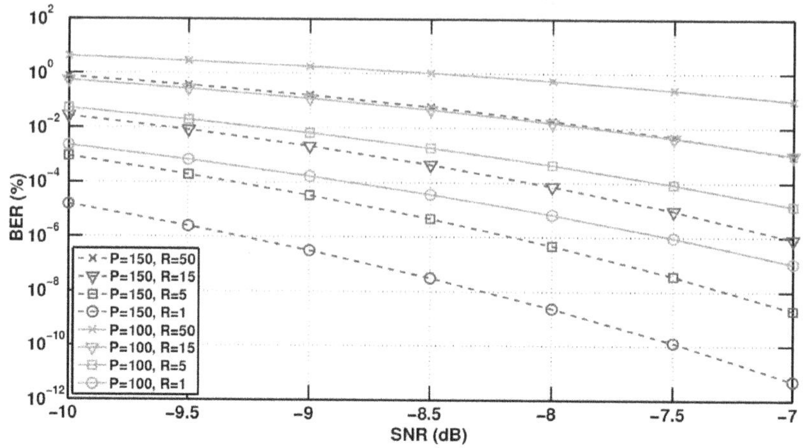

Figure 19.4 BER curves for the DPSK modulation with different values of P and R.

Theoretically, the performance degradation that results from a larger R can be compensated for by adopting a spreading waveform with a larger P, as is also verified by Figure 19.4. A larger P, however, decreases the data rate for both the orthogonal and DPSK modulation schemes; also, the time-varying nature of the UWA channel constrains the maximum P value that we can use. This constraint leads us to considering a performance enhancement method that does not require increasing P.

19.3.2 RAKE reception based on the principal arrival

Many channels, including UWA communication channels, can be considered to be sparse, as they consist of only a few dominant delay and Doppler taps [Carbonelli et al. 2007][Ling et al. 2011]. This sparsity feature, however, has not been addressed in our discussion so far. In UWA communication over a sparse frequency-selective channel, the channel paths associated with the nulls in the CIR contain no information about the transmitted signal, and the finger outputs on these paths comprise only noise. The EGC stage of the RAKE receiver does not distinguish these defective paths from those carrying the signal power. Indeed, the finger outputs are equally weighted and summed. Consequently, after EGC, the noise contaminates also the reliable finger outputs with higher SNR, and a poor BER performance is expected.

A natural way to alleviate this problem is to discard the finger outputs associated with the nulls in the CIR and instead combine only those outputs with higher SNR [Proakis 2001]. In [Sozer et al. 1999][Blackmon et al. 2002][Hursky et al. 2006], a threshold module is cascaded after each finger output, and an output whose energy is below that threshold is not included when performing EGC. However, the resulting performance becomes sensitive to the threshold value whose selection, to obtain optimal performance, is not straightforward.

An alternative method to the threshold-RAKE reception is to instead make a decision by using only the finger output along the principal arrival (i.e., the tap with the strongest power, see [Stojanovic et al. 1994]) that enjoys the highest SNR. The feasibility of this idea is demonstrated by the fact that, in general, a dominant channel tap can be associated in relatively benign shallow water environments with the principal arrival [Stojanovic et al. 1994][Yang & Yang 2008]. The key problem then is how to identify this principal arrival. In a strict sense, the attempt to identify the principal arrival incorporates partial channel information, which basically violates the assumed noncoherent nature of the scheme. On the other hand, this is not coherent reception either, since the receiver does not undo the phase shift introduced by the UWA channel during propagation. We will show shortly that the BER performance of this hybrid reception scheme is considerably enhanced as long as the principal arrival is successfully identified and its corresponding path sufficiently dominates the channel power.

Identification of the principal arrival requires that a known probing sequence be first transmitted. The probing sequence helps not only to synchronize the transmitted information [Sozer et al. 1999], but also to identify the principal arrival. With binary DPSK modulation for example, a probing sequence can be constructed by repeating $\tilde{\mathbf{x}}$, say α times, where $\tilde{\mathbf{x}}$ is constructed by prefixing the PeCAN waveform, as in (19.15). We still use the receiver structure in Figure 19.3, but only the finger outputs $\{d_i^{(r)}\}_{r=1}^{R}$ $(i = 1, \ldots, \alpha)$ are of interest in the probing mode. Based on (19.20), the sample mean along the rth RAKE finger, i.e., $\frac{1}{\alpha}\sum_{i=1}^{\alpha} d_i^{(r)}$, has a $\mathcal{CN}(Ph(r), P\sigma^2/\alpha)$ distribution, where $r = 1, \ldots, R$. Statistically, a larger α decreases the variance of the sample mean, and centralizes the sample mean to $Ph(r)$. Consequently, a larger α favors the identification of the principal arrival: we simply pick up the finger path that generates the largest norm of the sample mean. However, a larger α decreases the net data rate. This method can also be applied to the orthogonal modulation case by sending $\mathbf{x}_1$ α times and, during the probing mode, projecting the received information only onto $\mathbf{x}_1$. Note that if the channel were time-invariant, a single long WeCAN or PeCAN waveform would have been used as the probing waveform. To address the piecewise time-invariant case, several shorter waveforms could be cascaded to avoid using a single probing waveform.

We can now derive the BER expressions when the RAKE receiver uses only the finger output along the principal arrival. For binary orthogonal modulation, by setting $R = 1$, (19.12) reduces to

$$P_{\text{BER}} = \frac{1}{2}\exp\left(-\frac{\eta P \times \text{SNR}}{2}\right), \tag{19.27}$$

where $\eta = |h(r)|^2/\|\mathbf{h}\|^2 \in [0, 1]$ is the ratio of the power of the principal arrival to the total channel power. Similarly, for DPSK modulation, by setting $R = 1$, (19.26) reduces to

$$P_{\text{BER}} = \frac{1}{2}\exp\left(-\eta P \times \text{SNR}\right). \tag{19.28}$$

Note that in either case, a factor of $1 - \eta$ of the signal power is lost by using the principal arrival only. By comparing (19.27) and (19.28), we can see, once again, that to achieve the same BER performance, orthogonal modulation requires twice the SNR

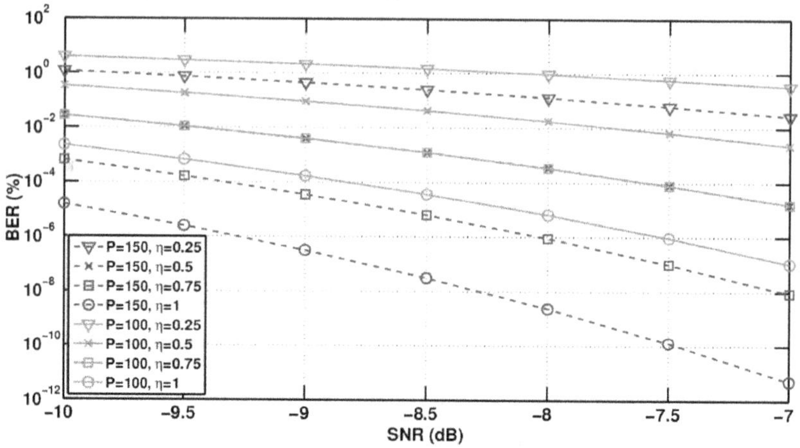

Figure 19.5 BER curves with different values of P and η formed by using only the principal arrival for the DPSK modulation.

that DPSK does. The BER curves versus SNR for different η and P values are plotted in Figure 19.5 for the DPSK scenario only. As evidenced, if the principal arrival occupies only $\eta = 0.25$ of the total channel power, the resultant BER performance for $P = 100$ is comparable to that shown in Figure 19.4 with $R = 50$. In general, however, a principal arrival with $\eta = 0.25$ is quite a conservative assumption. Larger values of η promising better BER performance (see Figure 19.5) occur in applications.

19.4 Numerical examples

In this section we assess the BER performance achieved by using the Multi-WeCAN and PeCAN waveforms. The covertness of the communication is also addressed. We start by considering binary orthogonal modulation, and then we consider the binary DPSK model.

19.4.1 Binary orthogonal modulation

Consider a simulated time-invariant frequency-selective channel represented by $R = 50$ resolved taps, as shown in Figure 19.6. To address sparsity, 34 out of 50 taps are nulls. The second tap $h(2)$ is the principal arrival with $\eta = 54.27\%$. Two Multi-WeCAN spreading waveforms $\{\mathbf{x}_i\}_{i=1}^{2}$, each with a chip length $P = 127$, are used to spread the information symbols (bits).

The P value for the Multi-WeCAN spreading waveforms can be arbitrary but is chosen as 127 to meet the length constraint imposed on the Gold sequence (to ensure a proper comparison in the forthcoming analysis).

Using the available Multi-WeCAN spreading waveforms, we first investigate the performance of the principal arrival identification by transmitting a probing sequence over

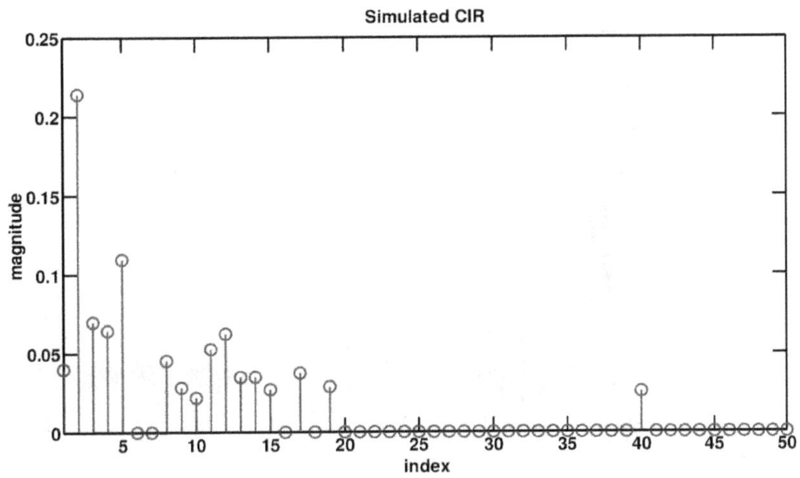

Figure 19.6 The magnitude of the simulated CIR. Among the $R = 50$ taps, 34 are nulls. The dominant tap is $h(2)$ with $\eta = 54.27\%$.

the simulated channel represented by the CIR shown in Figure 19.6. Note that this simulated channel is time-invariant. Consequently, a single long probing sequence could be employed. To be consistent with the approach developed in the previous section, we will apply a probing sequence constructed via repeating $\mathbf{x}_1$ α times. At the receiver side, the incoming measurements are projected onto $\{\mathbf{x}_1^{(r)}\}_{r=1}^{50}$ (see Figure 19.1). The norms of the sample mean along each RAKE finger are plotted in Figure 19.7 using 10 Monte-Carlo trials. We fix SNR = −10 dB and use two different α values, viz, 5 and 30. We can see that by using a larger α value, the curves in Figure 19.7(b) show less variance than those in Figure 19.7(a), which favors the identification of the principal arrival, an observation in line with the remark made in the previous section. In both cases, the principal arrival $h(2)$ is successfully identified.

We are interested in comparing the BER performance by using Multi-WeCAN waveforms and Gold waveforms. To be exact, since only two waveforms are required, Gold waveforms in this case reduce to a preferred pair of m-sequences [Gold 1968]. The preferred generator polynomials are [7 3 0] and [7 3 2 1 0], respectively. We investigate the correlation levels of the two types of waveforms before proceeding with the evaluation of the BER performance. The auto- and cross-correlation level of the Multi-WeCAN waveforms and of the Gold waveforms are plotted in Figures 19.8 and 19.9 respectively, where the correlation level is defined as follows:

$$\text{correlation level} = 20\log_{10}\left|\frac{r_{\tilde{i}\tilde{j}}(k)}{P}\right| \text{ dB}, \quad \tilde{i},\tilde{j} = 1,2, \quad k = 0,1,\ldots,P-1, \tag{19.29}$$

and $r_{\tilde{i}\tilde{j}}(k)$ is defined in (19.7). Note that the correlation levels of the Gold sequences are zero at certain lags, and these points are represented by −60 dB in Figure 19.9. We observe that overall the Multi-WeCAN waveforms give lower correlation levels over the lag ranges of interest than the Gold sequences.

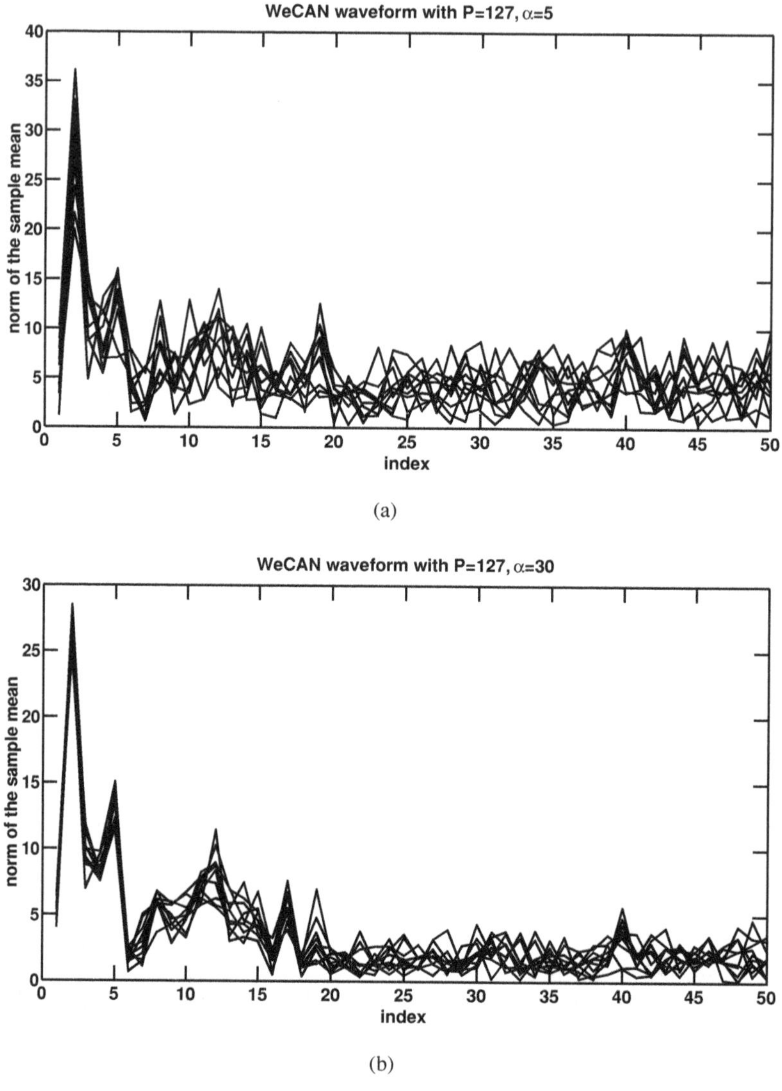

Figure 19.7 Principal arrival identification using a probing sequence with SNR = −10 dB. Ten Monte-Carlo trials are shown. (a) $\alpha = 5$ and (b) $\alpha = 30$.

Next, the BER performance is evaluated. The selected information sequence consists of 1000 symbols (bits), and each symbol is mapped to one of the waveforms. The incoming measurements are constructed according to (19.1), with the noise vector $\mathbf{e} \sim \mathcal{CN}(0, \sigma^2\mathbf{I})$. The frequency-selective channel, as shown in Figure 19.6, is considered here. We address two reception schemes. The first scheme implements an EGC over $R = 50$ RAKE fingers, as shown in Figure 19.1, and the other relies on the principal arrival only, which we assume has been successfully identified in the probing mode. The resulting empirical BER curves are shown in Figure 19.10, along with the theoretical BER given by (19.12) and (19.27). Each point is averaged over 1000 independent

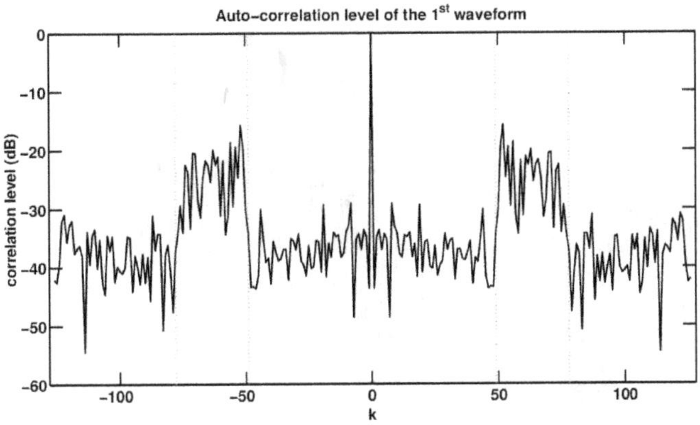

(a)

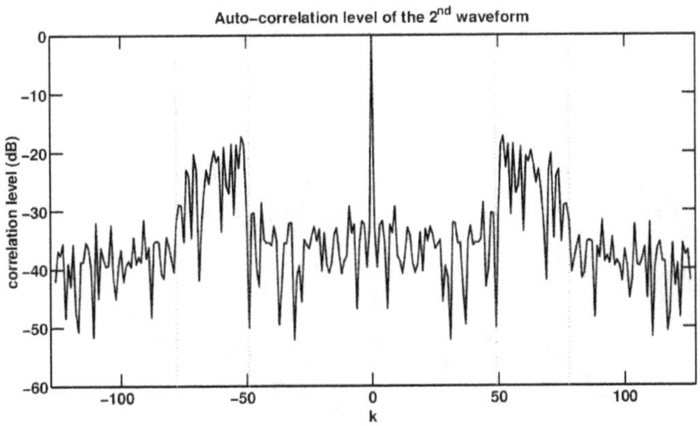

(b)

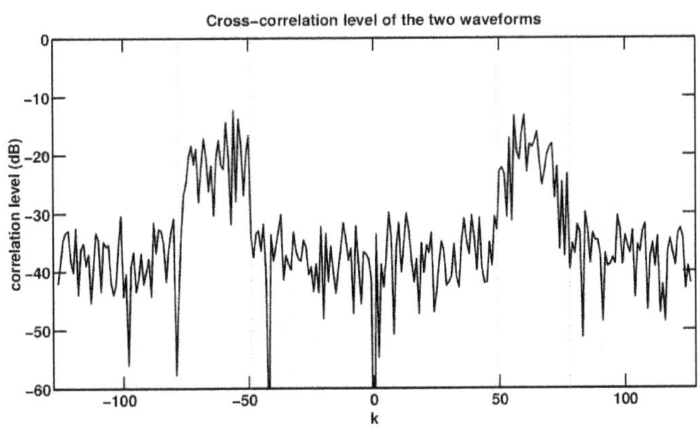

(c)

Figure 19.8 Correlation levels of the Multi-WeCAN sequences with $P = 127$. The lag intervals of interest are $[-126, -78] \cup [-49, 49] \cup [78, 126]$, over which we want to suppress the correlation levels.

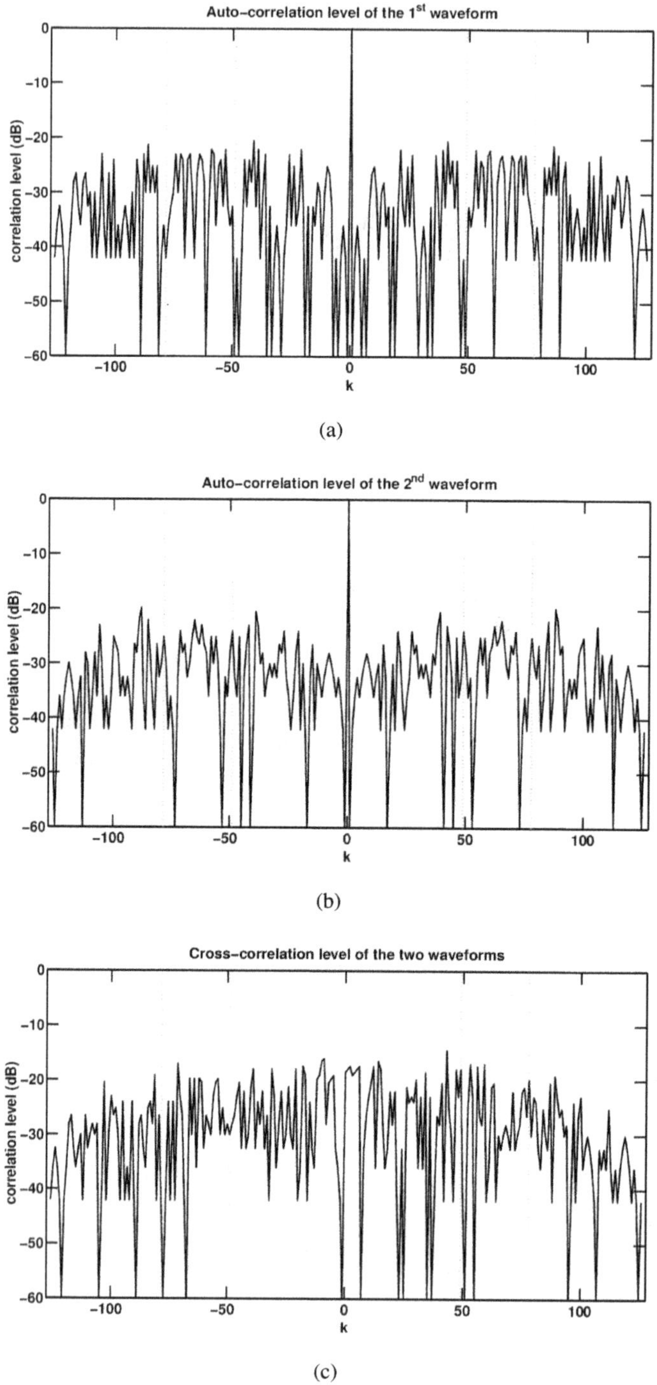

Figure 19.9 Correlation levels of the Gold sequences with $P = 127$. The lag intervals of interest are $[-126, -78] \cup [-49, 49] \cup [78, 126]$, over which we want to suppress the correlation levels.

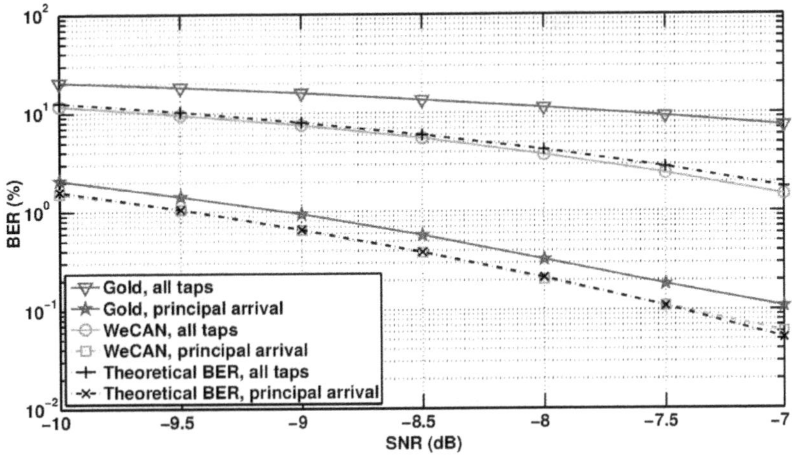

Figure 19.10 The empirical BER curves corresponding to Multi-WeCAN sequences and Gold sequences with $P = 127$, with the theoretical values plotted alongside. Two reception schemes are considered: one uses EGC over all the $R = 50$ finger outputs, and the other uses the principal arrival only.

Monte-Carlo runs. Since the principal arrival occupies $\eta = 54.27\%$ of the total channel power, a considerable BER improvement by using only the RAKE finger along this path is achieved. Note the good agreement between the theoretical BER curves and those derived by adopting Multi-WeCAN waveforms. This can be explained by the very low correlation levels at the lags of interest for the Multi-WeCAN waveforms. On the other hand, in this particular example, the BER curves of the Gold waveforms deviate from the theoretical curves due to their higher sidelobe levels within the lag ranges of interest. To enhance the BER performance, we can either increase P if the channel permits or apply a sophisticated channel coding scheme. This comes at the price of a reduced data rate.

Finally, we consider the covertness of the communication scheme. We assume that, except for precise knowledge of the spreading waveforms $\{\mathbf{x}_i\}_{i=1}^2$, an eavesdropper has the same information about the communication details as an intended receiver, such as the value of $P = 127$, the location of the principal arrival, the package structure and modulation scheme, etc. The eavesdropper tries to detect the transmitted information by generating a pair of spreading waveforms, whose chips have independently and randomly generated phase values. The resulting BER performance is plotted in Figure 19.11 by conducting 1000 independent Monte-Carlo runs. We can see that the BER is 50% on average. The error lowerbound or upperbound shown in the figure is not produced by a specific trial. Instead, they are the minimum and maximum values over the entire 1000 trials. From Figure 19.11, we can see that the detection performance achieved by generating spreading waveforms in a random manner is on the average the same as that of a uniformed guess. Consequently, the Multi-WeCAN waveforms have desirable LPI/LPD properties. For Gold sequences with length P, on the other hand, the

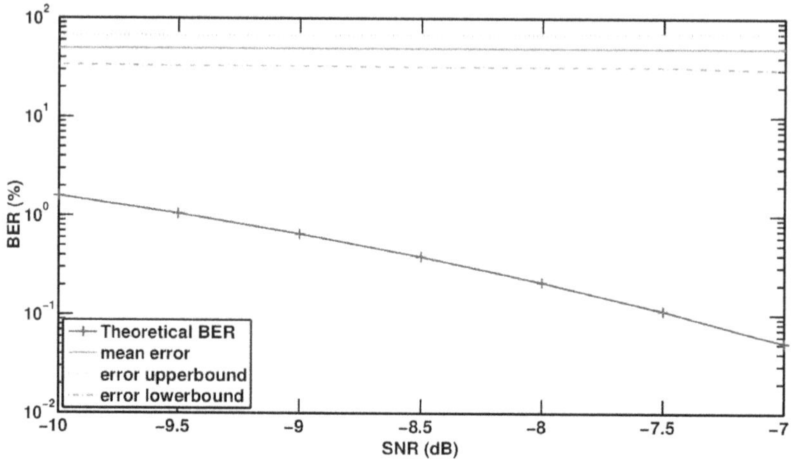

Figure 19.11 BER performance achieved by generating unimodular waveforms in a random manner. The BER is 50% on average, implying desired LPI or LPD properties for the Multi-WeCAN waveforms.

eavesdropper can exhaust all pairs of Gold waveforms by using all possible preferred pairs of P-length m-sequences.

19.4.2 DPSK modulation

Now we shift our focus to the DPSK modulation. We start by investigating the periodic auto-correlation property of the PeCAN waveform.

Figure 19.12 shows the periodic auto-correlation levels of two different PeCAN waveforms with P equal to 100 or 150. The two PeCAN waveforms have practically zero correlation sidelobes over the nonzero lags: the sidelobe level shown in Figure 19.12 is about −320 dB, i.e., 10^{-16}, which is the smallest number that can be properly handled in MATLAB and can thus be considered as "zero".

Next we consider the performance of the principal arrival identification. The simulated CIR shown in Figure 19.6 is also used here. We fix SNR = −10 dB, and the probing sequence is constructed by repeating five times the spreading waveform $\tilde{\mathbf{x}}$, with a chip length equal to $P + 49$. At the receiver side, after removing the 49 prefix chips from each $\tilde{\mathbf{x}}$, the norms of the sample mean along each RAKE finger are shown in Figure 19.13 for 10 independent Monte-Carlo trials. In both scenarios, the principal arrival $h(2)$ is successfully identified.

The original information sequence contains 1000 binary symbols $\{b_i\}_{i=1}^{1000}$, and the corresponding encoded symbols $\{a_i\}_{i=0}^{1000}$ are constructed based on (19.14). The encoded symbols then phase-modulate $\tilde{\mathbf{x}}$ before transmission over a frequency-selective channel represented by $R = 50$ resolved taps, as shown in Figure 19.6. Again, two reception schemes are considered. The first uses a conventional EGC approach as shown in Figure 19.2, and the other is based only on the principal arrival, which is assumed to be

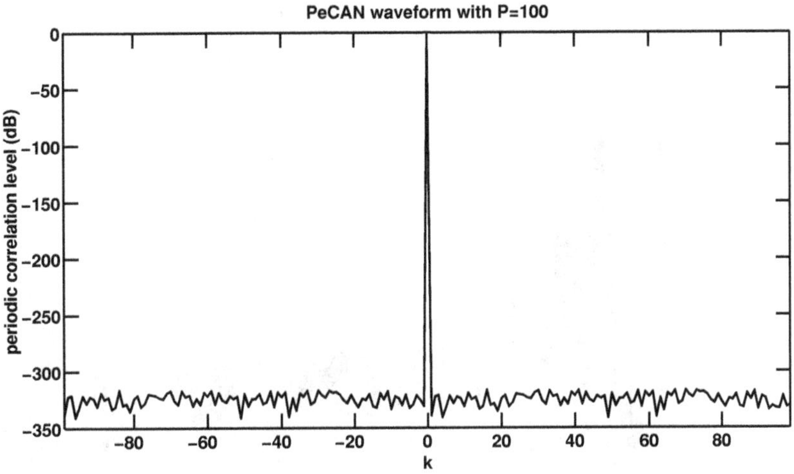

(a)

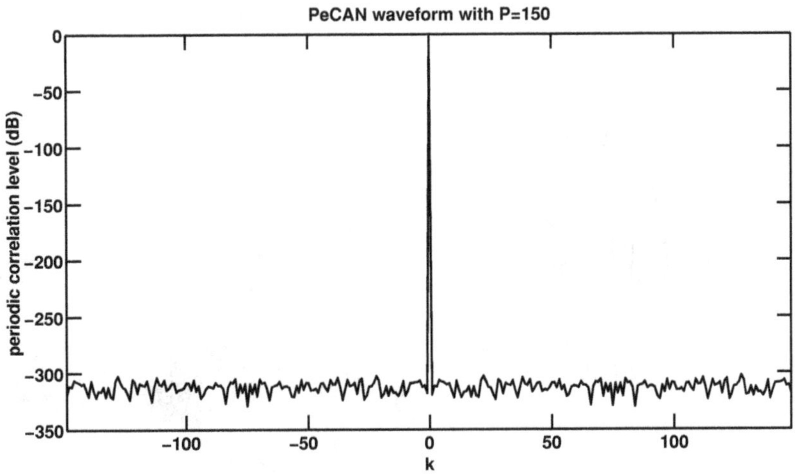

(b)

Figure 19.12 The periodic auto-correlation levels of the PeCAN waveform. (a) $P = 100$ and (b) $P = 150$.

correctly identified in the probing mode. The resulting empirical BER performance, along with the theoretical values, are shown in Figure 19.14. Each point in the figure is averaged over 2×10^6 independent Monte-Carlo runs. We can see that due to the (almost) zero periodic auto-correlation sidelobes of the PeCAN waveform, the empirical BER curves show perfect agreement with the theoretical values. Due to the same reason, the average performance of a Frank sequence with $P = 100$ also coincides with the theoretical BER value for $P = 100$, and therefore the corresponding curve is not included in the figure. Note that there is no Frank sequence corresponding to $P = 150$, and moreover a Frank sequence can be guessed once P is known.

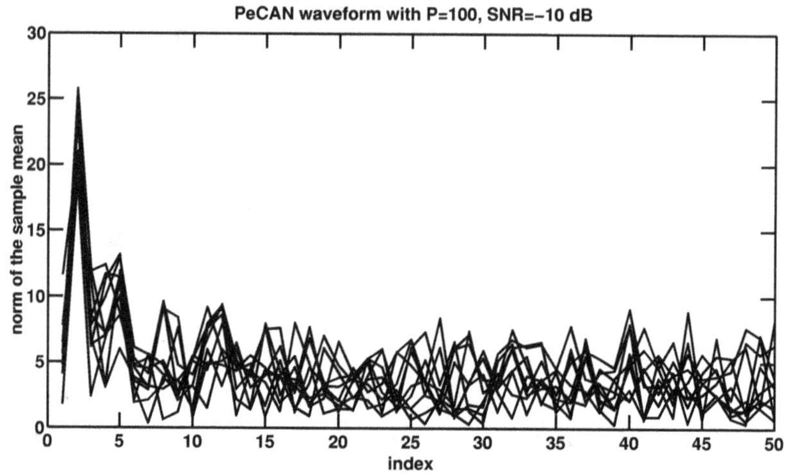

(a)

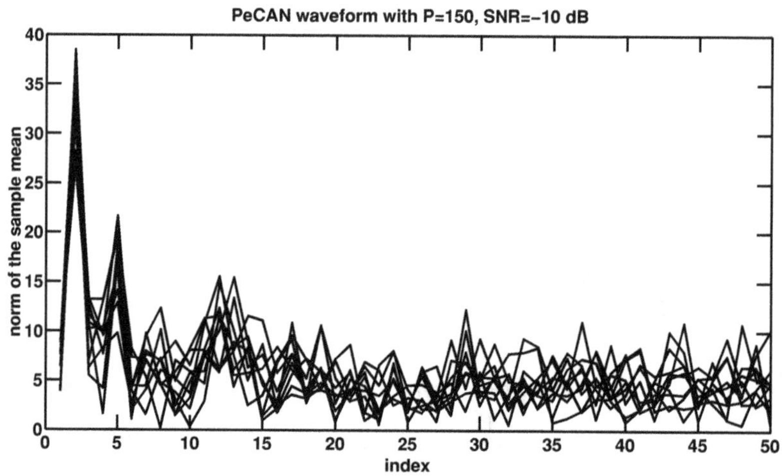

(b)

Figure 19.13 Principal arrival identification using probing waveforms with SNR = −10 dB. Ten Monte-Carlo trials are shown. (a) $P = 100$ and (b) $P = 150$.

Finally, we consider the covertness of the communication scheme. We assume that, except for precise knowledge of the spreading waveforms, an eavesdropper has the same information about the communication details as an intended receiver. The resulting BER performance, plotted in Figure 19.15, is obtained by conducting 1000 independent Monte-Carlo runs. The observations made for Figure 19.11 are equally applicable in this case.

We now relax the assumptions by allowing the eavesdropper to use even the PeCAN algorithm for waveform generation. Consequently, the initial sequence (see step 0 in Table 9.1) becomes the only information that the eavesdropper does not have. Again we

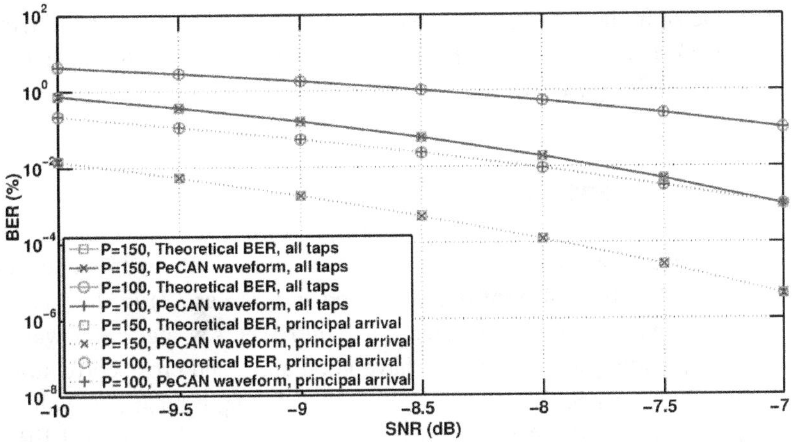

Figure 19.14 The empirical BER curves corresponding to two different PeCAN waveforms with $P = 100$ and 150, with the theoretical values plotted alongside. Two reception schemes are considered: one uses EGC over all the $R = 50$ finger outputs, and the other uses the principal arrival only. Each point is averaged over 2×10^6 independent Monte-Carlo trials.

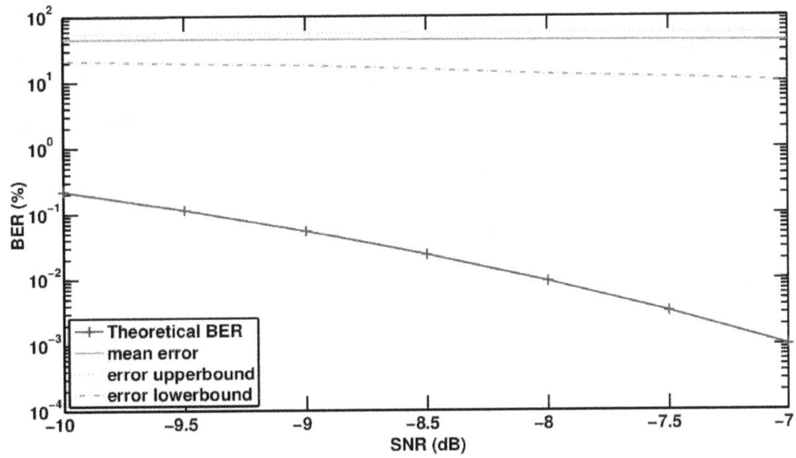

Figure 19.15 BER performance achieved by generating the unimodular waveform chips in a random manner. The BER is 50% on average, implying desired LPI or LPD properties for the PeCAN waveform.

conduct 1000 independent Monte-Carlo runs by randomly generating initial sequences, and we perform the detection by using the resulting PeCAN spreading waveforms. Interestingly, the so-obtained BER results are almost identical to those given in Figure 19.15. This fact suggests that the different PeCAN waveforms obtained by using different initial random sequences are almost uncorrelated to one another, which is a desired feature from a LPI/LPD point of view. Note that we could have conducted a similar simulation for the Multi-WeCAN waveforms by initializing the algorithm with an independent waveform, and then running the Multi-WeCAN algorithm to generate eavesdropper's waveforms. However, for the Multi-WeCAN algorithm, such an experiment turns out to be too time consuming.

19.5 Conclusions

We have considered covert UWA communications with noncoherent schemes (as opposed to the coherent one discussed in Chapter 18) in this chapter. For the binary orthogonal modulation scheme, the Multi-WeCAN algorithm was used to synthesize two spreading waveforms that have good auto- and cross-correlation properties over a certain lag range. For the binary DPSK scheme, by using a cyclic prefix, we demonstrated how a PeCAN sequence can be used to enhance the BER performance. The utility of the aforementioned Multi-WeCAN and PeCAN waveforms for covert communications is illustrated using numerical examples.

References

Abramovich, Y. I. & Frazer, G. J. [2008], 'Bounds on the volume and height distributions for the MIMO radar ambiguity function', *IEEE Signal Processing Letters* **15**, 505–508.

Ackroyd, M. H. & Ghani, F. [1973], 'Optimum mismatched filters for sidelobe suppression', *IEEE Transactions on Aerospace and Electronic Systems* **9**(2), 214–218.

Barker, R. H. [1953], "Group synchronizing of binary digital systems" *Communication Theory,* W. Jackson (ed.), Butterworths, London.

Benedetto, J. J., Konstantinidis, I. & Rangaswamy, M. [2009], 'Phase-coded waveforms and their design: The role of the ambiguity function', *IEEE Signal Processing Magazine* **26**(1), 22–31.

Blackmon, F., Sozer, E. M., Stojanovic, M. & Proakis, J. [2002], 'Performance comparison of RAKE and hypothesis feedback direct sequence spread spectrum techniques for underwater communication applications', *MTS/IEEE OCEANS'02* pp. 594–603.

Bliss, D. W. & Forsythe, K. W. [2003], 'Multiple-input multiple-output (MIMO) radar and imaging: degrees of freedom and resolution', *37th Asilomar Conf on Signals, Systems and Computers,* Pacific Grove, CA pp. 54–59.

Blunt, S. D. & Gerlach, K. [2006], 'Adaptive pulse compression via MMSE estimation', *IEEE Transactions on Aerospace and Electronic Systems* **42**(2), 572–584.

Bonami, A., Garrigos, G. & Jaming, P. [2007], 'Discrete radar ambiguity problems', *Applied and Computational Harmonic Analysis* **23**, 388–414.

Borwein, P. & Ferguson, R. [2005], 'Polyphase sequences with low autocorrelation', *IEEE Transactions on Information Theory* **51**(4), 1564–1567.

Boyd, S. & Vandenberghe, L. [2004], *Convex Optimization*, Cambridge University Press, Cambridge.

Brennan, D. [1959], 'Linear diversity combining techniques', *Proc. IRE* (6), 1075–1102.

Brenner, A. R. [1998], 'Polyphase Barker sequences up to length 45 with small alphabets', *Electronics Letters* **34**(16), 1576–1577.

Capon, J. [1969], 'High resolution frequency-wavenumber spectrum analysis', *Proceedings of the IEEE* **57**, 1408–1418.

Carbonelli, C., Vedantam, S. & Mitra, U. [2007], 'Sparse channel estimation with zero tap detection', *IEEE Transactions on Wireless Communications* **6**(5), 1743–1753.

Cardone, G., Cincotti, G. & Pappalardo, M. [2002], 'Design of wide-band arrays for low side-lobe level beam patterns by simulated annealing', *IEEE Transactions on Ultrasonics, Ferroelectrics and Frequency Control* **49**(8).

Chen, C.-Y. & Vaidyanathan, P. P. [2008], 'MIMO radar ambiguity properties and optimization using frequency-hopping waveforms', *IEEE Transactions on Signal Processing* **56**(12), 5926–5936.

Chu, D. C. [1972], 'Polyphase codes with good periodic correlation properties (correspondence)', *IEEE Transactions on Information Theory* **18**(4), 531–532.

Costas, J. P. [1984], 'A study of a class of detection waveforms having nearly ideal range-doppler ambiguity properties', *Proceedings of the IEEE* **72**(8), 996–1009.

DeLong Jr., D. F. & Hofstetter, E. M. [1967], 'On the design of optimum radar waveforms for clutter rejection', *IEEE Transactions on Information Theory* **13**(3), 454–463.

DeLong Jr., D. F. & Hofstetter, E. M. [1969], 'The design of clutter-resistant radar waveforms with limited dynamic range', *IEEE Transactions on Information Theory* **15**(3), 376–385.

Deng, H. [2004], 'Polyphase code design for orthogonal netted radar systems', *IEEE Transactions on Signal Processing* **52**(11), 3126–3135.

Diaz, V., Urena, J., Mazo, M., Garcia, J., Bueno, E. & Hernandez, A. [1999], 'Using Golay complementary sequences for multi-mode ultrasonic operation', *IEEE 7th International Conf on Emerging Technologies and Factory Automation,* UPC Barcelona, Spain .

Diederich, C. J. & Hynynen, K. [1999], 'Ultrasound technology for hyperthermia', *Ultrasound in Medicine and Biology* **25**(6), 871–887.

Dolph, C. L. [1946], 'A current distribution for broadside arrays which optimizes the relationship between beam width and side-lobe level', *Proceedings of the IRE* **34**(6), 335–348.

Elliott, R. [1975], 'Design of line source antennas for narrow beamwidth and asymmetric low sidelobes', *IEEE Transactions on Antennas and Propagation* **23**(1), 100–107.

Falk, M. H. & Issels, R. D. [2001], 'Hyperthermia in oncology', *International Journal of Hyperthermia* **17**(1), 1–18.

Fan, P. & Darnell, M. [1997], 'Construction and comparison of periodic digital sequence sets', *IEE Proceedings Communications* **144**(6), 361–366.

Fan, P. & Mow, W. H. [2004], 'On optimal training sequence design for multiple-antenna systems over dispersive fading channels and its extensions (correspondence)', *IEEE Transactions on Vehicular Technology* **53**(5), 1623–1626.

Fan, P., Suehiro, N., Kuroyanagi, N. & Deng, X. [1999], 'Class of binary sequences with zero correlation zone', *IET Electronics Letters* **35**(10), 777–779.

Fenn, A. J., Sathiaseelan, V., King, G. A. & Stauffer, P. R. [1996], 'Improved localization of energy deposition in adaptive phased-array hyperthermia treatment of cancer', *The Lincoln Laboratory Journal* **9**(2), 187–195.

Fienup, J. R. [1982], 'Phase retrieval algorithms: a comparison', *Applied Optics* **21**(15), 2758–2769.

Fishler, E., Haimovich, A., Blum, R., Chizhik, D., Cimini, L. & Valenzuela, R. [2004], 'MIMO radar: an idea whose time has come', *IEEE Radar Conf,* Philadelphia, PA pp. 71–78.

Fishler, E., Haimovich, A., Blum, R., Cimini, L., Chizhik, D. & Valenzuela, R. [2004], 'Performance of MIMO radar systems: advantages of angular diversity', *38th Asilomar Conf on Signals, Systems and Computers,* Pacific Grove, CA pp. 305–309.

Fishler, E., Haimovich, A., Blum, R., Cimini, L., Chizhik, D. & Valenzuela, R. [2006], 'Spatial diversity in radars - models and detection performance', *IEEE Transactions on Signal Processing* **54**(3), 823–838.

Fletcher, R. [1970], 'A new approach to variable metric algorithms', *The Computer Journal* **13**(3), 317–322.

Forsythe, K. W. & Bliss, D. W. [2005], 'Waveform correlation and optimization issues for MIMO radar', *39th Asilomar Conf on Signals, Systems and Computers,* Pacific Grove, CA pp. 1306–1310.

Frazer, G., Abramovich, Y. & Johnson, B. [2007], 'Spatially waveform diverse radar: Perspectives for high frequency OTHR', *IEEE Radar Conf,* Boston, MA, USA pp. 385–390.

Freedman, A. & Levanon, N. [1994], 'Properties of the periodic ambiguity function', *IEEE Transactions on Aerospace and Electronic Systems* **30**(3), 938–941.

Freedman, A., Levanon, N. & Gabbay, S. [1995], 'Perfect periodic correlation sequences', *Signal Processing* **41**(2), 165–174.

Friedlander, B. [2007], 'Waveform design for MIMO radars', *IEEE Transactions on Aerospace and Electronic Systems* **43**(3), 1227–1238.

Friese, M. & Zottmann, H. [1994], 'Polyphase Barker sequences up to length 31', *Electronics Letters* **30**(23), 1930–1931.

Fuhrmann, D. R. & San Antonio, G. [2004], 'Transmit beamforming for MIMO radar systems using partial signal correlation', *38th Asilomar Conf on Signals, Systems and Computers,* Pacific Grove, CA pp. 295–299.

Fuhrmann, D. R. & San Antonio, G. [2008], 'Transmit beamforming for MIMO radar systems using signal cross-correlation', *IEEE Transactions on Aerospace and Electronic Systems* **44**(1), 1–16.

Gerchberg, R. W. & Saxton, W. O. [1972], 'A practical algorithm for the determination of the phase from image and diffraction plane pictures', *Optik(Stuttgart)* **35**, 237–246.

Getz, B. & Levanon, N. [1995], 'Weight effects on the periodic ambiguity function', *IEEE Transactions on Aerospace and Electronic Systems* **31**(1), 182–193.

Gianfelice, D., Khiat, A., Amara, M., Belblidia, A. & Boulanger, Y. [2003], 'MR imaging-guided focused US ablation of breast cancer: histopathologic assessment of effectiveness-initial experience', *Radiology* **227**, 849–855.

Gladkova, I. & Chebanov, D. [2004], 'On the synthesis problem for a waveform having a nearly ideal ambiguity functions', *International Conf on Radar Systems,* Toulouse, France .

Gold, R. [1967], 'Optimal binary sequences for spread spectrum multiplexing', *IEEE Transactions on Information Theory* **IT-13**, 619–621.

Gold, R. [1968], 'Maximal recursive sequences with 3-valued recursive cross-correlation functions', *IEEE Transactions on Information Theory* **14**(1), 154–156.

Golub, G. H. & Van Loan, C. F. [1984], *Matrix Computations*, Johns Hopkins University Press, Baltimore, MD.

Guey, J. C. & Bell, M. R. [1998], 'Diversity waveform sets for delay-doppler imaging', *IEEE Transactions on Information Theory* **44**(4).

Guo, B. & Li, J. [2008], 'Waveform diversity based ultrasound system for hyperthermia treatment of breast cancer', *IEEE Transactions on Biomedical Engineering* **55**(2), 822–826.

Guo, B., Wang, Y., Li, J., Stoica, P. & Wu, R. [2006], 'Microwave imaging via adaptive beamforming methods for breast cancer detection', *Journal of Electromagnetic Waves and Applications* **20**(1), 53–63.

Han, Y. K. & Yang, K. [2009], 'New m-ary sequence families with low correlation and larger size', *IEEE Transactions on Information Theory* **55**(4), 1815–1823.

Haykin, S. [2006], 'Cognitive radar: a way of the future', *IEEE Signal Processing Magazine* **23**(1), 30–40.

Headrick, J. M. & Skolnik, M. I. [1974], 'Over-the-horizon radar in the HF band', *Proceedings of the IEEE* **62**(6), 664–673.

Høholdt, T. [2006], "The merit factor problem for binary sequences." *Applied Algebra, Algebraic Algorithms and Error-Correcting Codes,* M. Fossorier, H. Imai, S. Lin, and A. Poli, (eds.), Vol. 3857, Lecture Notes in Computer Science, pp. 51–59, Springer-Verlag, Heidelberg.

Hui, H. T. [2007], 'Decoupling methods for the mutual coupling effect in antenna arrays: a review', *Recent Patents on Engineering* **1**(2), 187–193.

Hursky, P., Porter, M. B. & Siderius, M. [2006], 'Point-to-point underwater acoustic communications using spread-spectrum passive phase conjugation', *Journal of the Acoustical Society of America* **120**(1), 247–257.

Iltis, R. A. & Fuxjaeger, A. W. [1991], 'A digital DS spread-spectrum receiver with joint channel and Doppler shift estimation', *IEEE Transactions on Communications* **39**, 1255–1267.

Jakowatz Jr., C. V., Wahl, D. E., Eichel, P. H., Ghiglia, D. C. & Thompson, P. A. [1996], *Spotlight-Mode Synthetic Aperture Radar: A Signal Processing Approach*, Kluwer Academic Publishers, Norwell, MA.

Jedwab, J. [2005], 'A survey of the merit factor problem for binary sequences' *Sequences and Their Applications – SETA 2004,* T. Helleseth, D. Sarwate, H. Y. Song, and K. Yang, Eds., Vol. 3486, Lecture Notes in Computer Science, pp. 30–55, Springer-Verlag, Heidelberg.

Kasami, T. [1966], 'Weight distribution formula for some class of cyclic codes', *Report 0475236 at Coordinated Science Lab, University of Illinois at Urbana-Champaign* .

Katsibas, T. K. & Antonopoulos, C. S. [2004], 'A general form of perfectly matched layers for three-dimensional problems of acoustic scattering in lossless and lossy fluid media', *IEEE Transactions on Ultrasonics, Ferroelectrics, and Frequency Control* **51**(8), 964–972.

Kay, S. [2007], 'Optimal signal design for detection of Gaussian point targets in stationary Gaussian clutter/reverberation', *IEEE Journal of Selected Topics in Signal Processing* **1**(1), 31–41.

Khan, H. A., Zhang, Y., Ji, C., Stevens, C. J., Edwards, D. J. & O'Brien, D. [2006], 'Optimizing polyphase sequences for orthogonal netted radar', *IEEE Signal Processing Letters* **13**(10), 589–592.

Kilfoyle, D. B. & Baggeroer, A. B. [2000], 'The state of the art in underwater acoustic telemetry', *IEEE Journal of Oceanic Engineering* **25**, 4–27.

Kremkau, F. W. [1993], *Diagnostic Ultrasound: Principles and Instruments*, W. B. Saunders, Philadelphia, PA.

Kretschmer Jr., F. F. & Gerlach, K. [1991], 'Low sidelobe radar waveforms derived from orthogonal matrices', *IEEE Transactions on Aerospace and Electronic Systems* **27**(1), 92–102.

Kruger, R. A., Kopecky, K. K., Aisen, A. M., Reinecke, D. R., Kruger, G. A. & Kiser, W. L. [1999], 'Thermoacoustic CT with radio waves: a medical imaging paradigm', *Radiology* **211**(1), 275–278.

Kumar, P. V. & Moreno, O. [1991], 'Prime-phase sequences with periodic correlation properties better than binary sequences', *IEEE Transactions on Information Theory* **37**(3), 603–616.

Lebret, H. & Boyd, S. [1997], 'Antenna array pattern synthesis via convex optimization', *IEEE Transactions on Signal Processing* **45**(3), 526–532.

Levanon, N. & Mozeson, E. [2004], *Radar Signals*, Wiley, NY.

Li, J. & Stoica, P. [1996], 'An adaptive filtering approach to spectral estimation and SAR imaging', *IEEE Transactions on Signal Processing* **44**(6), 1469–1484.

Li, J. & Stoica, P. [2009], *MIMO Radar Signal Processing*, John Wiley & Sons, Inc., Hoboken, NJ.

Li, J., Stoica, P., Xu, L. & Roberts, W. [2007], 'On parameter identifiability of MIMO radar', *IEEE Signal Processing Letters* **14**(12), 968–971.

Li, J., Stoica, P. & Zheng, X. [2008], 'Signal synthesis and receiver design for MIMO radar imaging', *IEEE Transactions on Signal Processing* **56**(8), 3959–3968.

Li, J., Xie, Y., Stoica, P., Zheng, X. & Ward, J. [2007], 'Beampattern synthesis via a matrix approach for signal power estimation', *IEEE Transactions on Signal Processing* **55**(12), 5643–5657.

Lin, Z. [1988], 'Wideband ambiguity function of broadband signals', *Acoustical Society of America* **83**(6), 2108–2116.

Lindenfeld, M. J. [2004], 'Sparse frequency transmit and receive waveform design', *IEEE Transactions on Aerospace and Electronic Systems* **40**(3).

Ling, J., He, H., Li, J., Roberts, W. & Stoica, P. [2010], 'Covert underwater acoustic communications: Transceiver structures, waveform designs and associated performances', *OCEANS Conf,* Seattle, WA, USA .

Ling, J., Tan, X., Yardibi, T., Li, J., He, H. & Nordenvaad, M. L. [2009], 'Enhanced channel estimation and efficient symbol detection in MIMO underwater acoustic communications', *43th Asilomar Conf on Signals, Systems and Computers,* Pacific Grove, CA .

Ling, J., Yardibi, T., Su, X., He, H. & Li, J. [2009], 'Enhanced channel estimation and symbol detection for high speed multi-input multi-output underwater acoustic communications', *Journal of the Acoustical Society of America* **125**(5), 3067–3078.

Ling, J., Zhao, K., Li, J. & Nordenvaad, M. [2011], 'Multi-input multi-output underwater communications over sparse and frequency modulated acoustic channels', *Journal of the Acoustical Society of America* **130**(1), 249–262.

Mailloux, R. J. [1982], 'Phased array theory and technology', *Proceedings of the IEEE* **70**(3), 246–291.

Meaney, P. M., Fanning, M. W., Li, D., Poplack, S. P. & Paulsen, K. D. [2000], 'A clinical prototype for active microwave imaging of the breast', *IEEE Transactions on Microwave Theory and*

Techniques **48**(11), 1841–1853.

Moore, I. C. & Cada, M. [2004], 'Prolate spheroidal wave functions, an introduction to the slepian series and its properties', *Applied and Computational Harmonic Analysis* **16**(3), 208–230.

Oppermann, J. & Vucetic, B. S. [1997], 'Complex spreading sequences with a wide range of correlation properties', *IEEE Transactions on Communications* **45**(3), 365–375.

Orsi, R., Helmke, U. & Moore, J. B. [2004], 'A Newton-like method for solving rank constrained linear matrix inequalities', *43rd IEEE Conf on Decision and Control* pp. 3138–3144.

Palmese, M., Bertolotto, G., Pescetto, A. & Trucco, A. [2007], 'Spread spectrum modulation for acoustic communication in shallow water channel', *MTS/IEEE OCEANS'07* pp. 1–4.

Patton, L. K. & Rigling, B. D. [2008], 'Modulus constraints in adaptive radar waveform design', *IEEE Radar Conf,* Rome, Italy .

Proakis, J. G. [2001], *Digital Communications, Fourth edition*, McGraw-Hill Inc.

Pursley, M. B. [1977], 'Performance evaluation for phase-coded spread-spectrum multiple-access communication–part I: System analysis', *IEEE Transactions on Communications* (8), 795–799.

Ritcey, J. A. & Griep, K. R. [1995], 'Coded shift keyed spread spectrum for ocean acoustic telemetry', *MTS/IEEE OCEANS'95* pp. 1386–1391.

Roberts, W., Stoica, P., Li, J., Yardibi, T. & Sadjadi, F. [2010], 'Iterative adaptive approaches to MIMO radar imaging', *IEEE Journal of Selected Topics in Signal Processing: Special Issue on MIMO Radar and Its Applications* **4**(1), 5–20.

Robey, F. C., Coutts, S., Weikle, D. D., McHarg, J. C. & Cuomo, K. [2004], 'MIMO radar theory and experimental results', *38th Asilomar Conf on Signals, Systems and Computers,* Pacific Grove, CA pp. 300–304.

Rummler, W. D. [1967], 'A technique for improving the clutter performance of coherent pulse train signals', *IEEE Transactions on Aerospace and Electronic Systems* **3**(6), 898–906.

Salzman, J., Akamine, D. & Lefevre, R. [2001], 'Optimal waveforms and processing for sparse frequency UWB operation', *IEEE Radar Conf,* Atlanta, GA pp. 105–110.

San Antonio, G. & Fuhrmann, D. R. [2005], 'Beampattern synthesis for wideband MIMO radar systems', *1st IEEE International Workshop on Computational Advances in Multi-Sensor Adaptive Processing,* Puerto Vallarta, Mexico pp. 105–108.

Sarwate, D. V. [1979], 'Bounds on crosscorrelation and autocorrelation of sequences', *IEEE Transactions on Information Theory* **IT-25**(6), 720–724.

Sarwate, D. V. [1999], 'Meeting the Welch bound with equality', *Sequences and Their Applications (Proceedings of SETA '98)* pp. 79–102, Springer, London.

Sarwate, D. V. & Pursley, M. B. [1980], 'Crosscorrelation properties of pseudorandom and related sequences', *Proceedings of the IEEE* **68**(5), 593–619.

Scholnik, D. R. & Coleman, J. O. [2000], 'Formulating wideband array-pattern optimizations', *Proceeding of IEEE International Conf. on Phased Array Systems and Technology* pp. 489–492.

Schotten, H. D. & Lüke, H. D. [2005], 'On the search for low correlated binary sequences', *International Journal of Electronics and Communications* **59**(2), 67–78.

Sharma, R. [2010], 'Analysis of MIMO radar ambiguity functions and implications on clear region', *IEEE International Radar Conf,* Washington DC .

Simon, M. & Alouini, M. [1998], 'A unified approach to the performance analysis of digital communication over generalized fading channels', *Proceedings of the IEEE* **86**(9), 1860–1877.

Skolnik, M. I. [2008], *Radar Handbook, Third Edition*, McGraw-Hill, New York, NY.

Smadi, M. A. & Prabhu, V. K. [2004], 'Analysis of partially coherent PSK system in wireless channels with equal-gain combining', *IEEE Canadian Conf Electrical Computer Engineering (CCECE'04)* Ontario, Canada pp. 99–102.

Sozer, E. M., Proakis, J. G., Stojanovic, M., Rice, J. A., Benson, A. & Hatch, M. [1999], 'Direct sequence spread spectrum based modem for under water acoustic communication and channel measurements', *MTS/IEEE OCEANS'99,* Seattle, WA pp. 228–233.

Spafford, L. J. [1968], 'Optimum radar signal processing in clutter', *IEEE Transactions on Information Theory* **14**(5), 734–743.

Stein, S. [1981], 'Algorithms for ambiguity function processing', *IEEE Transactions on Acoustics, Speech and Signal Processing* **29**(3).

Stimson, G. W. [1998], *Introduction to Airborne Radar*, SciTech Publishing, Mendham, NJ.

Stoica, P. & Ganesan, G. [2002], 'Maximum-SNR spatial-temporal formatting for MIMO channels', *IEEE Transactions on Signal Processing* **50**(12), 3036–3042.

Stoica, P., He, H. & Li, J. [2009*a*], 'New algorithms for designing unimodular sequences with good correlation properties', *IEEE Transactions on Signal Processing* **57**(4), 1415–1425.

Stoica, P., He, H. & Li, J. [2009*b*], 'On designing sequences with impulse-like periodic correlation', *IEEE Signal Processing Letters* **16**(8), 703–706.

Stoica, P., Jakobsson, A. & Li, J. [1998], 'Capon, APES and matched-filterbank spectral estimation', *Signal Processing* **66**(1), 45–59.

Stoica, P., Li, H. & Li, J. [1999], 'A new derivation of the APES filter', *IEEE Signal Processing Letter* **6**(8), 205–206.

Stoica, P., Li, J. & Xie, Y. [2007], 'On probing signal design for MIMO radar', *IEEE Transactions on Signal Processing* **55**(8), 4151–4161.

Stoica, P., Li, J. & Xue, M. [2008], 'Transmit codes and receive filters for radar', *IEEE Signal Processing Magazine* **25**(6), 94–109.

Stoica, P., Li, J. & Zhu, X. [2008], 'Waveform synthesis for diversity-based transmit beampattern design', *IEEE Transactions on Signal Processing* **56**(6), 2593–2598.

Stoica, P. & Moses, R. L. [2005], *Spectral Analysis of Signals*, Prentice-Hall, Upper Saddle River, NJ.

Stojanovic, M., Catipovic, J. & Proakis, J. [1994], 'Phase coherent digital communications for underwater acoustic channels', *IEEE Journal of Oceanic Engineering* **19**(1), 100–111.

Stojanovic, M. & Freitag, L. [2000], 'Hypothesis-feedback equalization for direct-sequence spread-spectrum underwater communications', *MTS/IEEE OCEANS'00* pp. 123–128.

Stojanovic, M. & Freitag, L. [2004], 'MMSE acquisition of DSSS acoustic communications signals', *MTS/IEEE OCEANS'04* pp. 14–19.

Stojanovic, M., Proakis, J. G., Rice, J. A. & Green, M. D. [1998], 'Spread spectrum underwater acoustic telemetry', *MTS/IEEE OCEANS'98* pp. 650–654.

Sturm, J. F. [1999], 'Using SeDuMi 1.02, a MATLAB toolbox for optimization over symmetric cones', *Optimization Methods and Software Online* **11-12**, 625–653. Available: http://www2.unimaas.nl/ sturm/software/sedumi.html.

Stutt, C. A. & Spafford, L. J. [1968], 'A 'best' mismatched filter response for radar clutter discrimination', *IEEE Transactions on Information Theory* **14**(2), 280–287.

Suehiro, N. [1994], 'A signal design without co-channel interference for approximately synchronized CDMA systems', *IEEE Journal on Selected Areas in Communications* **12**(5), 837–841.

Sussman, S. [1962], 'Least-square synthesis of radar ambiguity functions', *IEEE Transactions on Information Theory* **8**(3), 246–254.

Svantesson, T. [1999], 'Modeling and estimation of mutual coupling in a uniform linear array of dipoles', *1999 IEEE International Conf on Acoustics, Speech, and Signal Processing,* Phoenix, Arizona .

Szabo, T. [2004], *Diagnostic Ultrasound Imaging: Inside Out*, Elsevier Academic Press, Burlington, MA.

Tang, X. H., Fan, P. Z. & Matsufuji, S. [2000], 'Lower bounds on correlation of spreading sequence set with low or zero correlation zone', *IEEE Electronic Letters* **36**(6), 551–552.

Tang, X. & Mow, W. H. [2008], 'A new systematic construction of zero correlation zone sequences based on interleaved perfect sequences', *IEEE Transactions on Information Theory* **54**(12), 5729–5734.

Torii, H., Nakamura, M. & Suehiro, N. [2004], 'A new class of zero-correlation zone sequences', *IEEE Transactions on Information Theory* **50**(3), 559–565.

Trefethen, L. N. & Bau, D. [1997], *Numerical Linear Algebra*, Society for Industrial and Applied Mathematics, Philadelphia, PA.

Tropp, J. A., Dhillon, I. S., Heath, R. W. & Strohmer, T. [2005], 'Designing structured tight frames via an alternating projection method', *IEEE Transactions on Information Theory* **51**(1), 188–209.

Tse, D. & Viswanath, P. [2005], *Fundamentals of Wireless Communication*, Cambridge University Press, New York, NY.

Tsimenidis, C. C., Hinton, O. R., Adams, A. E. & Sharif, B. S. [2001], 'Underwater acoustic receiver employing direct-sequence spread spectrum and spatial diversity combining for shallow-water multi-access networking', *IEEE Journal of Oceanic Engineering* **26**, 594–603.

Turin, G. L. [1960], 'An introduction to matched filters', *IRE Transactions on Information Theory* **6**, 311–329.

Van Trees, H. L. [2002], *Optimum Array Processing: Part IV of Detection, Estimation, and Modulation Theory*, John Wiley & Sons, New York, NY.

Vernon, C. C., Hand, J. W., Field, S. B., Machin, D., Whaley, J. B., Van Der Zee, J., Van Putten, W. L. J., Van Rhoon, G. C., Van Dijk, J. D. P., González, D. G., Liu, F., Goodman, P. & Sherar, M. [1996], 'Radiotherapy with or without hyperthermia in the treatment of superficial localized breast cancer: results from five randomized controlled trials', *International Journal of Radiation Oncology Biology Physics* **35**(4), 731–744.

Wang, G. & Lu, Y. [2011], 'Designing single/multiple sparse frequency transmit waveforms with sidelobe constraint', *IET Radar Sonar and Navigation* **5**(1), 32–38.

Wang, J. & Moeneclaey, M. [1992], 'DS-SSMA star network over indoor radio multipath Rician fading channels', *Military Communications Conf* **3**, 841–845.

Ward, D. B., Kennedy, R. A. & Williamson, R. C. [1996], 'FIR filter design for frequency invariant beamformers', *IEEE Signal Processing Letters* **3**(3), 69–71.

Weiss, A. J. & Picard, J. [2008], 'Maximum-likelihood position estimation of network nodes using range measurements', *IET Signal Processing* **2**(4), 394–404.

Welch, L. R. [1974], 'Lower bounds on the maximum correlation of signals', *IEEE Transactions on Information Theory* **IT-20**(3), 397–399.

Wolf, J. D., Lee, G. M. & Suyo, C. E. [1968], 'Radar waveform synthesis by mean-square optimization techniques', *IEEE Transactions on Aerospace and Electronic Systems* **5**(4), 611–619.

Woodward, P. M. [1957], *Probability and Information Theory with Applications to Radar*, Pergamon, New York, NY.

Xu, L., Li, J. & Stoica, P. [2006], 'Radar imaging via adaptive MIMO techniques', *14th European Signal Processing Conf,* (invited), Florence, Italy .

Xu, L., Li, J. & Stoica, P. [2008], 'Target detection and parameter estimation for MIMO radar systems', *IEEE Transactions on Aerospace and Electronic Systems* **44**(3), 927–939.

Yang, T. C. & Yang, W. B. [2008], 'Performance analysis of direct-sequence spread-spectrum underwater acoustic communications with low signal-to-noise-ratio input signals', *Journal of the Acoustical Society of America* **123**(2), 842–855.

Yang, Y. & Blum, R. S. [2007*a*], 'MIMO radar waveform design based on mutual information and minimum mean-square error estimation', *IEEE Transactions on Aerospace and Electronic Systems* **43**(1), 330–343.

Yang, Y. & Blum, R. S. [2007*b*], 'Minimax robust MIMO radar waveform design', *IEEE Journal of Selected Topics in Signal Processing* **1**(1), 147–155.

Yardibi, T., Li, J., Stoica, P., Xue, M. & Baggeroer, A. B. [2010], 'Source localization and sensing: A nonparametric iterative adaptive approach based on weighted least squares', *IEEE Transactions on Aerospace and Electronic Systems* **46**(1), 425–443.

Yuan, X., Borup, D., Wiskin, J., Berggren, M. & Johnson, S. A. [1999], 'Simulation of acoustic wave propagation in dispersive media with relaxation losses by using FDTD method with PML absorbing boundary condition', *IEEE Transactions on Ultrasonics, Ferroelectrics, and Frequency Control* **46**(1), 14–23.

Zeng, X., Li, J. & McGough, R. [2010], 'A waveform diversity method for optimizing 3-D power depositions generated by ultrasound phased arrays', *IEEE Transactions on Biomedical Engineering* **57**(1), 41–47.

Zhang, N. & Golomb, S. W. [1993], 'Polyphase sequence with low autocorrelations', *IEEE Transactions on Information Theory* **39**(3), 1085–1089.

Zoraster, S. [1980], 'Minimum peak range sidelobe filters for binary phase-coded waveforms', *IEEE Transactions on Aerospace and Electronic Systems* **16**(1), 112–115.

Index

www.ingramcontent.com/pod-product-compliance
Lightning Source LLC
Chambersburg PA
CBHW081957110325
23302CB00015B/243
* 9 7 8 1 1 0 7 0 1 9 6 9 0 *